A COLLEGE COURSE ON RELATIVITY AND COSMOLOGY

A College Course on Relativity and Cosmology

Ta-Pei Cheng

OXFORD
UNIVERSITY PRESS

OXFORD
UNIVERSITY PRESS

Great Clarendon Street, Oxford, OX2 6DP,
United Kingdom

Oxford University Press is a department of the University of Oxford.
It furthers the University's objective of excellence in research, scholarship,
and education by publishing worldwide. Oxford is a registered trade mark of
Oxford University Press in the UK and in certain other countries

Published in the United States of America by Oxford University Press
198 Madison Avenue, New York, NY 10016, United States of America

British Library Cataloguing in Publication Data
Data available

Library of Congress Control Number: 2014957469

ISBN 978–0–19–969340–5 (hbk.)
ISBN 978–0–19–969341–2 (pbk.)

Printed and bound by
CPI Group (UK) Ltd, Croydon, CR0 4YY

Links to third party websites are provided by Oxford in good faith and
for information only. Oxford disclaims any responsibility for the materials
contained in any third party website referenced in this work.

To
Brian Benedict

Preface

Physics students, including undergraduates, are interested in learning Einstein's general theory of relativity and its application to modern cosmology. This calls for an introduction to the subject that is suitable for upper-level physics and astronomy students in a college. Given the vast number of topics in this area, most general relativity books tend to be rather lengthy. While an experienced instructor can select appropriate parts from such texts to cover in an undergraduate course, it may be desirable to have a book that is written, from the start, as an accessible and concise introduction for undergraduates.

My aim is to present an overview of general relativity and cosmology. While not an exhaustive survey, it has sufficient details so that even a mathematics-shy undergraduate can grasp the fundamentals of the subject, appreciate the beauty of its structure, and place a reported cosmological discovery in its proper context.

The reader is assumed to have a mathematical preparation at the calculus level, plus some familiarity with matrices. As far as physics preparation is concerned, some exposure to intermediate mechanics and electromagnetism as well as special relativity (as part of an introductory modern physics course) will be helpful, although not a prerequisite.

The emphasis of the book is pedagogical. Following the historical development, the implications of Einstein's principle of equivalence (between gravity and inertia) are presented in detail, before deducing them from the more difficult framework of Riemannian geometry. The necessary mathematics is introduced gradually. The full tensor formalism of general relativity is postponed to the last chapter of the book, since much of Einstein's theory can be understood with a metric description of spacetime. Many applications, including black holes and cosmology, can be discussed at this more accessible level. The book also has some other features that a reader should find useful:

- A bullet list of topics at the beginning of each chapter serves as the chapter's abstract, giving the reader a foretaste of upcoming material.

- Review questions at the end of each chapter should help a beginning student to focus on the key elements of the chapter; the answers to these questions are provided at the back of the book. (The practice of frequent quizzes based on these questions can be an effective means of ensuring that each student is keeping up with the progress of the class.)

- Calculational details, peripheral topics, and historical tidbits are segregated in boxes that can be skipped over, depending on a reader's interest.

- Exercises (with solution hints) are dispersed throughout the text.

This book is the undergraduate edition of my previous work, *Relativity, Gravitation and Cosmology: A Basic Introduction,* second edition (2010), published as part of the Oxford Master Series. This college edition streamlines the presentation by concentrating on the core topics and omitting, or placing in marked boxes, more peripheral material. Still, there being varieties of undergraduate courses (some for physics or for astronomy students, some assuming much familiarity with special relativity, others less so, etc.), enough material is presented in this book to allow an instructor to choose different chapters to emphasize. Most undergraduate courses will probably omit the last chapter on tensor formalism leading to the Riemannian curvature tensor. It is included for those wishing to have a glimpse of the full tensor formalism needed to set up Einstein's field equation (and for them to realize that it is not so difficult after all). This last chapter serves another purpose: a class may be a combined course for undergraduates and beginning graduate students. This mathematical material in the last chapter may serve as an extra requirement for the advanced students.

I am grateful to Brian Benedict for his contribution to this book. In his careful reading of the entire manuscript (twice over) Brian made numerous suggestions for improvement: not only in the presentation but also in the substance of the physics. His input has been invaluable. I also thank my editors, Sonke Adlung, Jessica White, and Ania Wronski, for their advice and encouragement. The book grew from a course that I taught several times at the Physics Departments of the University of Missouri – St. Louis and Portland State University; their continual support is gratefully acknowledged. The illustrations in the book were prepared by Cindy Bertram and Wendy Allison. To them I express my thanks.

As always, I will be glad to receive readers' comments and spotted errors at tapeicheng@gmail.com. An updated list of corrections, as well as information about the online presentation of the book material, can be found at the website (http://www.umsl.edu/~chengt/GR3.html). Instructors interested in obtaining the solution manual are invited to its entry link included in the Oxford University Press's book site.

St. Louis

T.P.C.

Contents

Introduction

- Relativity means that it is physically impossible to detect absolute motion. This can be stated as a symmetry in physics: physics equations are unchanged under transformations among reference frames in relative motion.

- Special relativity (SR) is the symmetry with respect to coordinate transformations among inertial frames, general relativity (GR) the symmetry among more general frames, including accelerating coordinate systems or systems in which gravity is present.

- Tensor equations are covariant under coordinate transformation. The familiar case of rotation symmetry illustrates how coordinate symmetry follows automatically when the equations of physics are written in tensor form.

- Newtonian mechanics is covariant under Galilean transformations among inertial frames of reference. Maxwell's equations lack this Galilean symmetry.

- The apparent contradiction between constancy of light speed and classical (i.e., Galilean and Newtonian) relativity led most physicists to assume the presence of a unique reference frame, an electromagnetic ether. Einstein took a different approach by developing a new conception of time, leading to a completely new kinematics. Although both viewpoints would lead independently to the Lorentz transformation, their physics contents are fundamentally different.

- The equivalence principle between acceleration and gravity played an important role in Einstein's progress from SR to GR, which is a geometric theory of gravitation.

Relativity and quantum theory are the two great advances of physics in the twentieth century. Special relativity (SR) teaches us that the arena of physics is four-dimensional spacetime, and general relativity (GR) that gravity is the structure of this 4D spacetime. GR is the classical field theory of gravitation. Quantum field theory, the quantum description of fields, is based on the unification of quantum mechanics with special relativity. It culminates in the Standard Model of particle physics, which gives a precise and successful

A College Course on Relativity and Cosmology. First Edition. Ta-Pei Cheng.
© Ta-Pei Cheng 2015. Published in 2015 by Oxford University Press.

description of electromagnetic, weak, and strong interactions. A major focus of twenty-first-century physics is the effort to unify quantum theory with general relativity to discover the quantum theory of gravitation. This book gives a basic introduction to the relativity theories of Albert Einstein (1879–1955). We start with Einstein's special relativity, together with its geometric formulation in Minkowski spacetime. A detailed discussion of the principle of equivalence of inertia and gravitation then leads to his general theory with its notable application to cosmology.

Einstein's theory of gravitation encompasses and surpasses Newton's theory, which is seen to be valid only for particles moving with velocities much less than that of light, and only in a weak and static gravitational field. Although the effects of GR are often small in the terrestrial and solar domains, its predictions have been impressively verified whenever high-precision observations have been performed. GR is indispensable to situations involving strong gravity: large systems such as in cosmology, or compact stellar objects. Einstein's theory predicts the existence of black holes, objects whose gravity is so strong that even light cannot escape from them. GR, with its fundamental feature of a dynamical spacetime, offers a natural conceptual framework for the cosmology of an expanding universe. Furthermore, GR can accommodate the possibility of a constant vacuum energy density, giving rise to a repulsive gravitational force. Such an agent is the key ingredient of modern cosmological theories of the big bang (the inflationary cosmology) and of the accelerating universe (driven by so-called dark energy) in the present epoch.

1.1   Relativity as a coordinate symmetry

Creating new theories to describe phenomena that are not easily observed on earth poses great challenges. There are few existing experimental results one might synthesize to develop new theoretical content—the way, for example, that led to the formulation of the electromagnetic theory by James Clerk Maxwell (1831–1879). Einstein pioneered the elegant approach of using physics symmetries as a guide to new theories that would be relevant to realms yet to be explored. As we shall explain below, relativity is a coordinate symmetry. Symmetries impose restrictions on the equations of physics. The requirement that any new theory should reduce to known physics in the appropriate limit often narrows one's focus further to a manageably few possibilities, whose consequences can then be explored in detail.

Symmetries in physics

Symmetry in physics means that physics equations remain the same after some definite changes, a symmetry transformation. The symmetry of relativity is a symmetry of physics equations under coordinate transformations. A change of the coordinate system is equivalent to a change of the observer's reference frame. Special relativity asserts that physics is the same for all inertial observers, that there is no physics observation one can make to distinguish between inertial frames of reference in uniform rectilinear motion. General relativity extends the symmetry to include non-inertial observers in reference frames that may be accelerating with respect to one another or that may include gravity.

Inertial frames of reference

Recall the first law of Newton: In an inertial frame of reference, an object, if not acted upon by an external force, will continue its constant-velocity motion (including the state of rest). What is an inertial reference frame? It is actually defined by the first law.[1] The physics content of the first law can then be interpreted as saying that physical processes are described most simply using inertial frames, and that such reference frames exist in nature. This is one of the most important lessons taught to us by Galileo Galilei (1564–1642) and Isaac Newton (1643–1727). In reality, it has been found that the inertial frames are coordinate frames moving with constant velocity with reference to distant galaxies (the so-called fixed stars).[2] In Chapter 4, we will extend the notion of inertial frames to mean frames from which gravity is absent.

1.1.1 Coordinate transformations

For a given coordinate system with a set of orthonormal basis vectors $\{\mathbf{e}_i\}$ in three dimensions (so $i = 1, 2, 3$), a vector—for example, the vector $\mathbf{A}$—can be represented by its components $A_1, A_2,$ and A_3, with $\{A_i\}$ being the coefficients of expansion of $\mathbf{A}$ with respect to the basis vectors:

$$\mathbf{A} = \sum_{i=1}^{3} A_i \mathbf{e}_i = A_1 \mathbf{e}_1 + A_2 \mathbf{e}_2 + A_3 \mathbf{e}_3. \tag{1.1}$$

With a change of the coordinate basis $\{\mathbf{e}_i\} \to \{\mathbf{e}_i'\}$, the same vector would have another set of components (Fig. 1.1):

$$\mathbf{A} = \sum_{i=1}^{3} A_i' \mathbf{e}_i' = A_1' \mathbf{e}_1' + A_2' \mathbf{e}_2' + A_3' \mathbf{e}_3'. \tag{1.2}$$

Thus a coordinate change would bring about the transformation[3] of the vector components as

$$\begin{pmatrix} A_1 \\ A_2 \\ A_3 \end{pmatrix} \longrightarrow \begin{pmatrix} A_1' \\ A_2' \\ A_3' \end{pmatrix}. \tag{1.3}$$

Inertial frames are related to each other by rotation and boost transformations. We shall first discuss the rotation transformation.

[1] Namely, an inertial frame of reference is defined as one in which an object, if not acted upon by an external force, will continue its constant-velocity motion. Similarly, Newton's second law can be interpreted as giving the definition of force (mass times acceleration). The important lesson these two laws impart is that description of physical processes is simplest if one carries them out in inertial frames and concentrates on this product of mass and acceleration.

[2] That this connection has a physical basis was named (by Einstein) Mach's principle: the inertia of a body is determined by the large-scale distribution of matter in the universe. This principle can be regarded as partially realized in the GR theory of gravity.

[3] Such a transformation is called a passive coordinate transformation, because the axes are rotated while the physical entities themselves are unchanged. An active transformation, on the other hand, leaves the axes fixed as the physical objects move. Thus the rotation in (1.4) is a passive rotation, and the boosts in (1.9) and (1.16) are passive boosts. For these simple cases, the active transformations differ from the passive ones just by a sign change of the respective transformation parameters: $\theta \to -\theta$ and $v \to -v$.

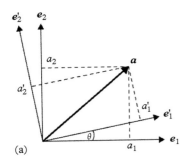

(a)

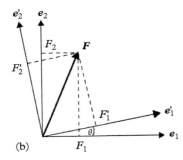

(b)

Figure 1.1 *Coordinate change of a vector under rotation. Components of different vectors, whether the vector is acceleration **a** as in (a) or the force **F** as in (b), all transform in the same way, as in (1.7).*

Rotation transformation

Take for example the coordinate transformation that describes a rotation of the coordinate axes by an angle θ around the z axis. The new position components are related to the original ones by relations that can be worked out geometrically from Fig. 1.1:

$$\begin{aligned} A'_1 &= (\cos\theta)A_1 + (\sin\theta)A_2, \\ A'_2 &= -(\sin\theta)A_1 + (\cos\theta)A_2, \\ A'_3 &= A_3. \end{aligned} \tag{1.4}$$

This set of equations can be written compactly as a matrix, the **rotation transformation matrix**, acting on the original vector components to yield the components in the new coordinate frame:

$$\begin{pmatrix} A'_1 \\ A'_2 \\ A'_3 \end{pmatrix} = \begin{pmatrix} \cos\theta & \sin\theta & 0 \\ -\sin\theta & \cos\theta & 0 \\ 0 & 0 & 1 \end{pmatrix} \begin{pmatrix} A_1 \\ A_2 \\ A_3 \end{pmatrix}. \tag{1.5}$$

Expressed in component notation, this matrix equation becomes

$$A'_i = \sum_{j=1}^{3} [R]_{ij}A_j = [R]_{i1}A_1 + [R]_{i2}A_2 + [R]_{i3}A_3, \tag{1.6}$$

where i and j are the row and column indices, respectively. Namely, the components of any vector should transform under a rotation of coordinate axes in the manner of (1.6). The relation given in (1.4) is a specific example of a rotational transformation, i.e., around a specific axis (z) by a specific amount (θ) with $[R]_{11} = \cos\theta$ and $[R]_{12} = \sin\theta$, etc.

Rotational invariance is a prototype of a physics symmetry. Having rotational symmetry means that the physics is unchanged under a rotation of coordinates by a fixed angle (N.B., not a rotating coordinate system, in which the coordinate axes have a rotational velocity). Take, for example, the equation of Newton's second law $F_i = ma_i$ ($i = 1, 2, 3$), which is the familiar $\mathbf{F} = m\mathbf{a}$ equation written in vector-component form. These components are projections onto the coordinate axes. In order for the equation to exhibit rotational symmetry, the validity of $F_i = ma_i$ in a system O must imply the validity of $F'_i = m'a'_i$ in any other system O' that is related to O by a rotation. Since mass m is a scalar (unchanged under a transformation) whereas the vector components of acceleration a_i and force F_i transform according to (1.6) (as displayed in Fig. 1.1),

$$m' = m, \qquad a'_i = \sum_j [R]_{ij}a_j, \qquad F'_i = \sum_j [R]_{ij}F_j, \tag{1.7}$$

the validity of $F'_i - m'a'_i = 0$ follows from $F_i - ma_i = 0$, because the non-singular transformation matrix $[R]$ is the **same** for each set of vector components, $\{F_i\}$ and $\{a_i\}$:

$$F'_i - m'a'_i = \sum_j [R]_{ij}(F_j - ma_j) = 0. \qquad (1.8)$$

Under a transformation, the components of force and acceleration change values. However, since the corresponding components of force $\{F_i\}$ and acceleration $\{a_i\}$, both being vector components, transform in the same way, $F_j - ma_j = 0$ implies $F'_i - m'a'_i = 0$, and so the physics equation keeps the same form and their relation is not altered. The resultant equation is said to be **covariant** because all terms in the equation transform in the same way.

Scalars and vectors are the simplest examples of tensors, having no or one index, respectively. A tensor in general is a mathematical object having definite transformation properties under coordinate transformation. Each term in a tensor equation must transform in the same way. A covariant equation maintains its form under transformation. Therefore, if a physics equation can be written in tensor form, it is covariant and automatically respects coordinate (e.g., rotation) symmetry.

Boost transformation

In classical mechanics, the transformation from one inertial frame O with coordinates x_i to another inertial frame O' with coordinates x'_i is the **Galilean transformation**. If the relative velocity of the two frames is a constant **v** and their relative orientations are specified by three angles α, β, and γ, the new coordinates are related to the original ones by $x_i \longrightarrow x'_i = [R]_{ij}x_j - v_it$, where $[R] = [R(\alpha, \beta, \gamma)]$ is the rotation matrix discussed above. Here we are mainly interested in coordinate transformations among inertial frames with the same orientation, $[R(0,0,0)]_{ij} = \delta_{ij}$ (see Fig. 1.2). Such a transformation is called a (Galilean) **boost**:[4]

$$x'_i = x_i - v_it, \qquad (1.9)$$
$$t' = t.$$

In Newtonian physics, the time coordinate is assumed to be absolute; i.e., it is the same in every coordinate frame. As a direct corollary, we have the **velocity addition rule**, with the velocity components $u_i = dx_i/dt$ and $u'_i = dx'_i/dt'$ related by

$$u'_i = u_i - v_i, \qquad (1.10)$$

which is obtained, because $t' = t$, by differentiation on both sides of (1.9) by $d/dt' = d/dt$.

[4] N.B., we are actually discussing the relation for the **intervals** between two events, say, $A = (x_A, t_A)$ and $B = (x_B, t_B)$. This relation, simplified to the case of one spatial dimension, should be understood as $\Delta x' = \Delta x - v\Delta t$, with $\Delta x = x_A - x_B$ and $\Delta t' = \Delta t = t_A - t_B$. Thus, the x's written in (1.9) are understood to be the coordinate displacements between position point x and the coordinate origin: $(x_A = x, x_B = 0)$. Similarly, the x''s are the coordinate displacements between position point x' and the origin O'. It is also understood that the coordinate origins of the two systems O and O' coincide at $t = 0$.

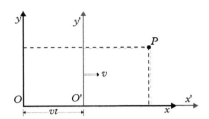

Figure 1.2 *The point P is located at* (x, y) *in the O system and at* (x', y') *in the O' system, which is moving with velocity v in the x direction. The coordinate transformations are* $x' = x - vt$ *and* $y' = y$.

Exercise 1.1 Graphic representation of Galilean transformation

Draw a (2D) vector diagram showing the Galilean relation $\mathbf{r}' = \mathbf{r} - \mathbf{v}t$ *for a general relative velocity, instead of for the restricted case* $\mathbf{v} = v\hat{\mathbf{x}}$ *as shown in Fig. 1.2.*

1.1.2   The principle of relativity

The idea of relativity can be traced back at least to the 1632 book *Dialogue Concerning the Two Chief World Systems* by Galileo in his defense of the Copernican system against the traditional Ptolemaic worldview. Responding to the criticism that one lacks the sensation of speed although the earth is in high-speed motion, Galileo discussed a thought experiment of an observer below decks on a ship. The observer is not able to tell whether the ship is docked or is moving smoothly through the water. This idea was subsequently incorporated in Newton's laws of mechanics.

Newtonian relativity (i.e., Galilean symmetry) says that the laws of physics are unchanged under the Galilean transformation (1.9). This implies that no mechanical experiment can detect any intrinsic difference between two inertial frames. It is easy to check that a physics equation, such as the one describing the Newtonian gravitational acceleration ($\mathbf{a} = -G_N M\,\hat{\mathbf{r}}/r^2$ with G_N being Newton's constant) of a test particle due to a point source with mass M, does not change its form under the Galilean transformation.

Exercise 1.2 Galilean covariance of Newton's law

Demonstrate that the law of universal gravitational attraction,

$$\mathbf{F} = -G_N \frac{m_A m_B}{r_{AB}^3} \mathbf{r}_{AB}, \tag{1.11}$$

is unchanged under Galilean transformation. In the above equation, we have two mass points, m_A *and* m_B, *located at positions* $\mathbf{r}_A$ *and* $\mathbf{r}_B$, *separated by* $\mathbf{r}_{AB} = \mathbf{r}_A - \mathbf{r}_B$. *The force on either mass due to the other is related to the acceleration by Newton's second and third laws,* $\mathbf{F} = -m_A \mathbf{a}_A = m_B \mathbf{a}_B$.

Exercise 1.3 Galilean covariance of Newtonian momentum conservation

Consider the two-particle collision $A + B \longrightarrow C + D$. *Demonstrate that if momentum conservation holds in one frame O,*

$$m_A \mathbf{u}_A + m_B \mathbf{u}_B = m_C \mathbf{u}_C + m_D \mathbf{u}_D, \tag{1.12}$$

it also holds in another frame O' in relative motion $(\mathbf{v})$, *provided the total mass is also conserved:* $m_A + m_B = m_C + m_D$.

Galilean vs. Lorentz symmetry

While Newtonian mechanics has Galilean symmetry, Maxwell's equations are not covariant under the Galilean transformation. The easiest way to see this is by recalling the fact that the propagation speed of electromagnetic waves is a constant

$$c = \sqrt{\frac{1}{\epsilon_0 \mu_0}} \simeq 3 \times 10^8 \text{ m/s}, \qquad (1.13)$$

where ϵ_0 and μ_0, the permittivity and permeability of free space, are the respective constants appearing in Coulomb's and Ampère's laws. For the speed of light c to be the same in all inertial frames violates the Galilean velocity addition rule (1.10).

The ether theory　For nineteenth-century physicists (including Maxwell himself), the most natural interpretation of light speed's constancy was that the Newtonian relativity principle should not apply. It was thought that, like all mechanical waves, electromagnetic waves must have an elastic medium, called ether, for their propagation. Maxwell's equations were thought to be valid only in the rest frame of this ether medium.[5] The constant c was interpreted to be the speed of light waves relative to this ether—the frame of absolute rest.

Einstein's new kinematics　Einstein proposed a different explanation of light speed's constancy: Maxwell's equations do obey the principle of relativity, but the relationship between inertial frames is not correctly given by the Galilean transformation. Hence the velocity addition rule (1.10) is invalid; the correct velocity relation should be such that c can be the same in every inertial frame. The modification of the velocity addition rule is related to a new conception of a coordinate-dependent time, $t' \neq t$. This turned out to be the correct interpretation. We shall discuss this in Section 1.2.

As nineteenth-century physicists generally believed in the existence of the ether and in the validity of Maxwell's equations only in the rest frame of that ether, a key problem in physics was to understand motion in the ether frame. This was often referred to[6] as "the electrodynamics of moving bodies." The most developed dynamical ether theory was that of Hendrik Lorentz (1853–1928). In this 1890 theory, Lorentz introduced a mathematical construct (i.e., not a physical entity), called local time, to describe objects moving with respect to the ether:

$$t' = t - \frac{v}{c^2}x. \qquad (1.14)$$

He then demonstrated that if one allowed as well for the change (as measured by moving observers) of electromagnetic fields $(E_i, B_i) \rightarrow (E_i', B_i')$, it then followed that the speed of light was essentially unchanged in a moving frame $c' = c + O(v^2/c^2)$. However, the interferometry experiment by Albert Michelson (1852–1931) and E.W. Morley (1838–1923) showed that in different moving frames[7] the speed of light remained the same, even to an accuracy $O(v^2/c^2)$.

[5] Many physicists were nonetheless unsettled by the idea of an all-permeating, yet virtually undetectable, substrate. The ether must not impede the motion of any object; yet it must be extremely stiff, as the wave speed should be proportional to the rigidity of the medium in which the wave propagates.

[6] Einstein's 1905 relativity paper was titled "The electrodynamics of moving bodies."

[7] If there had existed an ether medium, the motion of the earth would have brought about an ether wind, which would have caused the speed of light to vary with its direction of propagation.

[8] George FitzGerald (1851–1901) independently proposed this hypothesis of length contraction. But his proposal was not made in the context of any comprehensive theory as was Lorentz's.

This famous null result (not finding a different light speed) then compelled Lorentz to introduce the hypothesis of length contraction:[8]

$$l' = \frac{l}{\gamma}, \quad \text{with} \quad \gamma = \frac{1}{\sqrt{1 - \frac{v^2}{c^2}}}. \tag{1.15}$$

An object of length l would shrink to the length l' when moving with speed v in the ether medium. When this length contraction, together with local time, was formally incorporated into Lorentz's theory, the effective change in coordinates for a frame moving with speed v in the x direction was represented by the Lorentz transformation[9]

[9] Equation (1.16), published in 1904, was later called by Henri Poincaré (1854–1912) the "Lorentz transformation." All this was unknown to Einstein, who was then working in the Swiss Patent Office, outside the mainstream of academia. Poincare' was the first one to state publicly the principle of relativity. For an accessible account of Poincaré's contribution to special relativity, see (Logunov 2001).

$$x' = \gamma(x - vt), \quad y' = y, \quad z' = z, \quad t' = \gamma\left(t - \frac{v}{c^2}x\right). \tag{1.16}$$

It was then discovered that under such a transformation, when supplemented with the necessary change of $(E_i, B_i) \rightarrow (E_i', B_i')$, Maxwell's equations were unchanged[10]—to all orders of v/c. Thus this Lorentz transformation (1.16) was first written down within the framework of ether theory in 1904. As we shall discuss below, Einstein independently arrived at this transformation in 1905 by a completely differently route, devising a fundamentally new kinematics and dispensing completely with any ether medium.

[10] In particular, the light velocity remains the same in every inertial frame. See Exercise (1.4) for a discussion of the velocity addition rule implied by Lorentz transformation.

The γ factor (called the Lorentz factor) in (1.15) and (1.16) ranges from unity for $v = 0$ to infinity as $v \rightarrow c$. We note that the spatial transformation in the direction of the frame's motion (1.16) is just the Galilean transformation (1.9) multiplied by the γ factor, so it reduces to (1.9) in the low-velocity limit ($v/c \rightarrow 0$, hence $\gamma \rightarrow 1$).

1.2   Einstein and relativity

Looking over Einstein's work, it is quite clear that his motivation for new investigation was seldom related to any desire to account for this or that experimental puzzle. His approach often started with fundamental principles that would have the widest validity. He was always motivated by a keen sense of aesthetics in physics. His approach to the problem of the "electrodynamics of moving bodies" was to take the principle of relativity as an axiom. This symmetry approach turned out to be a powerful and productive method in the discovery of new physics[11].

[11] For an overview of Einstein's physics, see, for example, (Pais 2005) and (Cheng 2013).

1.2.1   The new kinematics

The relativity principle understood prior to Einstein's was the Galilean symmetry, but its application to electrodynamics immediately ran into a contradiction. Maxwell's equations lack Galilean symmetry. Most notably, a constant light speed cannot obey the Galilean velocity addition rule (1.10). Einstein struggled with this contradiction for a long time. The breakthrough came when he noted that the velocity addition rule was based on the Newtonian conception of an absolute time ($t' = t$). Einstein's philosophical reading helped him to realize that physical concepts should be based on definite measurement, and that absolute time lacked such a physical basis. In particular, a comparison of time intervals ultimately involves the notion of simultaneity. But when signal exchange cannot be instantaneous, two events that take place at the same time in one reference frame may not be simultaneous to another observer moving with respect to the first reference frame. This led Einstein to a new conception of time—that time, just like space, can be coordinate-dependent. From this new understanding, the Lorentz transformation follows immediately,[12] and Maxwell's equations are relativistic since they are Lorentz-symmetric.

[12] Einstein's derivation of the Lorentz transformation will be presented in Section 2.2.1.

Exercise 1.4 The SR velocity addition rule

For motion in one spatial dimension (x only), the space and time coordinates transform according to the Lorentz transformation (1.16). From its differential form,

$$dx' = \gamma\,(dx - v\,dt), \qquad dt' = \gamma\left(dt - \frac{v}{c^2}\,dx\right), \qquad (1.17)$$

prove this new SR velocity addition rule, replacing the familiar relation (1.10) with

$$u' = \frac{u - v}{1 - \dfrac{uv}{c^2}}, \qquad (1.18)$$

where $u = dx/dt$ and $u' = dx'/dt'$ are the velocities of a particle as measured in two reference frames in relative motion (v). This addition rule exhibits the feature that $u = c$ implies $u' = c$; light speed is invariant under Lorentz transformation. Equation (1.18) reduces to the familiar relation $u' = u - v$ in the $v/c \to 0$ limit.

Einstein's revolutionary idea was that time and space should be treated on a similar footing; they are both coordinate-dependent. Hermann Minkowski (1864–1909) then proposed an instructive way to highlight this symmetry between space and time. He united them in a single mathematical structure, a four-dimensional manifold called **spacetime**. Such a geometric formulation of special relativity does not by itself bring out any new SR physics results, but it makes the relativistic formulation much more straightforward. All we need to do is to write physics equations as tensor equations with respect to this 4D Minkowski spacetime. They will automatically be covariant under Lorentz transformation. Most importantly, as we will discuss in Chapters 4 and 5, Minkowski spacetime facilitated Einstein's discovery of the general theory of relativity, the physics symmetry with respect to more general coordinate systems, including gravity and accelerated coordinates.

1.2.2   GR as a field theory of gravitation

Soon after the discovery of SR, Einstein started on his search for a relativistic theory of gravitation. He then recalled the famous result of Galileo that all objects fall with the same acceleration in a given gravitational field. To focus on this important result (in order to generalize it to physics beyond mechanics), Einstein elevated it to a principle: the equivalence between inertia (i.e., resistance to acceleration) and gravity. This led him to the idea that an accelerated frame is just an inertial frame with gravity. This became an important handle that allowed Einstein to generalize the principle of relativity for inertial frames to accelerated frames. Furthermore, this connection of acceleration and gravity shows that the theory of general relativity is a field theory of gravitation. In this geometric formulation, gravity is simply the structure of spacetime; a curved spacetime is identified as the gravitational field.

This new theory of gravitation extends Newton's theory to the wider realm of particles moving with relativistic velocities in gravitational fields that may be strong and/or time-dependent. For instance, GR allows cosmology theory to treat the entire universe as a physical system. Other surprising physical possibilities suggested by GR are black holes and gravitational waves.

Review questions

1. What is relativity? What is the principle of special relativity? What is general relativity?

2. What is a symmetry in physics? Explain how the statement that no physical measurement can detect a particular physical feature (e.g., the orientation or the constant velocity of a lab) can be phrased as a statement about a symmetry in physics. Illustrate your explanation with the examples of rotation and boost symmetries.

3. In general terms, what is a tensor? Explain how a physics equation, when written in terms of tensors, automatically displays the relevant coordinate symmetry.

4. What are inertial frames of reference? Answer this in three ways.

5. That light always propagates with the same speed (as indicated by Maxwell's equations and supported by the Michelson–Morley measurement) contradicts (Galilean) relativity's velocity addition rule: $u' = u - v$. What are the two distinct theoretical approaches to resolve this conflict?

6. The equations of Newtonian physics are unchanged when we change coordinates from one inertial frame to another. What is this coordinate transformation? The equations of electrodynamics are unchanged under another set of coordinate transformations. How are these two transformations related? (Just give their names and a qualitative description of their relation.)

Special Relativity: The New Kinematics

2

- Einstein created the new theory of special relativity (SR) by combining the principle of relativity with the principle of constancy of light speed.

- These two postulates appear to be contradictory: how can light travel at the same speed in two different reference frames that are in relative motion? Einstein resolved this paradox by realizing that simultaneity is a relative concept and that time runs at different rates in different inertial frames.

- Once different coordinate times are allowed, one can derive the Lorentz transformation, $(t, \mathbf{r}) \rightarrow (t', \mathbf{r}')$, in a straightforward manner. If frame O' is moving with speed v in the $+x$ direction relative to frame O,

$$\Delta x' = \gamma(\Delta x - v\Delta t), \quad \Delta y' = \Delta y, \quad \Delta z' = \Delta z,$$

$$\Delta t' = \gamma\left(\Delta t - \frac{v\Delta x}{c^2}\right), \quad \text{with} \quad \gamma = \left(1 - \frac{v^2}{c^2}\right)^{-1/2} > 1.$$

- Physical consequences follow immediately. A moving clock runs slow (time dilation); a moving object contracts (length contraction). This underscores the profound change in our conception of space and time brought about by SR.

- Nearly all the counterintuitive relativistic effects spring from this new conception of time. Length contraction is one such effect. To obtain the length of an object, one must measure its front and back ends at specific times. For a moving object, observers will not agree on the times of these measurements.

- Just as light speed is absolute ($c' = c$) in relativity, it follows that a particular combination of space and time intervals is invariant under coordinate transformation: $\Delta s' = \Delta s$, where $\Delta s^2 \equiv \Delta x^2 + \Delta y^2 + \Delta z^2 - c^2 \Delta t^2$.

- The electromagnetic fields also change when observed in different frames. Einstein derived their transformation, $(\mathbf{E}, \mathbf{B}) \rightarrow (\mathbf{E}', \mathbf{B}')$, by requiring that Maxwell's equations be covariant under Lorentz transformation.

- We present two worked examples: the twin paradox and the pole-and-barn paradox. These scenarios are counterintuitive puzzles shedding light on several basic concepts in relativity: time dilation, length contraction, reciprocity of relativity, as well as relativity of event order.

A College Course on Relativity and Cosmology. First Edition. Ta-Pei Cheng.
© Ta-Pei Cheng 2015. Published in 2015 by Oxford University Press.

2.1 Einstein's two postulates and Lorentz transformation

While most nineteenth-century investigations of electrodynamics used the device of the ether medium, Einstein concluded that there was no physical evidence of the existence of such an ether. Physics would be much simpler and more natural if ether was dispensed with altogether. From the modern perspective, what Einstein did was to introduce a principle of symmetry, the relativity principle, as one of the fundamental laws of nature.

In Chapter 1, we discussed Lorentz's derivation of the Lorentz transformation. He tried to use the electromagnetic theory of the electron to show that matter composed of electrons, when in motion, would behave in such a way as to make it impossible to detect the effect of this motion on the speed of light. One of his aims was to explain the persistent failure to detect any difference in the speed of light in different inertial reference frames. Einstein, by contrast, took as a fundamental axiom that light speed is the same to all observers. The key feature of his new kinematics was a new conception of time that allowed the same light speed, even viewed by observers in relative motion.

These were the two basic postulates for Einstein's new theory of relativity:

- Principle of relativity: Physics equations must be the same in all inertial frames of reference.
- Constancy of light speed: The speed of light is the same in all coordinate frames, regardless of the motion of its emitter or receiver.

Before deriving the Lorentz transformation from these two postulates, we can already obtain (in Box 2.1) the famous result of energy/matter equivalence: $E = mc^2$.

Box 2.1 A first look at $E = mc^2$

Einstein's original derivation (1905b) of this famous result (that a particle has rest energy $E = mc^2$) involved a somewhat complicated Lorentz transformation of electromagnetic radiation.[1] Here we present a much simpler deduction he gave many years later (Einstein 1946). Besides the two basic postulates, it uses only the familiar energy and momentum conservation constraints. We will discuss more generally the relativistic energy of a particle in Section 3.2.2. Consider an atom that radiates two identical back-to-back pulses of light (two photons) along the y axis. Figure 2.1(a) shows what an observer at rest with respect to the atom would see, before and after this emission. We

[1] This involves the transformation of the radiation amplitude as well as that of the volume filled by the radiation.

allow that the atom's mass may be different before and after the event (as justified below). Figure 2.1(b) shows what an observer moving to the left (call it the $-x$ direction) with speed v would see: the atom moving to the right with speed v and two photons moving at an angle θ with respect to the x axis. Although each photon's speed is still c, each photon's velocity acquires an x component equal to v. Thus, each photon's momentum p_γ has an x component $p_\gamma \cos\theta = p_\gamma v/c$, as shown in Fig. 2.1(c).

Along the x axis, we have the momentum conservation relation

$$Mv = mv + 2p_\gamma \frac{v}{c}. \tag{2.1}$$

At this stage, we do not need to assume any relativistic result—only the classical electromagnetic relation between light's momentum p_γ (i.e., the Poynting vector) and energy $E_\gamma = p_\gamma c$. Canceling the common factor of v in each term of (2.1) leads to the energy relation

$$(M - m)c^2 = 2E_\gamma, \tag{2.2}$$

showing that the mass of the atom must decrease after the radiation emission. Per energy conservation, the energy lost by the atom must equal the energy of the two photons. If we denote the change in atomic energy by ΔE, the energy conservation constraint for this emission process then reads $\Delta E = 2E_\gamma$. In this way, one gets

$$\Delta E = 2E_\gamma = Mc^2 - mc^2 = \Delta\left(mc^2\right). \tag{2.3}$$

Therefore, the atom's energy must contain a constant term, the rest energy of the particle, $E = mc^2$, which can be converted into other forms of energy—in this case, radiation energy.

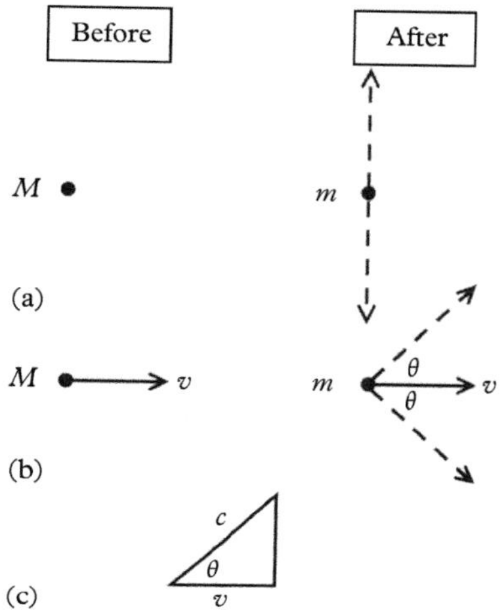

Before   After

$M \bullet$ m

(a)

$M \longrightarrow v$ m θ v

(b)

(c)

Figure 2.1 *(a) and (b) show two views of an atom, before and after it emits two back-to-back light pulses: (a) by an observer at rest with respect to the atom and (b) by an observer moving to the left. (c) To the moving observer, the photon's velocity acquires an x component equal to v, while its magnitude remains c; hence the propagation angle is given by $\cos\theta = v/c$. The mass of the atom is allowed to change with emission: $\Delta m = M - m \neq 0$.*

2.1.1   Relativity of simultaneity and the new conception of time

Einstein noted that at a fundamental level the measurement of time involves the matching of simultaneous events. Yet if signal exchange is not instantaneous, two events (at separate locations) that are simultaneous in one reference frame take place at different times according to a moving observer. An example is shown in Fig. 2.2. Two light pulses are sent from the midpoint of a moving railcar toward its front and back ends. To an observer on the railcar, the light pulses

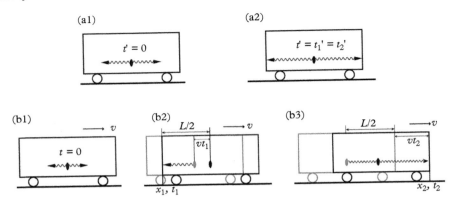

Figure 2.2 *Simultaneity is relative when signals cannot be transmitted instantaneously. Consider two events, (x'_1, t'_1) and (x'_2, t'_2), corresponding to light pulses (wavy lines) arriving at opposite ends of a moving railcar, after being emitted from the midpoint (a1, b1). The events are seen as simultaneous, $t'_1 = t'_2$, by an observer on the train as in (a2). But to another observer standing on the rail platform, these two events, (x_1, t_1) and (x_2, t_2), are not simultaneous, $t_1 \neq t_2$, because for this observer the light signals arrive at different times at the two ends of the moving railcar (b2, b3).*

arrive simultaneously at the front and back ends. However, to an observer standing on some stationary rail platform that the train passes, the light signals arrive at different times (at the back end before the front end). This leads the platform observer to conclude that these two events are not simultaneous.[2]

[2] In this same example, had the emission point not been at the midpoint of the railcar, so (for example) $t'_2 < t'_1$, it would still be possible to have $t_2 > t_1$. This shows that the order of events can be relative.

Exercise 2.1 Illustrating the relativity of equilocality

While the notion of the relativity of simultaneity may appear strange to us, we are all familiar with its analog in spatial coordinates—the relativity of equilocation. Two events that take place at the same location (but at different times) will be seen by a moving observer to have happened at different positions. This is a straightforward consequence of Galilean transformation. Use the setup as shown in Fig. 2.2 (i.e., light emissions at a fixed location on the railcar) to illustrate this phenomenon of relativity of equilocality.

Exercise 2.2 Calculating the nonsynchronicity of two events

In Fig. 2.2, the two events (light pulses arriving at the front and back ends of the railcar) are viewed as simultaneous in the O' frame: $t'_1 = t'_2$. (a) Work out the *nonsynchronicity $\Delta t = t_2 - t_1$ of these two events as viewed by an observer on the ground as the train passes by with speed v. (b) Show that simultaneity would be absolute, $\Delta t = \Delta t'$, had we followed the classical velocity addition rule (1.10), so that light signals would propagate forward with speed $c + v$ and backward with $c - v$.*

With Einstein's epiphany that time, like space, is coordinate-dependent, and the postulate of light speed's constancy, he could immediately demonstrate that the correct relation between space and time intervals measured in two inertial frames in relative motion is the Lorentz transformation.

2.1.2 Coordinate-dependent time leads to Lorentz transformation

From the above-stated postulates, Einstein (1905a) derived the relation between coordinates (x, y, z, t) of one inertial frame O and those (x', y', z', t') of another inertial frame O' that moves with a constant velocity v in the $+x$ direction. The origins of these two systems are assumed to coincide at the initial time.

1. Consider the transformation $(x, y, z, t) \rightarrow (x', y', z', t')$. Because the transverse length must be unchanged,[3] $y' = y$ and $z' = z$, we can simplify our equation display by concentrating on the two-dimensional problem of $(x, t) \rightarrow (x', t')$. The relevant transformation, like the rotation transformation (1.5), may be written in matrix form[4] as

$$\begin{pmatrix} x' \\ t' \end{pmatrix} = \begin{pmatrix} a_1 & a_2 \\ b_1 & b_2 \end{pmatrix} \begin{pmatrix} x \\ t \end{pmatrix}. \tag{2.4}$$

If one assumes that space is homogeneous and the progression of time is uniform, the transformation must be linear. That is, the elements (a_1, a_2, b_1, b_2) of the transformation matrix $[L]$ must be independent of the coordinates (x, t); we make the same coordinate transformation at every point. We call this a global transformation. Of course, these position- and time-independent factors can still depend on the relative speed v of the two coordinate frames.

2. The origin $(x' = 0)$ of the O' frame has the trajectory $x = vt$ in the O frame (cf. Fig. 1.2); thus, reading off from (2.4), we have $x' = 0 = a_1 x + a_2 t$ with $x = vt$, leading to

[3] Invariance of the transverse length: Consider three frames $O \xrightarrow{v} O' \xrightarrow{-v} O''$, with their relative velocities marked above the respective arrows. The transverse length could change by an overall factor, $y' = A(v)y$, with A depending possibly on the magnitude, but not the direction or the sense, of the relative velocity. The principle of relativity dictates that $y'' = Ay' = A^2 y$. But the O'' frame is in fact the O frame; $y'' = y$, so $A = \pm 1$. A must be a continuous function of v approaching unity as v approaches zero, so $A = 1$. Thus transverse length is invariant.

[4] Namely, (2.4) is just a compact way of writing $x' = a_1 x + a_2 t$ and $t' = b_1 x + b_2 t$.

$$a_2 = -va_1. \tag{2.5}$$

3. The origin ($x = 0$) of the O frame has the trajectory $x' = -vt'$ in the O' frame; from (2.4), we have (for $x = 0$) $x' = a_2 t$ and $t' = b_2 t$; thus $x'/t' = a_2/b_2 = -v$. Along with (2.5), this implies

$$b_2 = a_1. \tag{2.6}$$

Substituting the relations (2.5) and (2.6) into the matrix equation (2.4), we have $x' = a_1(x - vt)$ and $t' = b_1 x + a_1 t$. Taking their ratio, we get

$$\frac{x'}{t'} = \frac{a_1 \left(\dfrac{x}{t} - v \right)}{b_1 \dfrac{x}{t} + a_1}. \tag{2.7}$$

4. We now impose the constancy-of-c condition, $x/t = c = x'/t'$, on (2.7): $c(b_1 c + a_1) = a_1(c - v)$, leading to

$$b_1 = -\frac{v}{c^2} a_1. \tag{2.8}$$

5. Because of (2.5), (2.6), and (2.8), the whole transformation matrix $[L]$ has only one unknown constant, a_1:

$$[L] = a_1 \begin{pmatrix} 1 & -v \\ -v/c^2 & 1 \end{pmatrix}, \qquad [L^{-1}] = a_1 \begin{pmatrix} 1 & v \\ v/c^2 & 1 \end{pmatrix}. \tag{2.9}$$

In the above, we have also written down the inverse transformation matrix $[L^{-1}]$ by simply reversing the sign of the relative speed. That the transformations $[L]$ and $[L^{-1}]$ have the same coefficient a_1 follows from the principle of relativity: one cannot distinguish the transformation going from frame O to O' and the one from O' to O. The last unknown constant a_1 can then be fixed by the consistency condition $[L][L^{-1}] = [1]$:

$$a_1^2 \begin{pmatrix} 1 & -v \\ -v/c^2 & 1 \end{pmatrix} \begin{pmatrix} 1 & v \\ v/c^2 & 1 \end{pmatrix} = \begin{pmatrix} 1 & 0 \\ 0 & 1 \end{pmatrix}, \tag{2.10}$$

which implies that

$$a_1 = \sqrt{\frac{1}{1 - v^2/c^2}} \equiv \gamma. \tag{2.11}$$

This is the same Lorentz factor γ we encountered in Chapter 1.

This concludes Einstein's derivation of the Lorentz transformation:

$$x' = \gamma(x - vt), \quad y' = y, \quad z' = z$$
$$t' = \gamma\left(t - \frac{v}{c^2}x\right). \tag{2.12}$$

We note that the above steps 1–3 set up two coordinate frames in relative motion, step 4 imposed constancy of light speed[5] while step 5 is a consistency condition implied by the principle of relativity.

We can rewrite the Lorentz transformation in a more symmetric form by multiplying the time coordinate by a factor of c, so it will have the same dimension as the other coordinates:

$$\begin{pmatrix} x' \\ ct' \end{pmatrix} = \gamma \begin{pmatrix} 1 & -\beta \\ -\beta & 1 \end{pmatrix} \begin{pmatrix} x \\ ct \end{pmatrix}. \tag{2.13}$$

We can similarly write the inverse transformation as

$$\begin{pmatrix} x \\ ct \end{pmatrix} = \gamma \begin{pmatrix} 1 & \beta \\ \beta & 1 \end{pmatrix} \begin{pmatrix} x' \\ ct' \end{pmatrix}. \tag{2.14}$$

We have introduced the often-used dimensionless velocity parameter

$$\beta \equiv \frac{v}{c}. \tag{2.15}$$

We note that because $0 \leq \beta \leq 1$, the Lorentz factor $\gamma = (1 - \beta^2)^{-1/2}$ is always greater than unity: $\gamma \geq 1$; it approaches unity only in the low-velocity (nonrelativistic) limit $v \ll c$, and blows up when v approaches c.

[5] Had one imposed at step 4 the condition of absolute time, $t' = t$, the result would be the Galilean transformation.

Exercise 2.3 Lorentz transformation for a general relative velocity

The Lorentz transformation given in (2.12) is the special case in which the relative velocity **v** *of the two frames is along the direction of the x axis. Namely, the coordinate system is chosen such that the x axis is parallel to* **v**. *For the case where* **v** *is in a general direction, show that the position transformation may be written as*

$$\mathbf{r}' = \mathbf{r} + (\gamma - 1)\frac{\mathbf{r} \cdot \mathbf{v}}{v^2}\mathbf{v} - \gamma \mathbf{v} t. \tag{2.16}$$

Hint: The position vector can always be decomposed into two components $\mathbf{r} = \mathbf{r}_\parallel + \mathbf{r}_\perp$, *with the parallel part being* $\mathbf{r}_\parallel = (\mathbf{r} \cdot \mathbf{v})\mathbf{v}/v^2$.

Transformation of coordinate derivatives

We next present the Lorentz transformation of the coordinate derivatives $(\partial_x, \partial_0) \to (\partial_x', \partial_0')$. For this purpose, we introduce the commonly used notation[6]

$$x^0 \equiv ct, \quad x^{0\prime} \equiv ct', \quad \partial_0 \equiv \frac{\partial}{\partial x^0} = \frac{1}{c}\frac{\partial}{\partial t}, \quad \partial_0' \equiv \frac{\partial}{\partial x^{0\prime}} = \frac{1}{c}\frac{\partial}{\partial t'}. \tag{2.17}$$

Since $\partial_\mu x^\nu = \delta_\mu^\nu$ (for $\mu, \nu = 0, 1, 2, 3$) and the Kronecker delta[7] is unchanged under a Lorentz transformation, the derivatives must transform **oppositely** to the coordinates themselves, i.e., as their inverse, obtained by a simple sign change of β in (2.13):

$$\begin{pmatrix} \partial_x' \\ \partial_0' \end{pmatrix} = \gamma \begin{pmatrix} 1 & \beta \\ \beta & 1 \end{pmatrix} \begin{pmatrix} \partial_x \\ \partial_0 \end{pmatrix}. \tag{2.18}$$

[6] As we shall explain in Section 3.2.1, a tensor with superscript index is a contravariant component, while one with subscript index is a covariant component. For spacetime vectors in particular, $x^0 = -x_0 = ct$, while $x^i = x_i$ with spatial indices $i = 1, 2, 3$.

[7] The Kronecker delta is defined to have the values $\delta_\mu^\nu = 1$ if $\mu = \nu$ and $\delta_\mu^\nu = 0$ if $\mu \neq \nu$.

Exercise 2.4 The transformation of coordinate differentials via the chain rule

Given the transformation for the space and time coordinates, find the Lorentz transformation for the coordinate derivatives (2.18) by the chain rule of differentiation:

$$\frac{\partial}{\partial x'} = \frac{\partial}{\partial x}\frac{\partial x}{\partial x'} + \frac{\partial}{\partial t}\frac{\partial t}{\partial x'},$$
$$\frac{\partial}{\partial t'} = \frac{\partial}{\partial x}\frac{\partial x}{\partial t'} + \frac{\partial}{\partial t}\frac{\partial t}{\partial t'}. \tag{2.19}$$

The partial derivatives $(\partial x/\partial x'$, *etc.) can be read off from the differential form of (2.13) by interpreting it also as a chain rule equation:*

$$dx = \frac{\partial x}{\partial x'}dx' + \frac{\partial x}{\partial t'}dt',$$
$$dt = \frac{\partial t}{\partial x'}dx' + \frac{\partial t}{\partial t'}dt'. \tag{2.20}$$

That is, the transformation is a matrix of partial derivatives

$$
\begin{pmatrix} \partial'_x \\ \partial'_0 \end{pmatrix} = \begin{pmatrix} \dfrac{\partial x}{\partial x'} & \dfrac{\partial x^0}{\partial x'} \\ \dfrac{\partial x}{\partial x'^0} & \dfrac{\partial x^0}{\partial x'^0} \end{pmatrix} \begin{pmatrix} \partial_x \\ \partial_0 \end{pmatrix}. \tag{2.21}
$$

2.2 Physics implications of Lorentz transformation

Just about all the novel physics implications of SR are contained in the new coordinate transformation among inertial frames. We first work out the effect on space and time measurements. We then discuss in Box 2.2 the change of the measured electric and magnetic fields, $(\mathbf{E}, \mathbf{B}) \longrightarrow (\mathbf{E}', \mathbf{B}')$, that follows from the Lorentz transformation of derivatives (2.18) and the Lorentz covariance of Maxwell's equations.

2.2.1 Time dilation and length contraction

We discuss here two important physical implications of the postulates of SR: time dilation and length contraction:

a moving clock appears to run slow;

a moving object appears to contract.

These physical features[8] underscore the profound change in our conception of space and time brought about by relativity. We must give up our belief that measurements of distance and time give the same results for all observers. Special relativity makes the strange claim that observers in relative motion will have different perceptions of distance and time. This means that two identical watches worn by two observers in relative motion will tick at different rates and will not agree on the amount of time that has elapsed between two given events. It is not that these two watches are defective; rather, it is a fundamental statement about the nature of time. For a more elaborate example, see the twin paradox discussed in Box 2.3.

Time dilation Let the O' frame be the rest frame of a clock. Because the clock has no displacement in this frame, $\Delta x' = 0$, we have, from (2.13),

$$
\Delta t = \gamma \, \Delta t'. \tag{2.22}
$$

Because $\gamma > 1$, $\Delta t > \Delta t'$. This is time dilation: a moving clock appears to run slow. Again it is useful to illustrate this with a concrete example. In Fig. 2.3, time

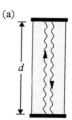

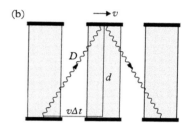

Figure 2.3 *Light-pulse clock with mirrors at the top and bottom of a vertical vacuum chamber: (a) at rest and (b) moving to the right with speed v. The comoving observer O' measures the time interval $\Delta t' = d/c$ as in (a). However, for the observer O with respect to whom the clock is moving with a velocity perpendicular to the light-pulse path, the light pulse traverses a diagonal distance D as in (b). This longer path requires a proportionally longer time interval $\Delta t = D/c = \sqrt{d^2 + v^2 \Delta t^2}/c$. Collecting Δt terms and using $\Delta t' = d/c$, we obtain the time dilation result $\Delta t = \gamma \Delta t'$.*

[8] It should be emphasized that in these expressions the word *appears* does not imply in any sense that these phenomena are illusionary. It simply means that a moving observer will find such results in measurements performed in her reference frame, compared with the measurements taken by another observer in the clock's (or object's) own frame of reference, its proper frame or rest frame.

[9] A basic clock rests on some physical phenomenon that has a direct connection to the underlying laws of physics. Different clocks—mechanical clocks, biological clocks, atomic clocks, or particle decays—simply represent different physical phenomena that can be used to mark time. They are all equivalent if their time intervals are related to the basic-clock intervals by correct physics relations. We note that while the familiar mechanical pendulum is a convenient basic clock for Newtonian physics, it is no longer so in relativity, because the dynamical equation for a pendulum must be modified to be compatible with SR. In short, time dilation holds for any clock, but it is easier to see in the case of a light clock as shown in Fig. 2.3. This explains why light signals are often used to demonstrate relativistic effects: the basic postulate of light speed's constancy can thus be easily incorporated.

[10] The setup is similar to that shown in Fig. 2.2, but with a clock fixed at a location on the rail platform.

[11] Operationally, an equivalent method is for the O' observer to read the same on-the-ground clock with the understanding that the clock is now in motion so she needs to translate the readings to her rest frame times by the time dilation formula.

[12] We often denote the rest frame as the O' system (be it a clock or some object whose length we are measuring) as in $\Delta t = \gamma \Delta t'$ or $\Delta x = \gamma^{-1} \Delta x'$. When using the results in (2.22) and (2.23), one must be certain which is the rest frame in the case being discussed and not blindly copy any written equation. For example, in the derivation here, we have $\Delta t' = \gamma \Delta t$ rather than the usual $\Delta t = \gamma \Delta t'$ as written in (2.22).

dilation shows up directly from the measurement by the most basic of clocks:[9] a light-pulse clock. It ticks away the time by having a light pulse bounce back and forth between two mirror ends separated by a fixed distance d.

Length contraction

The phenomenon of length contraction also follows directly from the Lorentz transformation. To measure the length of a moving object in the O frame, we can measure the two ends of the object simultaneously so $\Delta t = 0$; its measured length Δx is related to its rest-frame length $\Delta x'$ by (2.14):

$$\Delta x = \frac{\Delta x'}{\gamma}. \tag{2.23}$$

Because $1/\gamma < 1$, $\Delta x < \Delta x'$. This is the phenomenon of (longitudinal) length contraction of Lorentz and FitzGerald, (1.15).

Consider the specific example of length measurement of a moving railcar.[10] Let there be a clock attached to a fixed marker on the ground. A ground observer O, watching the train moving to the right with speed v, can measure the length L of the railcar by reading off the times when the front and back ends of the car pass this marker on the ground:

$$L = v(t_2 - t_1) \equiv v\Delta t. \tag{2.24}$$

An observer O' on the railcar can similarly deduce the length of the railcar in her reference frame from her own synchronized clocks located at the two ends of the railcar as they pass the marker on the ground,[11]

$$L' = v(t_2' - t_1') \equiv v\Delta t'. \tag{2.25}$$

These two time intervals in (2.24) and (2.25) are related by the above-considered time dilation (2.22): $\Delta t' = \gamma \Delta t$; the primes are flipped, because Δt is now the time recorded by a clock in its own rest frame, while $\Delta t'$ is the time recorded by an observer (with her own clocks) who sees the first clock in motion, so that events take place at two different (front and back) locations in her frame. From this, we immediately obtain

$$L = v\Delta t = \frac{v\Delta t'}{\gamma} = \frac{L'}{\gamma}, \tag{2.26}$$

which is the claimed result (2.23) of length contraction.[12]

We see from this example and from the derivation of (2.23) that length contraction is related to the relativity of simultaneity. This follows from the fact that in order to deduce the length of an object, one must make two separate measurements of the front and back ends of that object at specified times. Even the simplest way of making these two measurements (simultaneously) still requires, for observers in relative motion, a change of time, because simultaneity is relative. In fact, one finds that just about all of the counterintuitive results in SR are in one way or another ultimately related to the new conception of time. Thus one can conclude that the new conception of time is the key element of SR.

Exercise 2.5 Use a light-pulse clock to show length contraction

---

*In Fig. 2.3, we used a light-pulse clock to demonstrate the phenomenon of time dilation. This same clock can be used to demonstrate length contraction. Suppose the clock moves **parallel**, rather than perpendicular, to the bouncing light pulses. The length of the clock L' can be measured in the rest frame of the clock through the time interval $\Delta t'$ that it takes a light pulse to make the trip across the length of the clock and back: $c\left(\Delta t'_1 + \Delta t'_2\right) = c\Delta t' = 2L'$. In the moving reference frame, the length and time are measured to be L and Δt. A naive application of the time dilation formula $\Delta t = \gamma \Delta t'$ would suggest the incorrect effect of length elongation, $L = \gamma L'$. Demonstrate that a careful consideration of the light clock's operation in this setup does lead to the expected result $L = L'/\gamma$.*

Time dilation and length contraction appear counterintuitive to us, because our intuition is based on nonrelativistic physics ($v \ll c$), in which relativistic effects $O(v^2/c^2)$ are unobservably small to us. However, in high-energy particle physics experiments, accelerators push particles to speeds very close to the speed of light. As a result, particle physicists routinely encounter relativistic effects. Here we discuss such an example.

Example 2.1 The pion lifetime in the laboratory

In high-energy proton–proton collisions, a copious number of pions (a species of subatomic particles) are produced. Even though charged pions have a very short half-life $\tau_0 = 1.77 \times 10^{-8}$ s (they decay into a final state of a muon and a neutrino via the weak interaction), they can be collimated to form pion beams for other high-energy physics experiments. This is possible because the pions produced from proton–proton collisions have high kinetic energy and hence high velocity. Because of this high velocity, there is a special-relativistic time dilation effect. Consequently, the pions travel further, and their decay length is longer in the laboratory. For example, if a beam of pions has a speed of $0.99c$, what is the laboratory half-life of the pions? How far do they travel in that time? How far do the pions (or an observer in the pions' frame of reference) think they travel before half of them decay?

- **Time dilation in the laboratory frame:** The naive calculation of $\tau_0 c = (1.77 \times 10^{-8}\,\text{s}) \times (3 \times 10^8\,\text{m/s}) = 5.3\,\text{m}$ is incorrect, because the half-life $\tau_0 = 1.77 \times 10^{-8}$ s is the lifetime measured by a clock at rest with respect to the pion.

continued

Example 2.1 *continued*

The speed of $0.99c$ corresponds to a $\gamma = 7.1$. In the laboratory, the observer will see the pion decay time-dilated to $t = \gamma \tau_0 = 7.1 \times 1.77 \times 10^{-8}\,\text{s} = 1.26 \times 10^{-7}\,\text{s}$. Therefore, the decay length seen in the laboratory is

$$tv = \gamma \tau_0 v = 7.1 \times (1.77 \times 10^{-8}\,\text{s}) \times 0.99 \times (3 \times 10^8\,\text{m/s}) \simeq 38\,\text{m}.$$

- **Length contraction in the pion rest frame:** In the rest frame of the pion, this 38 m in the laboratory has a contracted length of 5.3 m, which, when divided by the particle speed of $0.99c$ yields its half-life of $\tau_0 = 1.77 \times 10^{-8}\,\text{s}$.

Exercise 2.6 Lorentz contraction of a moving sphere

A sphere of radius R is depicted as $x^2 + y^2 + z^2 = R^2$. *A moving observer O' with speed v will see this sphere as having the shape of an ellipsoid:*

$$\frac{x'^2}{X^2} + \frac{y'^2}{Y^2} + \frac{z'^2}{Z^2} = 1 \quad \text{with a volume} \quad V' = \frac{4\pi}{3}XYZ. \qquad (2.27)$$

How is this ellipsoidal volume related to the original spherical volume?

2.2.2   The invariant interval and proper time

In this new kinematics, time is no longer absolute (no longer an invariant under coordinate transformation). There is, however, a particular combination of space and time intervals[13] that is invariant:

$$ds^2 \equiv dx^2 + dy^2 + dz^2 - c^2\,dt^2. \qquad (2.28)$$

We will demonstrate this by an explicit calculation. Applying the Lorentz transformation (2.13) to (2.28) yields

$$ds'^2 = dx'^2 + dy'^2 + dz'^2 - c^2\,dt'^2$$

$$= \gamma^2 (dx - v\,dt)^2 + dy^2 + dz^2 - \gamma^2 c^2 \left(dt - \frac{v}{c^2}\,dx \right)^2 \qquad (2.29)$$

$$= \gamma^2 \left(1 - \frac{v^2}{c^2} \right) (dx^2 - c^2\,dt^2) + dy^2 + dz^2 = ds^2. \qquad (2.30)$$

What is the physical meaning of this interval? Why should one expect it to be an invariant? s is proportional to the time interval in the clock's rest frame, called the **proper time** (τ). In the rest frame, there is no displacement; hence $dx^2 + dy^2 + dz^2 = 0$, so we have

$$ds^2 = -c^2\,dt^2 \equiv -c^2\,d\tau^2. \qquad (2.31)$$

[13] Here we choose to display the infinitesimal invariant interval ds^2. In SR, we could just as well work with the finite interval $s'^2 = s^2$. However, the invariance of s holds only in SR—not in GR. Furthermore, in several applications to be discussed below, the infinitesimal form $ds'^2 = ds^2$ is directly relevant.

Since there is only one rest frame (hence only one proper time), all observers (i.e., all coordinate frames) can agree on this value. It must be an invariant.

We remember, of course, that the speed of light is a Lorentz invariant, and we should not be surprised by the equivalence of the statements "*c* is absolute" and "*s* is absolute." Namely, starting from either one of them, one can prove the other.

Exercise 2.7 Constancy of light speed in a general direction

In Chapter 1 (cf. (1.18)) we showed that if light travels in the same direction as the relative velocity of two observers, each observer sees the light propagate with the same speed, $u' = u = c$. Prove this result for a light pulse moving in an arbitrary direction. In principle, one can follow the same procedure and work out the three components $u_i' = dx_i'/dt'$ from the Lorentz transformation of the infinitesimal intervals and then show that for such a light pulse the magnitude of $\mathbf{u}$ is invariant. However, this approach involves a rather laborious calculation. Here you are asked to follow a much more efficient route by using the invariant interval ds^2, defined in (2.28), for your proof.

Exercise 2.8 *s* is absolute because *c* is absolute

The previous exercise showed that c is absolute because s is absolute. Here you are asked to prove the converse statement: the constancy of c leads to the invariance of s. Of course in (2.30), we already demonstrated this invariance by a direct application of the Lorentz transformation, which is based on c's constancy. You are now asked to demonstrate this directly without any detailed Lorentz transformation calculations.

Hint: From the vanishing invariant interval for light, $ds'^2 = ds^2 = 0$, and the fact that ds' and ds are infinitesimals of the same order, you can argue that the general intervals (not just for light) measured in two relative frames must be proportional to each other: $ds'^2 = P\,ds^2$, where P must be constant in space and time. From this, you can then show that the proportionality factor (which in principle can be velocity-dependent) must be the identity, $P = 1$, by considering three frames $O \xrightarrow{\mathbf{v}} O' \xrightarrow{-\mathbf{v}} O''$, where the symbols above the arrows indicate the relative velocities.

Box 2.2 The Lorentz transformation of electromagnetic fields

Since Maxwell's equations have the relativistic feature of a constant speed of light, we expect them to be covariant under Lorentz transformation. Einstein (1905a) used this expectation to derive the Lorentz transformation properties of electric and magnetic fields ($\mathbf{E}, \mathbf{B}$). From our basic knowledge of electromagnetism, we expect that these fields must change into each other when viewed in reference frames in relative motion. An electric charge at rest gives rise to an electric field, but no magnetic field. However, a moving observer sees a charge in motion, which produces both electric and

continued

Box 2.2 *continued*

magnetic fields. Different inertial observers will find different electric and magnetic field values, just as they would measure different position and time intervals.

How must the electromagnetic field change, $(\mathbf{E}, \mathbf{B}) \to (\mathbf{E}', \mathbf{B}')$, in order for the homogeneous Maxwell's equations to maintain the same form under a Lorentz transformation? In some inertial frame O, we have[14]

$$\mathbf{\nabla} \times \mathbf{E} + \frac{1}{c}\partial_t \mathbf{B} = 0, \qquad \mathbf{\nabla} \cdot \mathbf{B} = 0, \tag{2.32}$$

while in another frame O' moving in the $+x$ direction with speed v, we would still have

$$\mathbf{\nabla}' \times \mathbf{E}' + \frac{1}{c}\partial_t' \mathbf{B}' = 0, \qquad \mathbf{\nabla}' \cdot \mathbf{B}' = 0. \tag{2.33}$$

From this requirement, one can derive (part of) the transformation properties of $(\mathbf{E}, \mathbf{B})$. Let us start with the x component of Faraday's equation in the O' frame:

$$\partial_y' E_z' - \partial_z' E_y' + \partial_0' B_x' = 0. \tag{2.34}$$

Substituting in the Lorentz transformation of the derivatives (2.18), we then have

$$\partial_y E_z' - \partial_z E_y' + \gamma \partial_0 B_x' + \gamma\beta \partial_x B_x' = 0. \tag{2.35}$$

Likewise applying (2.18) to the no-monopole equation in (2.33) yields

$$\gamma \partial_x B_x' + \gamma\beta \partial_0 B_x' + \partial_y B_y' + \partial_z B_z' = 0. \tag{2.36}$$

Taking the linear combination (2.35)$-\beta$(2.36) and canceling the $\gamma\beta\partial_x B_x'$ terms, we get

$$\partial_y \left(E_z' - \beta B_y'\right) - \partial_z \left(E_y' + \beta B_z'\right) + \left(1 - \beta^2\right)\gamma \partial_0 B_x' = 0. \tag{2.37}$$

Multiplying this by a factor of γ and noting that $(1 - \beta^2)\gamma^2 = 1$, we have

$$\partial_y \left[\gamma \left(E_z' - \beta B_y'\right)\right] - \partial_z \left[\gamma \left(E_y' + \beta B_z'\right)\right] + \partial_0 B_x' = 0. \tag{2.38}$$

By comparing this equation with the x component of Faraday's equation (2.32) in the O frame, we can identify

$$B_x = B_x', \qquad E_y = \gamma \left(E_y' + \beta B_z'\right), \qquad E_z = \gamma \left(E_z' - \beta B_y'\right). \tag{2.39}$$

Starting with different components of Maxwell's equations, we can similarly show (see Exercise 2.9) that

$$E_x = E_x', \qquad B_y = \gamma \left(B_y' - \beta E_z'\right), \qquad B_z = \gamma \left(B_z' + \beta E_y'\right). \tag{2.40}$$

Here we have written fields in the O frame in terms of those in the O' frame; the inverse transformation is simply the same set of equations with a sign change for all the β factors.

[14] These equations in (2.32) are written in the Heaviside–Lorentz unit system. In this system, the measured parameter is the velocity of light c; in the more familiar SI system, $c = 1/\sqrt{\epsilon_0 \mu_0}$, with the measured permittivity ϵ_0 and the defined permeability μ_0 of free space. To go from SI to Heaviside–Lorentz units, one must scale the fields by $\sqrt{\epsilon_0} E_i \to E_i$, $\sqrt{1/\mu_0} B_i \to B_i$, and the charge and current densities by $\sqrt{1/\epsilon_0}(\rho, j_i) \to (\rho, j_i)$.

The field transformation rule displayed in (2.39) and (2.40) still seems mysterious. It appears quite different from that for the position and time coordinates. Why should it take this form? Is there a simple way to understand it? In Section 3.2.3 on Lorentz transformation of higher-rank tensors, this rule will indeed be explained.

Exercise 2.9 EM field transformation from the inhomogeneous Maxwell's equations

---

We derived the field transformation (2.39) by requiring the Lorentz covariance of the homogeneous Maxwell's equations (2.32). Now derive (2.40) from the inhomogeneous parts, Ampère's law and Gauss's law:

$$\nabla \times \mathbf{B} - \frac{1}{c}\partial_t \mathbf{E} = \frac{1}{c}\mathbf{j}, \qquad \nabla \cdot \mathbf{E} = \rho, \tag{2.41}$$

where $\mathbf{j}$ *is the electric current density and* ρ *is the charge density. For your proof, you will need to know that* $(\mathbf{j}, \rho)$ *have the same Lorentz transformation property as the space and time coordinates* $(\mathbf{r}, t)$.

2.3 Two counterintuitive scenarios as paradoxes

In Boxes 2.3 and 2.4, we present two worked examples illustrating the phenomena of time dilation and length contraction, as well as the issues of reciprocity of relativity and relativity of event order.

Box 2.3 The twin paradox

The twin paradox is a well-known reciprocity puzzle that sheds light on several basic concepts in relativity. We shall resolve the paradox by noting the involvement of noninertial frames.[15]

The paradox as a reciprocity puzzle

Two brothers, Al and Bill, have the same age. Al goes on a long journey in a spaceship at high speed; Bill stays at home. Al's biological clock appears to the stay-at-home Bill to run slow. When Al returns, he should be younger than Bill. Let us consider the case of Al traveling outward at speed $\beta = 4/5$ for 15 years, and then returning at the same speed $\beta = -4/5$ for 15 years.

[15] Here we follow the presentation by (Ellis and Williams 1988).

continued

Box 2.3 *continued*

These times are measured according to the clock in Al's rocketship. During both legs of the journey, the Lorentz factor is $\gamma = 5/3$. Thus, when Al and Bill meet again, Al should be younger. While Al has aged 30 years, Bill has aged $5/3 \times 30 = 50$ years according to SR time dilation. Letting the number of years Al has aged be A and the number of years Bill has aged be B, we have

$$A = 30, \quad B = 50 \qquad \text{(Bill's viewpoint)}. \qquad (2.42)$$

Of course, this SR prediction of asymmetric aging of the twins, while counterintuitive according to our low-velocity experience, is not paradoxical. It is just an example of time dilation, which is counterintuitive but real. However, there appears to be a reciprocity puzzle. If relativity is truly relative, we could just as well consider this separation and reunion from the viewpoint of Al, who sees Bill moving. According to Al, time dilation seems to imply that the stay-at-home Bill should be younger; while Al has aged 30 years, Bill should have aged only $(5/3)^{-1} \times 30$ years = 18 years.

$$A = 30, \quad B = 18 \qquad \text{(Al's viewpoint)}. \qquad (2.43)$$

Thus, Bill thinks he has aged 50 years, whereas Al thinks Bill should have aged only 18 years during the combined legs—a full 32-year difference in their perceptions. The twins cannot both be correct; they will meet again to compare their ages. Whose viewpoint is correct?

Checking theory by measurements

To answer this question, we can measure Bill's age as follows. Let the stay-at-home Bill celebrate his birthdays by setting off fireworks displays, which Al, with a powerful telescope aboard his spaceship, can always observe. Al can count the number of fireworks flashes he sees during his 30-year journey. If he sees 18 flashes, then Al's viewpoint is right; if 50, then Bill is right. During the outward-bound journey, Al sees flashes at an interval of $\Delta t_A^{(\text{out})}$, which differs from the interval $\Delta t_B = 1$ year at which Bill sets off fireworks. First of all, the time interval is dilated by a factor of γ. Moreover, because Al is moving away, the distance between the flashes increases by $v\Delta t_B$. Therefore, to reach Al in the spaceship, the light signals (i.e., the flashes) have to take an extra time interval of $v\Delta t_B/c$, which also has to be dilated by a factor of γ. Putting all this together,[16]

$$\Delta t_A^{(\text{out})} = \gamma(1 + \beta)\Delta t_B = \sqrt{\frac{1 + \beta}{1 - \beta}}\,\Delta t_B \qquad (2.44)$$

$$= \sqrt{\frac{1 + 4/5}{1 - 4/5}}\,(1 \text{ yr}) = 3 \text{ years}.$$

[16] Recall the relation $\gamma = (1 - \beta^2)^{-1/2}$. We chose to express the result in the form of a relativistic longitudinal Doppler shift, cf. (3.53), as the inverses of the time intervals are the produced and received firework frequencies.

Namely, during the 15-year outward-bound journey, Al sees Bill's birthday flashes every three years, for a total of 5 flashes. During the inward-bound journey, the relative velocity reverses sign ($\beta = -4/5$); Al sees flashes at an interval of

$$\Delta t_A^{(in)} = \sqrt{\frac{1-4/5}{1+4/5}}(1 \text{ yr}) = \frac{1}{3} \text{ years.} \qquad (2.45)$$

Thus, Al sees a total of 45 flashes on his 15-year inward-bound journey. Al sees a total of 50 flashes on his entire journey. This proves that Bill's viewpoint is correct. While the traveling twin Al has aged 30 years, the stay-at-home twin Bill has aged 50 years. But the following question remains: why is Bill's viewpoint correct, and not Al's?

The reciprocity puzzle is resolved when one realizes that while Bill remains an inertial frame of reference throughout Al's journey, Al does not— because Al must turn around! Thus Al's viewpoint cannot be represented by a single inertial frame of reference. This goes beyond SR, because noninertial frames are involved. Two different inertial frames must be used to describe Al's inbound and outbound legs. Moreover, the turnaround going from one inertial frame to the other requires acceleration.

Exercise 2.10 The twin paradox—the reciprocal measurement

We worked out the following measurement: the traveling Al, during his 30-year journey, sees 50 birthday fireworks set off by the stay-at-home Bill. You are now asked to work out explicitly the reciprocal measurement of Bill watching the birthday fireworks set off by Al. According to the time dilation calculation, during Bill's 50-year wait, he should observe 30 of Al's annual celebrations. This should be so, even though each twin sees the other's clock run slow.

Exercise 2.11 A twin-paradox puzzle resolved by the SR velocity addition rule

Just before the traveling twin (Al) turns around his rocketship (after traveling 15 years), his time dilation calculation tells him that the stay-at-home twin (Bill) has aged 9 years since his departure. Similarly, he also can conclude that on his 15-year inward-bound journey, a corresponding 9 years must elapse in Bills time. There seems to be a mismatch between this 18-year result (2.43) and Bill's 50 years that we confirmed by other means. How can we account for this missing 32 years?

Hint: The solution involves a comparison of clock readings just before and just after the turnaround point; to calculate this effect requires the SR velocity addition rule (1.18).

The relevant spacetime diagram will be discussed in Exercise 3.15.

Box 2.4 Relativity of event order: the pole-and-barn paradox

Imagine a runner rushing through a barn carrying a long pole horizontally. Let's say that the rest-frame lengths of the pole and barn are the same (L). From the perspective of the barn/ground observer, Fig. 2.4(a), the length contraction of the moving pole allows it be contained within the barn for a moment; the two ends of the pole, A and B, can be situated in-between the front (F) and rear (R) doors of the barn. On the other hand, from the perspective of the runner/pole, Fig. 2.4(b), the barn is moving toward the runner, so the Lorentz-contracted barn cannot possibly enclose the pole; the two ends of the pole would appear, for a moment, to protrude out from the front and rear doors of the barn. This appears to be another paradox. Does the pole fit in the barn or not? The barn observer believes he can simultaneously slam both doors and trap the pole inside. The runner believes this is impossible. Who is right?

We need to analyze carefully what we mean by the enclosure of the pole by the barn, or by the protrusion of the pole from the barn, in different coordinate frames. Let us list and precisely sequence the four distinct events that take place as the pole rushes past the barn: in the first event (call it AF), the front end of the pole A first reaches the front door of the barn F; in the last event (call it BR), the rear end of the pole B finally leaves the rear door of the barn R. Between these two occurrences, there are two other events: AR, in which the front end of the pole reaches the rear door, and BF, in which the back end of the pole leaves the front door of the barn. The temporal order of these in-between events, AR and BF, is different, depending on whether the barn completely contains the pole, or the pole protrudes out of the barn. In the first sequence, as seen by the ground/barn observer in Fig. 2.4(a), BF happens before AR; while in the second sequence, as seen by the runner in Fig. 2.4(b), AR happens before BF. Thus, we resolve the pole-and-barn paradox by noting that the temporal order of events can be changed in different coordinate frames. (Recall the prior example of relativity of event order mentioned in Sidenote 2.)

Let the event time for BF in the ground/barn's rest frame be $t_{BF} = 0$, and let the event AR take place an interval $\Delta t = t_{AR}$ later. Similarly, in the pole/runner's rest frame, let $t'_{BF} = 0$, and $t'_{AR} = \Delta t'$. Based on the above discussion, we

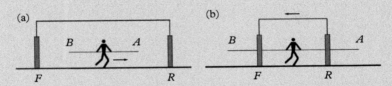

Figure 2.4 *A pole-carrying runner passes through a barn with a front door F and a rear door R. (a) The ground/barn observer sees the sequence of events as AF, BF, AR, BR. (b) The runner sees the sequence as AF, AR, BF, BR.*

expect Δt and $\Delta t'$ to have opposite signs. Since the two situations should be completely symmetric, they should have the same magnitude; hence $\Delta t = -\Delta t'$. In the O frame, where the pole can be completely contained inside the barn, the time interval for A to reach the rear door after the event of B passing the front door is just the time interval for the pole to cover the difference in length between of the barn and the (contracted) pole: $\Delta t = t_{AR} - t_{BF} = (L/v)(1 - \gamma^{-1})$.

We can plug this into the Lorentz transformation, (2.12) with $\Delta x = L$, in order to verify the relation between Δt and $\Delta t'$:

$$\Delta t' = \gamma \left(\Delta t - vL/c^2\right) = \gamma \left(L/v\right)\left[\left(1 - \gamma^{-1}\right) - v^2/c^2\right] = -(L/v)\left(1 - \gamma^{-1}\right) = -\Delta t. \tag{2.46}$$

Indeed, these two time measurements are equal and opposite, as expected.

So the barn observer may well slam the doors simultaneously at time $t_{\text{slam } FR}$ around the pole at a time after the tail end of pole passes through the front door BF but before the tip reaches the rear door AR; thus $t_{BF} < t_{AR}$. From the barn's viewpoint,

$$t_{AF} < t_{BF} < t_{\text{slam } FR} < t_{AR*} < t_{BR*}, \tag{2.47}$$

where t_{AR*} and t_{BR*} denote the times when the A and B ends of the pole passing (smashing $*$) through the closed rear barn door. But from the runner's viewpoint, $t'_{AR} < t'_{BF}$; by the time the B end reaches the front door, the A end of the pole has already smashed through the rear door. The pole can never be shut inside the barn. How do we account for the simultaneous slamming of doors observed in the barn's frame and the A and B ends smashing through the closed rear barn door at times t_{AR*} and t_{BR*} while the front doors must remain open when they pass through it? There is no contradiction, because simultaneity is relative; the runner will see the slamming of the front and back doors as two nonsimultaneous events: $t'_{\text{slam } R} < t'_{\text{slam } F}$. From the runner's viewpoint,

$$t'_{AF} < t'_{\text{slam } R} < t'_{AR*} < t'_{BF} < t'_{\text{slam } F} < t'_{BR*}. \tag{2.48}$$

Review questions

1. What does it mean that simultaneity is relative? Besides the familiar relativity of spatial equilocality, SR teaches us to expect the relativity of temporal simultaneity. What is the general lesson we can learn from this?

2. What are the two postulates of SR?

3. Two inertial frames are moving with respect to each other with velocity $\mathbf{v} = v\hat{\mathbf{x}}$. Write out the Lorentz transformation of the coordinates $(t, x, y, z) \longrightarrow (t', x', y', z')$. Show that in the low-velocity limit it reduces to the Galilean transformation. Write the inverse transformation $(t', x', y', z') \longrightarrow (t, x, y, z)$, as well as the transformation of the space and time derivatives $(\partial_t, \partial_x, \partial_y, \partial_z) \longrightarrow (\partial'_t, \partial'_x, \partial'_y, \partial'_z)$.

4. From the Lorentz transformation, perform the algebraic derivation of the time dilation and length contraction results. Pay particular attention to the input conditions assumed in these derivations.

5. Use the SR postulates and a simple light clock to derive the physical result of time dilation, $\Delta t = \gamma \Delta t'$. Consider a physical situation that allows you to deduce the result of length contraction, $\Delta x = \gamma^{-1} \Delta x'$, of an object in frame O' when viewed by an observer in frame O in relative motion. Do you need to use the physics of time dilation for this derivation of length contraction? Why?

6. To study the new kinematics of SR (clock synchronization, time dilation as illustrated by a light clock, etc.), one often employs light signals. Why is this so?

7. What combination of space and time intervals is invariant under Lorentz transformation? What is the physical significance of the invariant interval between two events? From this invariance, demonstrate that the speed of light is the same in all inertial frames regardless of the direction of propagation.

8. The Lorentz covariance of Maxwell's equations requires that electric and magnetic fields transform into each other under Lorentz transformation. Use the familiar laws of electromagnetism to explain why a static electric field due to an electric charge is seen by a moving observer to have a magnetic field component.

Special Relativity: Flat Spacetime

3

- The geometric description of SR interprets the relativistic invariant interval as the length in 4D Minkowski spacetime. The Lorentz transformation is a rotation in this pseudo-Euclidean manifold, whose metric is $\eta_{\mu\nu} = \text{diag}(-1, 1, 1, 1)$.

- Just about all the relativistic effects, such as time dilation, length contraction, and relativity of simultaneity, follow from the Lorentz transformation. Thus, the geometric formulation allows us to think of the metric $\eta_{\mu\nu}$ as embodying all of SR.

- A metric in general is a matrix whose diagonal elements are the magnitudes of the basis vectors and whose off-diagonal elements show their deviation from orthogonality. The metric of a space defines its inner product.

- Tensors in Minkowski space are quantities having definite transformation properties under Lorentz transformation. If a physics equation can be written in tensor form (as discussed in Chapter 1), it automatically respects the relativity principle.

- Tensors with no indices (scalars) are invariant under Lorentz transformation. They include the rest mass m of a particle, the proper time τ, and the speed of light c.

- Tensors with one index (4-vectors) include the position 4-vector $x^\mu = (ct, \mathbf{x})$, with components of time and 3D position, and the momentum 4-vector $p^\mu = (E/c, \mathbf{p}) = \gamma m(c, \mathbf{v})$, with components of relativistic energy $E = \gamma mc^2$ and 3D momentum $\mathbf{p} = \gamma m\mathbf{v}$. From the component expression of p^μ, we obtain the well-known energy–momentum relation $E^2 = p^2 c^2 + m^2 c^4$ and the conclusion that a massless particle ($E = pc$) always travels at speed c.

- Tensors with two indices (4-tensors of rank 2) include the symmetric energy–momentum–stress tensor $T_{\mu\nu} = T_{\nu\mu}$ (the source term for the relativistic gravitational field) and the antisymmetric electromagnetic field tensor $F_{\mu\nu} = -F_{\nu\mu}$, whose six components are the familiar electric and magnetic fields, $(\mathbf{E}, \mathbf{B})$. In terms of the field tensor, Maxwell's equations can be written compactly in tensor form, making their Lorentz covariance manifest.

A College Course on Relativity and Cosmology. First Edition. Ta-Pei Cheng.
© Ta-Pei Cheng 2015. Published in 2015 by Oxford University Press.

- The spacetime diagram is particularly useful in understanding the causal structure of relativity theory. It allows the simple visualization of various SR effects such as relativity of simultaneity and event order.
- In Box 3.6 we summarize various ideas and implications related to the spacetime formulation of SR. Such a description is indispensable for the eventual formulation of general relativity, a field theory in which curved spacetime is identified as the gravitational field.

3.1 Geometric formulation of relativity

Building on the prior work of Lorentz and Poincaré, Herman Minkowski, the erstwhile mathematics professor of Einstein at the ETH in Zürich, proposed in 1907 a geometric formulation of Einstein's SR. Minkowski concentrated on the invariance of the theory, emphasizing that the essence of SR is the proposition that time should be treated on equal footing with space. He highlighted this symmetry by uniting space and time in a single mathematical structure, now called **Minkowski spacetime**. The following are the opening words of an address he delivered at the 1908 Assembly of German National Scientists and Physicians held in Cologne:

> *The views of space and time which I wish to lay before you have sprung from the soil of experimental physics, and therein lies their strength. They are radical. Henceforth space by itself, and time by itself, are doomed to fade away into mere shadows, and only a kind of union of the two will preserve an independent reality.*

We treat position coordinates in three directions (x, y, z) on an equal footing by regarding them as the three components of a vector. They can be transformed into each other by rotations in a 3D space. Thus we can similarly express the symmetry between space and time by regarding time as the fourth coordinate of a 4D spacetime manifold (x, y, z, t). The different components can be transformed into each other by rotations in this 4D spacetime (i.e., Lorentz transformations). If physics equations are written as tensor equations in this spacetime, they are manifestly covariant under Lorentz transformation and are thus automatically relativistic. Einstein was not initially impressed by this new formulation, calling it "superfluous learnedness." He started using it only in 1912 for his geometric theory of gravity, in which he identified gravity as the structure of spacetime. Namely, Einstein adopted this geometric language and extended it—from flat Minkowski space to curved ones.

Minkowski spacetime

In Chapter 2, we showed that the Lorentz transformation leaves invariant the spacetime interval defined in (2.28):

$$s^2 = -c^2 t^2 + x^2 + y^2 + z^2. \qquad (3.1)$$

Minkowski pointed out that this may be viewed as the analog in a 4D pseudo-Euclidean space of 3D (squared) length. The Lorentz transformation is the general class of length-preserving transformations (rotations) in this space; see Box 3.1.

Box 3.1 Lorentz transformation as a rotation in 4D spacetime

The constancy of light speed requires[1] the invariance of the spacetime interval s^2 of (3.1). With the identification of s^2 as a (squared) length or magnitude, the Lorentz transformation is simply a rotation in spacetime. We can sharpen the analogy between Minkowski and Euclidean spaces by introducing an imaginary time coordinate, $w \equiv ict$. Concentrating on rotations in the subspace spanned by this time coordinate and one of the space coordinates (x), the interval $s^2 = w^2 + x^2$ (expressed as an ordinary Euclidean length) is invariant under ordinary rotation,[2] cf. (1.5), which implies

[1] See Exercise 2.8.

$$\begin{pmatrix} w' \\ x' \end{pmatrix} = \begin{pmatrix} \cos\theta & \sin\theta \\ -\sin\theta & \cos\theta \end{pmatrix} \begin{pmatrix} w \\ x \end{pmatrix}. \qquad (3.2)$$

[2] Here we ignore time and spatial parity flips to focus on proper Lorentz transformations.

Putting back $w = ict$, we have

$$ct' = \cos\theta\, ct - i\sin\theta\, x;$$
$$x' = -i\sin\theta\, ct + \cos\theta\, x. \qquad (3.3)$$

After re-parameterizing the rotation angle $\theta = -i\psi$ and using the identities[3] $\cos(i\psi) = \cosh\psi$ and $i\sin(-i\psi) = \sinh\psi$, we recognize (3.3) as the usual Lorentz transformation (2.13):

[3] Recall Euler's formula: $e^{i\theta} = \cos\theta + i\sin\theta$ and thus $\cos\theta = (e^{i\theta} + e^{-i\theta})/2$. By direct substitution, we have $\cos(i\psi) = (e^{-\psi} + e^{+\psi})/2 \equiv \cosh\psi$, etc.

$$\begin{pmatrix} ct' \\ x' \end{pmatrix} = \begin{pmatrix} \cosh\psi & -\sinh\psi \\ -\sinh\psi & \cosh\psi \end{pmatrix} \begin{pmatrix} ct \\ x \end{pmatrix} = \begin{pmatrix} \gamma & -\beta\gamma \\ -\beta\gamma & \gamma \end{pmatrix} \begin{pmatrix} ct \\ x \end{pmatrix}. \qquad (3.4)$$

To reach the last equality, we have used

$$\cosh\psi = \gamma \quad \text{and} \quad \sinh\psi = \beta\cosh\psi, \qquad (3.5)$$

with the standard notation

$$\beta = \frac{v}{c} \quad \text{and} \quad \gamma = \frac{1}{\sqrt{1-\beta^2}}. \qquad (3.6)$$

continued

Box 3.1 *continued*

The relations (3.5) between the rapidity[4] ψ and the relative velocity v can be derived by considering the motion of the $x' = 0$ origin in the O system. Plugging $x' = 0$ into the first equation in (3.4),

$$x' = 0 = -ct \sinh \psi + x \cosh \psi \quad \text{so} \quad \frac{x}{ct} = \tanh \psi. \qquad (3.7)$$

The $x' = 0$ coordinate origin moves with velocity $v = x/t$ along the x axis of the O system; thus $\beta = \tanh \psi$. From the identity $\cosh^2 \psi - \sinh^2 \psi = 1$, which may be written as $\cosh \psi \sqrt{1 - \tanh^2 \psi} = 1$, we find the first relation in (3.5).

3.2 Tensors in special relativity

As we discussed in Chapter 1, the simplest way to implement a coordinate symmetry (e.g., rotational symmetry) is to write physics equations in tensor form. Tensors in relativity are similar to the familiar tensors in 3D Euclidean space, except that they reside in a pseudo-Euclidean 4D Minkowski space. In order to extend the notion of tensors, we need to introduce generalized coordinates.

3.2.1 Generalized coordinates: bases and the metric

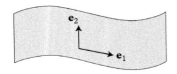

Figure 3.1 *Basis vectors for a 2D surface.*

To set up a coordinate system for 4D Minkowski space, we first need to choose a set of four basis vectors $\{\mathbf{e}_\mu\}$, where $\mu = 0, 1, 2, 3$. Each $\mathbf{e}_\mu$, for a definite index value μ, is a 4D vector (Fig. 3.1). In contrast to the Cartesian coordinate system in Euclidean space, this is not an orthonormal set: $\mathbf{e}_\mu \cdot \mathbf{e}_\nu \neq \delta_{\mu\nu}$. Nevertheless, we can represent the collection of scalar products of the basis vectors as a symmetric matrix, called the **metric**[5] or the **metric tensor**:

$$\mathbf{e}_\mu \cdot \mathbf{e}_\nu \equiv g_{\mu\nu}. \qquad (3.8)$$

We can display the metric of a 4D space as a 4×4 matrix whose elements are the dot products of the basis vectors:

$$[g] = \begin{pmatrix} g_{00} & g_{01} & \cdots \\ g_{10} & g_{11} & \cdots \\ \vdots & \vdots & \end{pmatrix} = \begin{pmatrix} \mathbf{e}_0 \cdot \mathbf{e}_0 & \mathbf{e}_0 \cdot \mathbf{e}_1 & \cdots \\ \mathbf{e}_1 \cdot \mathbf{e}_0 & \mathbf{e}_1 \cdot \mathbf{e}_1 & \cdots \\ \vdots & \vdots & \end{pmatrix}. \qquad (3.9)$$

Thus the diagonal elements are the (squared) magnitudes of the basis vectors, $|\mathbf{e}_0|^2, |\mathbf{e}_1|^2$, etc., while the off-diagonal elements represent their deviations from orthogonality. Any set of mutually perpendicular bases would be represented by a diagonal metric matrix.

The inverse basis vectors and the inverse metric   In Cartesian coordinate space, each basis vector is its own inverse: $\mathbf{e}_\mu \cdot \mathbf{e}_\nu = \delta_{\mu\nu}$. Namely, multiplying each basis vector by itself yields the identity, so the metric for Cartesian space is simply the identity matrix $[g] = [\mathbb{I}]$. Minkowski spacetime is non-Cartesian, so we must introduce a distinct set of *inverse basis vectors*, written as $\{\mathbf{e}^\mu\}$ with a superscript index, so that

$$\mathbf{e}_\mu \cdot \mathbf{e}^\nu = [\mathbb{I}]_\mu{}^\nu. \qquad (3.10)$$

Just as in (3.8), we can define the inverse metric tensor $g^{\mu\nu}$:

$$\mathbf{e}^\mu \cdot \mathbf{e}^\nu \equiv g^{\mu\nu} \quad \text{so that} \quad g_{\mu\nu}g^{\nu\lambda} = [\mathbb{I}]_\mu{}^\lambda, \qquad (3.11)$$

in which the last equality (where summation of the repeated ν indices is understood) follows from the direct product completeness relation $\mathbf{e}_\nu \otimes \mathbf{e}^\nu = \mathbb{I}$ and the orthogonality of the basis vectors (3.10).

Contravariant and covariant components

Because there are two sets of coordinate basis vectors, $\{\mathbf{e}_\mu\}$ and $\{\mathbf{e}^\mu\}$, there are two possible expansions for each vector $\mathbf{A}$:

Expansion of $\mathbf{A}$	Projections	Component names	
$\mathbf{A} = A^\mu \mathbf{e}_\mu$	$A^\mu = \mathbf{A} \cdot \mathbf{e}^\mu$	Contravariant components of $\mathbf{A}$	(3.12)
$\mathbf{A} = A_\mu \mathbf{e}^\mu$	$A_\mu = \mathbf{A} \cdot \mathbf{e}_\mu$	Covariant components of $\mathbf{A}$	

In (3.11) and (3.12), we have used the Einstein summation convention: when there is a matched pair of superscript and subscript indices (often called dummy indices), they are assumed to be summed over. Namely, we omit the summation sign (Σ):

$$A^\mu \mathbf{e}_\mu \equiv \sum_\mu A^\mu \mathbf{e}_\mu. \qquad (3.13)$$

The scalar product of any two vectors in terms of either contravariant or covariant components alone involves the metric matrices:

$$\mathbf{A} \cdot \mathbf{B} = g_{\mu\nu}A^\mu B^\nu = g^{\mu\nu}A_\mu B_\nu. \qquad (3.14)$$

One of the advantages of introducing these two types of tensor components is that when one uses both types, products are simplified; they can be written without the metrics:

$$\mathbf{A} \cdot \mathbf{B} = (A_\nu \mathbf{e}^\nu) \cdot (B^\mu \mathbf{e}_\mu) = A_\nu(\mathbf{e}^\nu \cdot \mathbf{e}_\mu)B^\mu = A_\mu B^\mu. \qquad (3.15)$$

The summation of a pair of superscript and subscript indices as in (3.13) is called a **contraction**. Because the summed dummy indices vanish, the total rank (number of indices) of the contracted tensor expression is reduced by two.[6] A comparison of (3.15) with (3.14) shows that the contravariant and covariant components of a vector are related as follows:

$$A_\mu = g_{\mu\nu}A^\nu, \qquad A^\mu = g^{\mu\nu}A_\nu. \tag{3.16}$$

We say that tensor indices can be lowered or raised through contractions with the metric tensor or inverse metric tensor.

Exercise 3.1 Product of symmetric and antisymmetric tensors

Use the fact that dummy indices can be freely relabeled to prove that the contraction of a symmetric tensor ($S_{\mu\nu} = +S_{\nu\mu}$) and an antisymmetric tensor ($A^{\mu\nu} = -A^{\nu\mu}$) vanishes: $S_{\mu\nu}A^{\mu\nu} = 0$.

Minkowski spacetime is pseudo-Euclidean

Consider a 4D space whose position coordinates are the contravariant components x^μ with the Greek index[7] $\mu \in \{0, 1, 2, 3\}$ so that

$$x^\mu = (x^0, x^1, x^2, x^3) = (ct, x, y, z). \tag{3.17}$$

$\{x^\mu\}$ are components of a 4-vector, as they satisfy the transformation property (3.4):

$$x'^\mu = [L]^\mu{}_\nu x^\nu, \qquad \text{with} \qquad [L]^\mu{}_\nu = \begin{pmatrix} \gamma & -\beta\gamma & 0 & 0 \\ -\beta\gamma & \gamma & 0 & 0 \\ 0 & 0 & 1 & 0 \\ 0 & 0 & 0 & 1 \end{pmatrix}. \tag{3.18}$$

The scalar product of a differential displacement with itself as in (3.14) and (3.15) is just its squared magnitude, the invariant interval:

$$g_{\mu\nu}\, dx^\mu\, dx^\nu = ds^2. \tag{3.19}$$

This equation relating the interval (ds^2) to the coordinates $\{dx^\mu\}$ is often taken as another (equivalent) definition of the metric. A simple comparison of (3.19) with the infinitesimal version of (3.1) leads to the identification of the metric elements for the Minkowski spacetime as

$$g_{\mu\nu} = \begin{pmatrix} -1 & 0 & 0 & 0 \\ 0 & 1 & 0 & 0 \\ 0 & 0 & 1 & 0 \\ 0 & 0 & 0 & 1 \end{pmatrix} \equiv \mathrm{diag}(-1, 1, 1, 1). \tag{3.20}$$

It can be shown that the geometry of a manifold is determined by its metric; geometric properties (relations between lengths, angles, and shapes) of the space are fixed by the metric.[8] The metric elements themselves can in turn be determined by length measurements, cf. (5.6). The metric (3.20) tells us that Minkowski spacetime is flat, because its elements are coordinate-independent; it is pseudo-Euclidean, since it differs from Euclidean space[9] by a sign change of the g_{00} element. This particular pseudo-Euclidean metric is often denoted by the specific symbol $\eta_{\mu\nu} \equiv \operatorname{diag}(-1,1,1,1)$. Its elements are obviously equal to those of its inverse: $\eta^{\mu\nu} \equiv \operatorname{diag}(-1,1,1,1)$, so $\eta^{\mu\nu}\eta_{\nu\lambda} = [\mathbb{I}]^\mu_\lambda$. Of course, this is not generally the case for metrics in curved spaces or even in other coordinate systems in flat space.

Position and position derivatives We have chosen in (3.17) to make the position elements contravariant components (with an upper index) of a vector. By lowering the index as in (3.16), we see that the covariant version of the position vector is $x_\mu = (-ct, x, y, z)$. While the position vector is naturally contravariant (i.e., there are no unnatural minus signs as on the covariant time coordinate), the closely related (4-) del operator is naturally covariant:

$$\partial_\mu \equiv \frac{\partial}{\partial x^\mu} = \left(\frac{1}{c}\frac{\partial}{\partial t} \quad \frac{\partial}{\partial x} \quad \frac{\partial}{\partial y} \quad \frac{\partial}{\partial z} \right), \tag{3.21}$$

so that $\partial_\mu x^\nu = [\mathbb{I}]^\nu_\mu$.

Because the contraction of a contravariant vector with a covariant one is a Lorentz scalar,[10] the contravariant and covariant vectors must transform oppositely (i.e., by a transformation matrix and its inverse, respectively). In particular, we have $dx'^\mu = [L]^\mu_\nu dx^\nu$, i.e.,

$$\begin{pmatrix} dx'_0 \\ dx'_1 \end{pmatrix} = \begin{pmatrix} \gamma & -\beta\gamma \\ -\beta\gamma & \gamma \end{pmatrix} \begin{pmatrix} dx_0 \\ dx_1 \end{pmatrix}, \tag{3.23}$$

whereas $\partial'_\mu = [L^{-1}]^\nu_\mu \partial_\nu$, i.e.,

$$\begin{pmatrix} \partial'_0 \\ \partial'_1 \end{pmatrix} = \begin{pmatrix} \gamma & \beta\gamma \\ \beta\gamma & \gamma \end{pmatrix} \begin{pmatrix} \partial_0 \\ \partial_1 \end{pmatrix}. \tag{3.24}$$

In general,[11] any contravariant vector A^μ transforms just like dx^μ, so

$$A'^\mu = [L]^\mu_\nu A^\nu, \tag{3.25}$$

while any covariant vector B_μ transforms just like ∂_μ, so

$$B'_\mu = [L^{-1}]^\nu_\mu B_\nu. \tag{3.26}$$

[8] For further discussion of the metric, see Section 5.1. Curved vs. flat spaces will be discussed in Section 6.1.

[9] The metric for a 4D Euclidean space with Cartesian coordinates is simply the 4D identity matrix, $\delta_{\mu\nu} \equiv \operatorname{diag}(1,1,1,1)$. The number of positive and negative elements of a diagonalized metric is its signature.

[10] Also, the contraction of the del operator with itself, just as in (3.19), is the Lorentz-invariant 4-Laplacian operator (called the D'Alembertian):

$$\square \equiv \partial^\mu \partial_\mu = -\frac{1}{c^2}\frac{\partial^2}{\partial t^2} + \nabla^2, \tag{3.22}$$

where the 3-Laplacian operator is $\nabla^2 = \partial^2/\partial x^2 + \partial^2/\partial y^2 + \partial^2/\partial z^2$. We can thereby write the relativistic wave equation as $\square\psi = -c^{-2}\partial^2\psi/\partial t^2 + \nabla^2\psi = 0$.

[11] In principle we should always refer the transformation of a tensor's components. But we shall adopt a simpler way by calling tensor components a tensor. Thus, by contravariant vector, we mean the contravariant components of a vector.

For tensors of higher rank, the transformation follows the same rule. For instance, a tensor with one contravariant and two covariant indices $T^{\mu}_{\nu\lambda}$ transforms like the direct product $A^{\mu}B_{\nu}C_{\lambda}$; hence

$$T'^{\mu}_{\nu\lambda} = [L]^{\mu}_{\alpha}[L^{-1}]^{\beta}_{\nu}[L^{-1}]^{\gamma}_{\lambda}T^{\alpha}_{\beta\gamma}. \tag{3.27}$$

That is, for every superscript index, there is a factor of $[L]$, and for every subscript index an $[L^{-1}]$. As in all tensor equations, free indices are matched on both sides and dummies contracted.

Exercise 3.2 The quotient theorem
_____

The quotient theorem states that in a tensor equation such as $A_{\mu\nu} = C_{\mu\lambda}B^{\lambda}_{\nu}$, if we know that $A_{\mu\nu}$ and B^{λ}_{ν} are tensors, then $C_{\mu\lambda}$ is also a tensor. Prove this.

Exercise 3.3 Lorentz invariance of the Minkowski metric
_____

In SR, the flat spacetime metric is invariant under Lorentz transformation: $g_{\mu\nu} = \eta_{\mu\nu} = g'_{\mu\nu}$. Show that this condition reduces to the orthogonality condition on the transformation matrix: $[L^{-1}] = [L]^{\top}$ in the Euclidean space. This is the usual length preserving orthogonality condition on a rotation matrix.

3.2.2   Velocity and momentum 4-vectors

Once we have identified the Lorentz transformation as a rotation in spacetime, all our knowledge of rotational symmetry can be applied to this coordinate symmetry of relativity. In particular, 4D spacetime tensor equations are automatically covariant under the $[L]$ transformation, and hence relativistic. This motivates our search for the 4-tensor expressions of physical quantities. Having already presented 4-vectors for position and position derivatives, we now seek 4-vectors for velocity and momentum.

4-velocity

The position 4-vector x^{μ} is given by (3.17). Its derivative with respect to the coordinate time, dx^{μ}/dt, is not a 4-vector, because t is not a Lorentz scalar. The derivative of x^{μ} that transforms as a 4-vector is the one with respect to scalar proper time τ. Proper time is directly related to the invariant spacetime interval $ds^2 = -c^2\,d\tau^2$, as shown in (2.31). Thus[12]

[12] Please keep in mind that $\dot{x}$ denotes differentiation with respect to proper time τ rather than coordinate time t.

$$\dot{x}^{\mu} \equiv \frac{dx^{\mu}}{d\tau} = \gamma\frac{dx^{\mu}}{dt} = \gamma\,(c, v_x, v_y, v_z), \tag{3.28}$$

where we have used the time dilation relation $dt = \gamma \, d\tau$ and $v_x = dx/dt$, etc. We can form its (squared) length by taking its scalar product with itself, (3.15):

$$\dot{x}^\mu \dot{x}_\mu = \gamma^2(-c^2 + v^2) = -c^2, \qquad (3.29)$$

which is invariant, as it should be.

Exercise 3.4 4-velocity transformation leads to the SR addition
rule for 3-velocities

With its transverse components suppressed, the 4-velocity of a particle is

$$\dot{x}^\mu = \gamma_u \begin{pmatrix} c \\ u \end{pmatrix}, \quad \text{with } u = \frac{dx}{dt} \text{ and } \gamma_u = \left(1 - \frac{u^2}{c^2}\right)^{-1/2}. \qquad (3.30)$$

As a 4-vector, $\dot{x}^\mu$ has the following boost (of velocity v parallel to u) transformation:

$$\gamma_u' \begin{pmatrix} c \\ u' \end{pmatrix} = \gamma_v \begin{pmatrix} 1 & -v/c \\ -v/c & 1 \end{pmatrix} \gamma_u \begin{pmatrix} c \\ u \end{pmatrix}. \qquad (3.31)$$

From this, derive the transformation of the gamma function,

$$\gamma_u' = \gamma_v \left(1 - \frac{uv}{c^2}\right) \gamma_u, \qquad (3.32)$$

as well as the velocity addition rule (1.18).

Exercise 3.5 Lorentz group property and SR velocity addition rule

---

(a) *Show that the Lorentz boost transformation, $[L(\psi)]$ with $\psi = \tanh^{-1} \beta$ for 1D space as displayed in (3.5), has the (mathematical) group property of*

$$[L(\psi_1)][L(\psi_2)] = [L(\psi_1 + \psi_2)]. \qquad (3.33)$$

(b) *From this group property, derive the velocity addition rule (1.18).*

Momentum 4-vector

We will naturally define the relativistic 4-momentum as the product of scalar mass and 4-velocity (3.28):

$$p^\mu \equiv m\dot{x}^\mu = \gamma m(c, \mathbf{v}), \qquad (3.34)$$

The spatial components of the relativistic 4-momentum p^μ are the components of the relativistic 3-momentum, $p^i = \gamma m v^i$, which reduces to $m v^i$ in the non-relativistic limit $\gamma = 1$. What then is the zeroth (timelike) component of the 4-momentum? Let us take its nonrelativistic limit ($v \ll c$):

$$p^0 = \gamma mc = mc\left(1 - \frac{v^2}{c^2}\right)^{-1/2}$$

$$\xrightarrow{\text{NR}} mc\left(1 + \frac{1}{2}\frac{v^2}{c^2} + \ldots\right) = \frac{1}{c}\left(mc^2 + \frac{1}{2}mv^2 + \ldots\right). \qquad (3.35)$$

The presence of the kinetic energy term $\frac{1}{2}mv^2$ in the nonrelativistic limit suggests that we interpret cp^0 as the relativistic energy E:

$$p^\mu = \left(\frac{E}{c}, p^i\right). \qquad (3.36)$$

The relativistic energy, $E = cp^0 = \gamma mc^2$, has a nonvanishing value mc^2, even when the particle is at rest ($\gamma = 1$).

Thus, for a particle with mass[13] the components of its relativistic 3-momentum and relativistic energy are

$$\mathbf{p} = \gamma m \mathbf{v} \quad \text{and} \quad E = \gamma mc^2. \qquad (3.37)$$

[13] The concept of a velocity-dependent mass $m^* \equiv \gamma m$ is sometimes used in the literature, so that $\mathbf{p} = m^*\mathbf{v}$ and $E = m^* c^2$. In our discussion, we will avoid this usage and restrict ourselves only to the Lorentz scalar mass m, which is equal to m^* in the rest frame of the particle $\left(m^*|_{v=0} = m\right)$; hence m is called the rest mass.

The ratio of a particle's momentum to its energy can be expressed as that of its velocity to c^2:

$$\frac{\mathbf{p}}{E} = \frac{\mathbf{v}}{c^2}. \qquad (3.38)$$

The momentum and energy transform into each other under the Lorentz transformation just as space and time do. According to (3.29) and (3.34), the invariant square of the 4-momentum must be $p_\mu p^\mu = -(mc)^2$. Plugging in (3.36), we obtain the important relativistic energy–momentum relation:

$$E^2 = (mc^2)^2 + (\mathbf{p}c)^2 = m^2 c^4 + \mathbf{p}^2 c^2. \qquad (3.39)$$

From (3.34) and (3.36), a particle's momentum and energy per unit mass can be written directly in terms of derivatives with respect to proper time:

$$\frac{\mathbf{p}}{m} = \dot{\mathbf{r}}, \qquad \frac{E}{mc} = c\dot{t}. \qquad (3.40)$$

These expressions will be useful when we discuss the SR limit of particle energy in general relativity when spacetime is curved.

Example 3.1 Relativistic momentum conservation

We have introduced the relativistic momentum through its tensor transformation property. Another approach is to show that momentum conservation in a collision process requires such an expression. Namely, it is easy to see that the conservation relation shown in (1.12) no longer holds in SR because it is not Lorentz-covariant. Here we shall demonstrate that for the 1D collision process of Exercise 1.3, the conservation law holds only for relativistic 3-momenta (γmu):

$$\gamma_A m_A u_A + \gamma_B m_B u_B = \gamma_C m_C u_C + \gamma_D m_D u_D, \tag{3.41}$$

with $\gamma_A = (1 - u_A^2/c^2)^{-1/2}$, etc., provided that relativistic energy (γmc^2) is also conserved. Our task is to show that if (3.41) holds in one frame, it also holds in another in relative motion (v):

$$(\gamma_A' m_A u_A' + \gamma_B' m_B u_B') - (\gamma_C' m_C u_C' + \gamma_D' m_D u_D') = 0. \tag{3.42}$$

Here are the algebraic steps. From the relativistic velocity addition rule (1.18), we have u_A', and from (3.32), we have the associated gamma factor γ_A'. This leads to the transformation

$$\gamma_A' u_A' = \gamma_v \gamma_A (u_A - v), \tag{3.43}$$

so that the left-hand side of (3.42) may be written as

$$\gamma_v \{ [(\gamma_A m_A u_A + \gamma_B m_B u_B) - (\gamma_C m_C u_C + \gamma_D m_D u_D)]$$
$$-v[(\gamma_A m_A + \gamma_B m_B) - (\gamma_C m_C + \gamma_D m_D)]\}, \tag{3.44}$$

which vanishes, because the first square bracket is zero by (3.41), and the second is zero by conservation of relativistic energy γmc^2. There are three equations for momentum and one for energy conservation. They can be written compactly in terms of 4-momenta:

$$p_A^\mu + p_B^\mu = p_C^\mu + p_D^\mu. \tag{3.45}$$

Being a tensor equation (with every term a 4-vector), it is manifestly covariant; and these relations are valid in every inertial frame.

We emphasize that relativistic momentum and energy are conserved quantities, and that they reduce to their classical counterparts in the appropriate limit.[14] We will use the same criteria when we discuss particle energy and momentum in general relativity.

[14] Here, the 3-momentum $p = \gamma mu$ reduces to mu in the nonrelativistic limit.

Massless particles always travel at speed c

When $m = 0$, we can no longer define 4-velocity as $\dot{x}^\mu$, because the concept of proper time does not apply (see the discussion below); thus the 4-momentum cannot be $p^\mu = m\dot{x}^\mu$. Nevertheless, we still assign a 4-momentum to such a particle, since a massless particle has energy and momentum just as on the right-hand side of (3.36). When $m = 0$, the relation (3.39) becomes

$$E = |\mathbf{p}|c. \tag{3.46}$$

[15] Gravitons are the quanta of the gravitational field, just as photons are the quanta of the electromagnetic field. They are massless, since they are the transmitters of long-range forces.

Plugging this into the ratio of (3.38), we obtain the well-known result that massless particles such as photons and gravitons[15] always travel at the speed of light, $v = c$. Hence there is no rest frame for massless particles. Stating this in another way: a massless particle has a 4-momentum with nontrivial components $p^\mu = (E/c, \mathbf{p})$ but null length $p^\mu p_\mu = 0$. The following exercise provides useful practice in using 4-momentum vectors for particles with and without mass.

Exercise 3.6 Compton scattering

---

[16] The low-frequency limit ($\hbar\omega \ll mc^2$) of Compton scattering, with no frequency shift $\omega' \simeq \omega$, is called Thomson scattering.

A photon with initial frequency ω scatters off a stationary electron (mass m) and moves off at an angle θ from its initial direction. Show that the frequency of the scattered photon[16] is

$$\omega' = \frac{\omega}{1 + \dfrac{\hbar\omega}{mc^2}(1 - \cos\theta)}. \qquad (3.47)$$

This problem provides practice in using 4-momentum, $p^\mu = (E/c, \mathbf{p})$, so that the energy–momentum conservation conditions can be imposed in a compact way: $k^\mu + p^\mu = k'^\mu + p'^\mu$, where (k^μ, k'^μ) are the initial and final photon 4-momenta and (p^μ, p'^μ) the electron 4-momenta. Recall that the photon has energy $E = \hbar\omega$. As a massless particle, its momentum magnitude is related to its energy by $|\mathbf{k}| \equiv k = E/c$.

Suggested steps: (a) Since $\omega' \sim k' \cdot p$, from 4-momentum conservation, show that $k' \cdot p' = k \cdot p$, so

$$k' \cdot p = k \cdot p - k \cdot k', \qquad (3.48)$$

where $k \cdot p \equiv k_\mu p^\mu$. (b) In the laboratory frame, the target electron is initially at rest (i.e., has zero 3-momentum); hence its initial 4-momentum is $p^\mu = (mc, \mathbf{0})$. Write out the scalar products of (3.48) in their components to obtain the result (3.47).

Exercise 3.7 Antiproton production threshold

Because of nucleon number conservation, the simplest reaction to produce an antiproton $\bar{p}$ in proton–proton scattering is $pp \to ppp\bar{p}$. The rest energy of a proton is $m_p c^2 = 0.94 \text{ GeV}$. Find the minimum kinetic energy a (projectile) proton must have in order to produce an antiproton after colliding with another (target) proton at rest. Let P_I^μ and P_F^μ be the respective initial and final total 4-momenta, so that the energy–momentum conservation conditions may be written simply as $P_I^\mu = P_F^\mu$. Since $P \cdot P$ is Lorentz-invariant, $P_I \cdot P_I = P_F \cdot P_F$, even if P_I^μ and P_F^μ are evaluated *in different coordinate frames. This observation can simplify the calculation of the initial projectile energy in the lab frame.*

Comment: *Another approach is to work in the center-of-mass frame: two protons with equal energy and opposite momenta collide into a stationary lump. You can use (3.32) to transform back to the lab frame. Comparing the lab-frame projectile energy with the center-of-mass frame energy explains why we might prefer colliders to fixed targets in high-energy experiments.*

Box 3.2 The wave vector

Here we discuss the wave 4-vector, which is closely related to the photon 4-momentum. Recall that for a dynamic quantity $A(\mathbf{x}, t)$ to be a solution to the wave equation, its dependence on the space and time coordinates must be in the combination $(\mathbf{x} - \mathbf{v}t)$, where $\mathbf{v}$ is the wave velocity. A harmonic electromagnetic wave is then proportional to $\exp[i(\mathbf{k} \cdot \mathbf{x} - \omega t)]$, where $k = |\mathbf{k}| = 2\pi/\lambda$ is the wavenumber, and $\omega = 2\pi/T$ is the angular frequency corresponding to a wave period T, so that the wave propagates with speed $\lambda/T = \omega/k = c$. The phase factor $(\mathbf{k} \cdot \mathbf{x} - \omega t)$, which measures the number of peaks and troughs of the wave (in radians), must be a frame-independent quantity (i.e., a Lorentz scalar). To make its scalar nature explicit, we write this phase in terms of the 4-vector $x^\mu = (ct, \mathbf{x})$ as

$$\mathbf{k} \cdot \mathbf{x} - \omega t = \begin{pmatrix} ct & \mathbf{x} \end{pmatrix} \begin{pmatrix} -1 & \\ & 1 \end{pmatrix} \begin{pmatrix} \omega/c \\ \mathbf{k} \end{pmatrix} \equiv x^\mu \eta_{\mu\nu} k^\nu = x_\nu k^\nu. \tag{3.49}$$

From our knowledge that x^μ is a 4-vector and $x_\nu k^\nu$ a scalar, we conclude[17] that ω and $\mathbf{k}$ must also form a 4-vector, the wave 4-vector $k^\mu = (\omega/c, k^i)$. Therefore, under a Lorentz boost in the $+x$ direction, we have

$$k'_x = \gamma \left(k_x - \beta \frac{\omega}{c} \right), \tag{3.50}$$

$$\omega' = \gamma (\omega - c\beta k_x) = \gamma (\omega - c\beta k \cos\theta), \tag{3.51}$$

where θ is the angle between the boost direction $\hat{\mathbf{x}}$ and the direction of wave propagation $\hat{\mathbf{k}}$. Since $ck = \omega$, we obtain the relativistic Doppler formula,

$$\omega' = \frac{1 - \beta \cos\theta}{\sqrt{1 - \beta^2}} \omega, \tag{3.52}$$

continued

[17] This follows from the quotient theorem, which is described in Exercise 3.2.

Box 3.2 *continued*

which can be compared with the nonrelativistic Doppler relation, $\omega' = (1 - \beta \cos\theta)\omega$. In the nonrelativistic limit, there is no Doppler shift ($\omega' = \omega$) in the transverse direction, $\theta = \pi/2$, whereas the relativistic transverse Doppler relation, $\omega' = \gamma\omega$, clearly demonstrates the relativistic time dilation. In the longitudinal direction, $\theta = 0$, we have the familiar longitudinal Doppler relation,

$$\frac{\omega'}{\omega} = \sqrt{\frac{1-\beta}{1+\beta}}, \tag{3.53}$$

which has the low-velocity ($v \ll c$, $\beta \ll 1$) approximation

$$\frac{\omega'}{\omega} \simeq 1 - \beta \quad \text{or} \quad \frac{\Delta\omega}{\omega} \simeq \frac{\Delta v}{c}. \tag{3.54}$$

Because $\omega = ck$,[18] the wave 4-vector $k^\mu = (k, k^i)$ has a null invariant magnitude, $\eta_{\mu\nu}k^\mu k^\nu = |\mathbf{k}|^2 - \omega^2/c^2 = 0$, compatible $E = |\mathbf{p}|c$ as shown in (3.46).

[18] The quantum mechanical relations $E = \hbar\omega = h\nu$ and $p = \hbar k = h/\lambda$ (of Planck and de Broglie, respectively) are, of course, consistent with (3.46) for a massless particle with speed c: $E/p = c = \omega/k = \nu\lambda$. They can be combined into a single 4-vector relation, $p^\mu = \hbar k^\mu$.

Box 3.3 Covariant force

Just as the ordinary velocity $\mathbf{v}$ has a complicated Lorentz property (hence our motivation to introduce 4-velocity), it is not easy to relate different components of the usual force vector $\mathbf{F} = d\mathbf{p}/dt$ in different moving frames. The notions of 4-velocity and 4-momentum naturally lead us to the definition of 4-force, or covariant force, as

$$K^\mu \equiv \frac{dp^\mu}{d\tau} = m\ddot{x}^\mu, \tag{3.55}$$

which, using (3.36), has components

$$K^\mu = \frac{dp^\mu}{d\tau} = \gamma\frac{d}{dt}(E/c, \mathbf{p}). \tag{3.56}$$

Next we show that the rate of energy change dE/dt is given, just as in nonrelativistic physics, by the dot product $\mathbf{F} \cdot \mathbf{v}$. Because $|\dot{x}^\mu|^2$ is a constant, its derivative vanishes:

$$0 = m\frac{d}{d\tau}(\eta_{\mu\nu}\dot{x}^\mu\dot{x}^\nu) = 2\eta_{\mu\nu}m\ddot{x}^\mu\dot{x}^\nu = 2\eta_{\mu\nu}K^\mu\dot{x}^\nu, \tag{3.57}$$

where we have used (3.55) to reach the last equality. Substituting in the components of K^μ and $\dot{x}^\nu$ from (3.56) and (3.28), we have

$$0 = K_\mu\dot{x}^\mu = \gamma^2\left(-\frac{dE}{dt} + \mathbf{F} \cdot \mathbf{v}\right); \tag{3.58}$$

thus $dE/dt = \mathbf{F} \cdot \mathbf{v}$ with $\mathbf{F} = d\mathbf{p}/dt$. We can then display the components of the covariant force (3.56) as

$$K^\mu = \gamma(\mathbf{F} \cdot \mathbf{v}/c, F^i). \tag{3.59}$$

This covariant force will be applied to the Lorentz force law of electromagnetism in Box 3.4, especially Exercise 3.10.

Exercise 3.8 Force components can have asymmetric appearances

---

While the relativistic force, when written as in (3.56) as a rate of change of relativistic momentum, has the same appearance in all three spatial components, this is not the case when it is expressed in terms of acceleration with respect to coordinate time. Show that the usual force component parallel to the motion is

$$F_x = \frac{dp_x}{dt} = \gamma^3 m \frac{d^2x}{dt^2}, \tag{3.60}$$

but the transverse components are

$$F_y = \frac{dp_y}{dt} = \gamma m \frac{d^2y}{dt^2} \quad and \quad F_z = \frac{dp_z}{dt} = \gamma m \frac{d^2z}{dt^2}. \tag{3.61}$$

They can be converted to covariant force by noting that $\mathbf{K} = \gamma \mathbf{F}$.

3.2.3 Electromagnetic field 4-tensor

The four spacetime displacement coordinates $(ct, \mathbf{r})$ form one 4-vector; the energy and momentum components of a particle $(E/c, \mathbf{p})$ form another. What sort of tensor can the six components of the electromagnetic fields, $\mathbf{E}$ and $\mathbf{B}$, form? As we have already seen (e.g., in Box 2.2), they transform into each other, and hence they must be components of the same tensor. This turns out to be an antisymmetric tensor of rank 2:

$$F_{\mu\nu} = -F_{\nu\mu}, \tag{3.62}$$

with the assignments

$$F_{0i} = -F_{i0} = -E_i \qquad F_{ij} = \varepsilon_{ijk}B_k, \tag{3.63}$$

where ε_{ijk} is the totally antisymmetric Levi-Civita symbol[19] with $\varepsilon_{123} = 1$, and the dummy index k is summed over. Writing out (3.63) explicitly, we have the covariant form of the electromagnetic field tensor:

$$F_{\mu\nu} = \begin{pmatrix} 0 & -E_1 & -E_2 & -E_3 \\ E_1 & 0 & B_3 & -B_2 \\ E_2 & -B_3 & 0 & B_1 \\ E_3 & B_2 & -B_1 & 0 \end{pmatrix}. \tag{3.64}$$

[19] The Levi-Civita symbol in an n-dimensional space is a quantity with n indices. Thus, in a 3D space, we have ε_{ijk}, whose indices run from 1 to 3 (or we could choose for them to run from 0 to 2, for example); in a 4D space, we have $\varepsilon_{\mu\nu\lambda\rho}$, whose indices run from 1 to 4 (or from 0 to 3, which is useful for relativity). The Levi-Civita symbol is totally antisymmetric; an interchange of any two adjacent indices results in a minus sign: $\varepsilon_{ijk} = -\varepsilon_{jik} = \varepsilon_{jki}$, etc. The symbol vanishes whenever any two indices are equal. For instance, ε_{122} must be zero because an interchange of the last two indices must result in a minus sign ($\varepsilon_{122} = -\varepsilon_{122}$), and the only way this can be true is if $\varepsilon_{122} = 0$. All the nonzero elements can be obtained by permutation of indices from $\varepsilon_{12} = \varepsilon_{123} = \varepsilon_{0123} \equiv 1$. One can use Levi-Civita symbols to express the cross product of vectors, $(\mathbf{A} \times \mathbf{B})_i = \varepsilon_{ijk}A_jB_k$.

Using the metric tensor $\eta^{\mu\nu}$ to raise the two indices, we can also write this field tensor in its contravariant form:

$$F^{\mu\nu} = \eta^{\mu\lambda} F_{\lambda\rho} \eta^{\rho\nu} = \begin{pmatrix} 0 & E_1 & E_2 & E_3 \\ -E_1 & 0 & B_3 & -B_2 \\ -E_2 & -B_3 & 0 & B_1 \\ -E_3 & B_2 & -B_1 & 0 \end{pmatrix}. \tag{3.65}$$

It is also useful to define the **dual field strength tensor**

$$\tilde{F}_{\mu\nu} \equiv -\frac{1}{2} \varepsilon_{\mu\nu\lambda\rho} F^{\lambda\rho}, \tag{3.66}$$

where $\varepsilon_{\mu\nu\lambda\rho}$ is the 4D Levi-Civita symbol with $\varepsilon_{0ijk} \equiv \varepsilon_{ijk}$ and thus $\varepsilon_{0123} = 1$. The elements of the dual field tensor are

$$\tilde{F}_{0i} = -\tilde{F}_{i0} = -B_i, \qquad \tilde{F}_{ij} = -\varepsilon_{ijk} E_k, \tag{3.67}$$

or, explicitly,

$$\tilde{F}_{\mu\nu} = \begin{pmatrix} 0 & -B_1 & -B_2 & -B_3 \\ B_1 & 0 & -E_3 & E_2 \\ B_2 & E_3 & 0 & -E_1 \\ B_3 & -E_2 & E_1 & 0 \end{pmatrix}. \tag{3.68}$$

Because $F_{\mu\nu}$ is a tensor of rank 2, under a Lorentz transformation it must change according to the rule (3.27) for higher-rank tensors—with a transformation matrix factor for each index. Thus we have

$$F_{\mu\nu} \longrightarrow F'_{\mu\nu} = [L^{-1}]^{\lambda}_{\mu} [L^{-1}]^{\rho}_{\nu} F_{\lambda\rho}. \tag{3.69}$$

Exercise 3.9 Lorentz transformation of the electromagnetic fields in the manifestly covariant formalism

Verify that the covariant transformation (3.69) is precisely the transformation relations (2.39) and (2.40) derived in Box 2.2. Carry out this comparison in two equivalent ways: (a) by direct substitution of (3.69) with the components of (3.63) and transformation relation (2.13) such as $E'_x = F'_{10}$, $B'_y = F'_{13}$, $[L^{-1}]^0_1 = \beta\gamma$, etc.; (b) by converting the right-hand side of (3.69) into a product of three 4×4 matrices.

Box 3.4 Manifestly covariant equations for electromagnetism

We now use the field tensor $F_{\mu\nu}$ to write the equations of electromagnetism in a form that clearly displays their Lorentz covariance.

Lorentz force law

Using (3.63), one can easily show (Exercise 3.10) that the Lorentz force, the electromagnetic force felt by a charge q moving with velocity $\mathbf{v}$,

$$\mathbf{F} = q \left(\mathbf{E} + \frac{1}{c}\mathbf{v} \times \mathbf{B} \right), \tag{3.70}$$

comes from the spatial part of the covariant force law,

$$K^\mu = \frac{q}{c} F^{\mu\nu} \dot{x}_\nu. \tag{3.71}$$

K^μ is the covariant force discussed in Box 3.3, $K^\mu = m\ddot{x}^\mu$, where the differentiation is with respect to the proper time and $\dot{x}^\mu$ is the 4-velocity of the charged particle.

Exercise 3.10 Covariant Lorentz force law

(a) In Box 3.3, we have identified the spatial part of the covariant force $\mathbf{K} = \gamma\mathbf{F}$ with the relativistic force $\mathbf{F} = d\mathbf{p}/dt$, where $\mathbf{p} = \gamma m\mathbf{v}$ is the relativistic momentum. Justify the identification of the force (which is usually taken as $m\mathbf{a}$) in (3.70) with this $\mathbf{F}$. (b) Check that the spatial $\mu = i$ components of (3.71) correspond to the familiar Lorentz force law (3.70). (c) Check that the $\mu = 0$ component has the correct interpretation as the time component of a covariant force, $K^0 = \gamma\mathbf{F} \cdot \mathbf{v}/c$, as required by (3.59).

Inhomogeneous Maxwell's equations Gauss's and Ampère's laws in (2.41),

$$\mathbf{\nabla} \cdot \mathbf{E} = \rho, \qquad \mathbf{\nabla} \times \mathbf{B} - \frac{1}{c}\frac{\partial \mathbf{E}}{\partial t} = \frac{1}{c}\mathbf{j}, \tag{3.72}$$

may be written as one covariant Maxwell's equation:

$$\partial_\mu F^{\mu\nu} = -\frac{1}{c} j^\nu, \tag{3.73}$$

where the electromagnetic current 4-vector[20] is given as

$$j^\mu = (j^0, \mathbf{j}) = (c\rho, \mathbf{j}). \tag{3.74}$$

The reader is invited to check in Exercise 3.11 that the four equations in (3.72) are just the four equations corresponding to different values of the index ν in (3.73).

[20] To show that j^μ is a bona fide 4-vector, we recall that the relation between charge-current density and charge density is $\mathbf{j} = \rho\mathbf{v}$, where $\mathbf{v}$ is the velocity field. Thus (3.74) may be written as $j^\mu = \rho(c, \mathbf{v})$. If we replace the density ρ by the rest-frame density ρ' (which is a Lorentz scalar) through the relation $\rho = \gamma\rho'$ (reflecting the usual Lorentz length/volume contraction), we can relate j^μ to the 4-velocity field (3.28): $j^\mu = \rho'\gamma(c, \mathbf{v}) = \rho'\dot{x}^\mu$. Since ρ' is a scalar and $\dot{x}^\mu$ is a 4-vector, this shows explicitly that j^μ is also a 4-vector.

Exercise 3.11 The inhomogeneous Maxwell's equations

(a) From the components of (3.65), work out the four equations of (3.73), showing that they are just the familiar Coulomb–Gauss law and Ampère's law of (3.72). (b) Use (3.27) to explicitly demonstrate that (3.73) is covariant under Lorentz transformation. (c) Prove that the charge conservation equation (3.81) to be discussed below follows directly from (3.73).

Homogeneous Maxwell's equations   Faraday's law and Gauss's law for magnetism in (2.32),

$$\mathbf{\nabla} \times \mathbf{E} + \frac{1}{c}\frac{\partial \mathbf{B}}{\partial t} = 0, \qquad \mathbf{\nabla} \cdot \mathbf{B} = 0 \tag{3.75}$$

correspond to the Bianchi identity of the electromagnetic field tensor,

$$\partial_\mu F_{\nu\lambda} + \partial_\lambda F_{\mu\nu} + \partial_\nu F_{\lambda\mu} = 0. \tag{3.76}$$

We note that this equation can be written (Exercise 3.12) in an alternative form as

$$\partial_\mu \tilde{F}^{\mu\nu} = 0. \tag{3.77}$$

Exercise 3.12 The homogeneous Maxwell's equations

Explicitly demonstrate that the two forms of the homogeneous equations (3.76) and (3.77) are equivalent.

Suggestion: Start with (3.77) with $\nu = 0$ to derive (3.76).

Electromagnetic duality   From (3.73) and (3.77), it is clear that Maxwell's equations in free space ($j^\mu = 0$) are invariant under the duality transformation of $F_{\mu\nu} \to \tilde{F}_{\mu\nu}$, which can be viewed as a rotation by 90° in the plane spanned by the perpendicular **E** and **B** axes: $\mathbf{E} \to \mathbf{B}$ and $\mathbf{B} \to -\mathbf{E}$. This is just a mathematical way to formalize the symmetry between electricity and magnetism that we all notice in Maxwell's theory. If $j^\mu \neq 0$, this duality symmetry can be restored by the introduction in (3.77) of the magnetic current $j_M^\mu = (c\rho_M, \mathbf{j}_M)$, where ρ_M is the magnetic charge (monopole) density and $\mathbf{j}_M$ is the magnetic charge-current density.

3.2.4 The energy–momentum–stress 4-tensor for a field system

We have already learned in Section 3.2.2 how to describe the energy and momentum of a single massive or massless particle; we now consider the case of a field system (having infinite degrees of freedom) with a continuous distribution of matter and energy. (Recall the energy and momentum carried by the electromagnetic fields.) For this purpose, we need to introduce another rank-2 tensor, the stress–energy–momentum tensor. We shall begin by way of an analogy with the current density (the flow of a charge) and its conservation law.

Structure of the charge 4-current density The charge density ρ and 3D current density components j_i can be transformed into each other under Lorentz transformation. This simply reflects the fact that a moving observer views a stationary charge as a current. Thus, together, they form the four components of a 4-current density $j^\mu = (\rho c, j_i)$. Recall that charge density is charge (q) per unit volume ($\Delta x \Delta y \Delta z$), and current density, say in the x direction, is the amount of charge moved over a cross-sectional area ($\Delta y \Delta z$) perpendicular to x in unit time Δt:

$$\rho = \frac{q}{\Delta x^1 \Delta x^2 \Delta x^3}, \qquad j_x = \frac{q}{\Delta t \Delta x^2 \Delta x^3} = \frac{cq}{\Delta x^0 \Delta x^2 \Delta x^3}. \tag{3.78}$$

We can express the meaning of all four components of j^μ more compactly:

$$j^\mu = \frac{cq}{\Delta S^\mu}, \tag{3.79}$$

where ΔS^μ is the Minkowski volume (3-surface) with x^μ held fixed. We also note that charge conservation of a field system is expressed in terms of a continuity equation,

$$\frac{d\rho}{dt} + \nabla \cdot \mathbf{j} = 0, \tag{3.80}$$

which can be compactly written in this 4-tensor formalism as a vanishing 4-divergence equation,

$$\partial_\mu j^\mu = 0. \tag{3.81}$$

Structure of the energy–momentum–stress tensor The charge 4-current j^μ describes the distribution and flow of a conserved scalar quantity, the charge q. However, if we wish to generalize this to energy and momentum, which are themselves components of a 4-vector p^μ, the relevant currents (generalizing (3.79)) must have two indices (hence rank 2):

$$T^{\mu\nu} = \frac{cp^\mu}{\Delta S^\nu}, \tag{3.82}$$

which is a 4×4 symmetric matrix $T^{\mu\nu} = T^{\nu\mu}$. The energy/momentum conservation of a system of continuous medium (i.e., a field system) can then be expressed by a vanishing divergence of $T^{\mu\nu}$:

$$\partial_\mu T^{\mu\nu} = 0. \tag{3.83}$$

The physical meaning of the ten independent components of $T^{\mu\nu}$ can be worked out from (3.82):

- T^{00} is the energy density of the system.
- $T^{0i} = T^{i0}$ are the three components of the momentum density or, equivalently, the energy current density.
- T^{ij} is a 3×3 symmetric matrix whose diagonal elements are the pressure (i.e., normal force per unit area) in the three directions, and whose off-diagonal elements are the shear (parallel) forces per unit area.

In short, the energy–momentum tensor, also called the stress–energy tensor, describes the distribution and flow of matter/energy in spacetime. As we shall see, the $T^{\mu\nu}$ tensor will appear as the inhomogeneous source term for gravity in Einstein's field equation in his general theory of relativity.

Exercise 3.13 The stress–energy tensor is symmetric

As part of the proof that the stress–energy tensor is symmetric, show that the equality of momentum density T_{i0} and energy density flux T_{0i} follows from the definition (3.82).

Box 3.5 Stress–energy tensor for an ideal fluid

The case of the energy–momentum tensor of an ideal fluid is a particularly important one. Here the fluid elements interact only through a perpendicular force (no shear). Thus the fluid can be characterized by two parameters: the mass density ρ and the isotropic pressure[21] p (no viscosity). In the comoving frame,[22] where each fluid element may be regarded as momentarily at rest, the stress tensor, according to (3.82), takes on a particularly simple form:

$$T^{\mu\nu} = \begin{pmatrix} \rho c^2 & & & \\ & p & & \\ & & p & \\ & & & p \end{pmatrix} = \left(\rho + \frac{p}{c^2}\right)\dot{x}^\mu \dot{x}^\nu + pg^{\mu\nu}, \tag{3.84}$$

[21] Do not confuse the symbol for pressure with that for the magnitude of 3-momentum, as both are denoted by p.

[22] The comoving frame is defined as the coordinate system in which the fluid elements themselves carry the position coordinate labels and clocks that are synchronized.

where $g^{\mu\nu}$ is the inverse metric tensor, which for a flat spacetime is the SR invariant $\eta^{\mu\nu}$. To reach the last equality, we have used the fact that in the comoving frame the fluid velocity field is simply $\dot{x}^{\mu} = (c, \mathbf{0})$. Since ρ and p are quantities in the rest (comoving) frame (hence Lorentz scalars), the right-hand side is a proper tensor of rank 2, so this expression should be valid in every frame (as Lorentz transformations will not change its form). One can easily check that the four energy–momentum conservation equations (3.83) are just the familiar continuity equation for mass conservation (3.80) and the Euler equation for hydrodynamics, which in the comoving frame takes on the simple form of $\nabla p = 0$.

The even simpler case when the pressure term is absent, $p = 0$, describes a swamp of noninteracting particles, i.e., a cloud of dust.

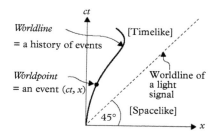

Figure 3.2 *Basic elements of a space-time diagram, with two spatial coordinates (y, z) suppressed, displaying only the region with $+x$ and $+t$.*

3.3 The spacetime diagram

Relativity brings about a profound change in the causal structure of space and time, which can be nicely visualized in terms of the **spacetime diagram**, a plot of event positions and times[23] (Figs. 3.2 and 3.4).

- Let us first recall the causal structure of space and time in pre-relativity physics (Fig. 3.3). For a given event P at a particular point in space and a particular instant in time, all the events that could in principle be reached by a particle starting from P are collectively labeled as the future of P, while all the events from which a particle can arrive at P form the past of point P. Those events that are in neither the future nor the past of the event P form a 3D set of events simultaneous with P. The events that are simultaneous with P are simultaneous with each other—simultaneity is absolute. The notion of simultaneous events allows one to discuss in pre-relativity physics all of the space at a given instant of time, and, as a corollary, allows one to study space and time separately.

- In relativistic physics, the events that fail to be causally connected to event P are much larger than a 3D space; they form a 4D subset of spacetime. As we shall see, all events outside the future and past lightcones are causally disconnected from the event P, which lies at the tip of the (future and past) lightcones in the spacetime diagram (cf. Fig. 3.4). However, events that are causally disconnected from P may be causally connected to one another. Thus simultaneity is relative, and there can be no absolute time labels in relativistic physics. Time and space coordinates are inextricably linked.

[23] To have the same length dimension for all coordinates, the temporal axis is represented by $x^0 = ct$.

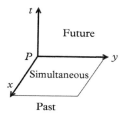

Figure 3.3 *Spacetime diagram (one dimension suppressed) showing the causal structure in pre-SR physics. Events that are neither in the future nor in the past are in the 3D space of simultaneous events with the same absolute time as the origin O.*

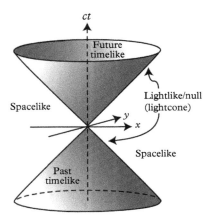

Figure 3.4 *Invariant regions in the spacetime diagram with one of the spatial coordinates suppressed. The regions within the top and bottom cones are respectively future- and past-timelike; the lightcones themselves are termed null/lightlike. The region outside the cones is spacelike.*

3.3.1 Invariant regions and causal structure

An event with coordinates (t, x, y, z) is represented by a worldpoint in the spacetime diagram. The history of a point object is a line of worldpoints called a worldline. In Fig. 3.2, the 3D space is represented by a 1D x axis. A light pulse travels with speed $c = \pm\Delta x/\Delta t$ (in one spatial dimension). Thus its history is represented by a straight line at a 45° angle with respect to the axes. Its invariant interval is $\Delta s^2 = \Delta x^2 - c^2 \Delta t^2 = 0$. Any path with constant velocity $v = \Delta x/\Delta t$ would be a straight line. Worldlines with $v = |\Delta x/\Delta t| < c$, corresponding to $\Delta s^2 < 0$, would make an angle greater than 45° with respect to the spatial axis (i.e., above the worldlines for a light ray). According to relativity, no worldline can have $v > c$. If there were such a line, it would have interval $\Delta s^2 > 0$ and would make an angle less than 45° (i.e., below the light worldline). Since Δs^2 is invariant, it is meaningful to divide the spacetime diagram into regions, as in Fig. 3.4.

$\Delta s^2 < 0$	timelike
$\Delta s^2 = 0$	lightlike
$\Delta s^2 > 0$	spacelike

- The lightlike region has all the events that are separated from the origin by interval $\Delta s^2 = 0$. These events can be connected to the origin by light signals. This 45°-inclined surface forms a lightcone. It has a slope of unity, because the speed of light is c. A displacement vector that connects an event in this surface to the origin is an example of a lightlike vector or a null vector, a nonzero 4-vector having zero invariant magnitude.

- The spacelike region has all the events that are separated from the origin by an interval of $\Delta s^2 > 0$. The displacement 4-vector from the origin to any point in this region is a spacelike vector—having a positive invariant magnitude. It would require a signal traveling at a speed greater than c to connect an event in this region to the origin. Thus, an event taking place at any point in this region cannot influence or be influenced causally (in the sense of cause-and-effect) by an event at the origin.

- The timelike region has all the events that are separated from the origin by an interval of $\Delta s^2 < 0$. (The displacement 4-vector from the origin to any point in this region is a timelike vector—having a negative invariant magnitude.) One can always find a frame O' such that a given event in this region takes place at the same spatial location as the origin, $x' = 0$, but at different time, $t' \neq 0$. This makes it clear that events in this region can be causally connected with the origin. In fact, all the worldlines passing through the origin will be confined to the region, inside (and, for massless particles, on) the lightcone.[24] The timelike (and the lightlike) region can be divided into past and future sections, where events can respectively influence and be influenced by events at the origin.

[24] The worldline of an inertial observer (i.e., one moving with constant velocity) passing through the origin must be a straight line inside the lightcone. This straight line is the time axis of the coordinate system in which the inertial observer is at rest.

Figure 3.4 displays the lightcone structure with respect to the origin of the spacetime coordinates. However, each point in a spacetime diagram has its own lightcone. The future lightcones of several worldpoints are shown in Fig. 3.5. If we consider a series of lightcones having vertices located along a given worldline, each subsequent worldline segment must lie within the lightcone of all prior points. It is then clear from Fig. 3.5 that any particle can only proceed (along a timelike path) in the direction of ever-increasing time (i.e., into the future)[25].

3.3.2 Lorentz transformation in the spacetime diagram

The nontrivial parts of the Lorentz transformation (3.18) of coordinate intervals are

$$\Delta x' = \gamma(\Delta x - \beta c \Delta t), \qquad c \Delta t' = \gamma(c \Delta t - \beta \Delta x) \qquad (3.85)$$

(taken, for example, with respect to the origin). We can represent these transformed axes in a spacetime diagram as shown in Fig. 3.6:

- The x' axis corresponds to the $c \Delta t' = \gamma(c \Delta t - \beta \Delta x) = 0$ line. Namely, the x' axis is a straight line in the (x, ct) plane with a slope of $c \Delta t / \Delta x = \beta$.
- The ct'-axis corresponds to the $\Delta x' = \gamma(\Delta x - \beta c \Delta t) = 0$ line. Namely, it is a straight line with a slope of $c \Delta t / \Delta x = 1/\beta$.

Depending on whether β is positive or negative, the new axes either close in or open up from the original perpendicular axes. Thus we have the **opposite-angle rule**: the two axes make opposite-signed rotations of $\pm\theta$ (Fig. 3.6). For $\beta > 0$, the x' axis rotates by $+\theta$ relative to the x axis, and the ct' axis rotates by $-\theta$ relative to the ct axis. The physical basis for this rule is that the lightcone has unit slope in every inertial frame. Therefore the lightcone bisects the angle between the space and time axes of every inertial frame. Since the lightcone itself is invariant under a boost transformation from one inertial frame to another, the axes must close in or open up around the lightcone by an equal amount.

Relativity of simultaneity, event order, and causality

It is instructive to use the spacetime diagram to demonstrate some of the physical phenomena we have discussed previously. In Fig. 3.7, we have two events, A and B, where A is the origin of both coordinate systems O and O': $(x_A = t_A = 0, x'_A = t'_A = 0)$. In Fig. 3.7(a), the events A and B are simultaneous, $t_A = t_B$, with respect to the O system. But in the O' system, we clearly have $t'_A > t'_B$. This shows the relativity of simultaneity.

[25] The well-known exception is the interior region of a black hole, where, in Schwarzschild coordinates, the roles of space and time are interchanged. In that case, a particle always proceeds toward the $r = 0$ singularity, cf. Fig 7.2.

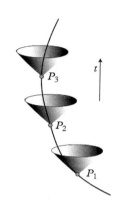

Figure 3.5 *Future lightcones of different worldpoints, P_1, P_2, etc. along a timelike worldline, which can only proceed in the direction of ever-increasing time, as each segment emanating from a given worldpoint must be contained within the future lightcone of that point.*

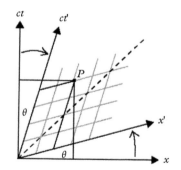

Figure 3.6 *Lorentz rotation in the spacetime diagram. The space and time axes rotate by the same amount but in opposite directions, so that the slope of the lightcone (the dashed line) remains unchanged. The shaded grid represents lines of fixed x' and t'.*

(a)

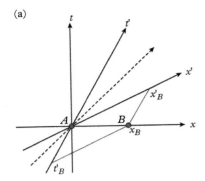

(b)

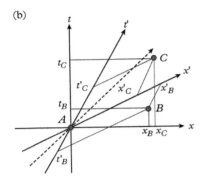

Figure 3.7 *(a) Relativity of simultaneity: $t_A = t_B$ but $t'_A > t'_B$. (b) Relativity of event order: $t_A < t_B$ but $t'_A > t'_B$. However, there is no change of event order with respect to A for all events located above the x' axis, such as the event C. This certainly includes the situation in which C is located in the forward lightcone of A (above the dashed line).*

---

[26] The (x', t') axes close in because the frame O' is moving with a positive β, and the (x'', t'') axes open out because the frame O'' is moving with a negative β with reference to the O frame.

In Fig. 3.7(b), we have $t_A < t_B$ in the O frame, but $t'_A > t'_B$ in the O' frame. Thus, the temporal order of events can be changed by a change of reference frames. However, this change of event order can take place only if event B is located in the region below the x' axis. This means that if we increase the relative speed between these two frames O and O' (so the x' axis moves ever closer to the lightcone), more and more events can have their temporal order (with respect to A at the origin) reversed as seen from the perspective of the moving observer. On the other hand, for all events above the x' axis, the temporal order is preserved. For example, with respect to event C, we have both $t_A < t_C$ and $t'_A < t'_C$. Of course the region above this x' axis includes the future lightcone of event A. This means that for two events that are causally connected (e.g., for A and any worldpoint in its future lightcone), their temporal order cannot be changed by a Lorentz transformation. The principle of causality is safe under SR. Namely, if A is the cause of C, there does not exist any coordinate frame in which C is the cause of A.

Exercise 3.14 Relativity of event order

We have already demonstrated in a spacetime diagram the relativity of event order. Now confirm this result algebraically via the Lorentz transformation of (3.85). (a) Given $t > 0$, what is the condition on (t, x) so that $t' < 0$? (b) Show that your derived condition will never lead to a violation of causality.

Exercise 3.15 Spacetime diagram for the twin paradox

Draw a spacetime diagram of the twin paradox discussed in Box 2.3. Let Bill's rest frame O have coordinates (x, t), outbound Al's frame O' have (x', t'), and inbound Al's frame O'' have (x'', t''). Draw your diagram so that the perpendicular lines represent the (x, t) axes.[26] (a) Mark the event when Al departs in the spaceship by the worldpoint O, the event when Al returns and reunites with Bill by the worldpoint Q, and the event when Al turns around (from outward-bound to inward-bound) by the point P. Thus, the stay-at-home Bill has worldline OQ. Al's worldline has two segments: OP for the outward-bound and PQ for the inward-bound parts of his journey. (b) On the t axis, which should coincide with Bill's worldline OQ, also mark the points M, P', and P'', which should be simultaneous with the turning point P as viewed in the coordinate frames of O, O', and O'', respectively. (c) Indicate the time values of $t_M, t_{P'}$, and $t_{P''}$ (i.e., the elapsed times since Al's departure at point O in the (x, t) coordinate system). In particular, show how changing the inertial frame from O' to O'' brings about a time change of 32 years in the O frame.

Exercise 3.16 Spacetime diagram for the pole-and-barn paradox

Draw a spacetime diagram for the pole-and-barn paradox discussed in Box 2.4. Let (x, t) be the coordinates for the ground (barn) observer and (x', t') for the rest frame of the runner (pole). Use $\beta = 4/5$ for the pole's/runner's speed. Show the worldlines for the front door (F) and rear door (R) of the barn, and the front end (A) and back end (B) of the pole. Your diagram should display the order-reversal phenomenon discussed in Box 2.4: $t_{AR} > t_{BF}$ and $t'_{AR} < t'_{BF}$. Your diagram should also demonstrate that the simultaneous slamming of two doors in the barn's frame does not lead to any contradiction when viewed in the runner's frame.

Box 3.6 The spacetime formulation of relativity—a reprise

Let us summarize the principal lessons we have learned from this geometric formulation of SR:

- The stage on which SR physics takes place is Minkowski spacetime, whose time coordinate is on an equal footing with its spatial ones. Space and time are thereby treated symmetrically. A spacetime diagram is often useful in clarifying ideas in relativity.

- Minkowski spacetime has a pseudo-Euclidean metric $\eta_{\mu\nu} = \text{diag}(-1, 1, 1, 1)$, corresponding to an invariant interval $s^2 = -c^2 t^2 + x^2 + y^2 + z^2$.

- The magnitude-preserving transformation in spacetime is the Lorentz transformation, from which all the special relativistic effects such as time dilation, length contraction, and relativity of simultaneity can be derived. Thus, in this geometric formulation, we can think of the metric $\eta_{\mu\nu}$ as embodying all of special relativity.

- The Lorentz transformation is a rotation in 4D spacetime. If we write physics equations as 4D tensor equations, they are automatically Lorentz-symmetric. This motivates us to seek 4-tensor expressions of physical quantities such as the scalars (rank-0 tensors) of mass and proper time, the vectors (rank-1 tensors) of momentum and current, and the rank-2 electromagnetic field and energy-momentum-stress tensors.

- Understanding SR as a theory of flat geometry in Minkowski spacetime is a crucial step in the progression toward general relativity, in which, as we shall see, this geometric formulation involves warped spacetimes. The corresponding metric must be position-dependent, $g_{\mu\nu}(x)$; this metric acts as the generalized gravitational potential.

- In our historical introduction, SR seems to be all about light; however, the speed c actually plays a broader role in relativity:

  - c is the universal conversion factor between space and time units that allows space and time to be treated symmetrically. It is just the speed that allows $\Delta s^2 = -c^2 \Delta t^2 + \Delta \mathbf{x}^2$ to be an invariant interval under coordinate transformations. This allows Δs^2 to be viewed as the (squared) length in spacetime.

  - c is the universal absolute maximum speed of signal transmission. Massless particles (e.g., photons and gravitons) travel at the speed c, while all massive particles move at slower speeds (so as not to have infinite energy).

Review questions

1. (a) Give the definition of the metric tensor in terms of the basis vectors. (b) When the metric is displayed as a square matrix, what is the interpretation of its diagonal and off-diagonal elements? (c) In terms of the Minkowski metric, what is the invariant interval (the magnitude of the 4-displacement) ds^2 between two neighboring events with coordinate separation $(c\,dt, dx, dy, dz)$ in flat spacetime?

2. We say the Lorentz transformation is a rotation in Minkowski spacetime. Explain how this is so.

3. Given that the position 4-vector changes as $x^\mu \to x'^\mu = [L]^\mu{}_\nu x^\nu$, what are the transformation rules for a contravariant vector $A^\mu \to A'^\mu$, a covariant vector $A_\mu \to A'_\mu$, a mixed rank-2 tensor $T^\mu{}_\nu \to T'^\mu{}_\nu$, and an inner (scalar) product $A^\mu B_\mu \to A'^\mu B'_\mu$?

4. What are the components of the displacement 4-vector x^μ? Why is dx^μ/dt not a 4-vector? How is it related to the velocity 4-vector $\dot{x}^\mu$? Why doesn't this definition of the 4-velocity apply to massless particles? Write out the contravariant components of $\dot{x}^\mu$ and the square magnitude $\dot{x}^\mu \dot{x}_\mu$ of the 4-velocity for a particle with mass.

5. What are the definitions (in terms of particle mass and velocity) of the relativistic energy E and momentum $\mathbf{p}$ of a particle? Write their nonrelativistic limits. What components of the momentum 4-vector p^μ are the (relativistic) energy E and 3-momentum $\mathbf{p}$? How are they related to each other? What is the $m = 0$ limit of this relation? How can we conclude from this that a massless particle must always travel at c?

6. What does one mean by saying that the inhomogeneous Maxwell's equation, $\partial_\mu F^{\mu\nu} = -j^\nu/c$, is manifestly covariant? Show that this equation implies electric charge conservation.

7. Use Minkowski index notation to express in a compact form the meaning of the components of the current 4-vector $j^\mu = (j^0, \mathbf{j})$. Use this answer to define the energy–momentum tensor $T^{\mu\nu}$ and state the meanings of its various components.

8. In the spacetime diagram, display the timelike, spacelike, and lightlike regions. Also, draw in several lightcones with their vertices located along a worldline for some inertial observer.

9. The coordinate frame O' is moving at a constant velocity v in the $+x$ direction with respect to the coordinate frame O. Draw the transformed axes (x', ct') in a 2D spacetime diagram with axes (x, ct). (You are not asked to solve the Lorentz transformation equations, but only to justify the directions of the new axes.)

10. In coordinate frame O, two events, A and B, are simultaneous ($t_A = t_B$), but not equilocal ($x_A \neq x_B$). Use a spacetime diagram to show that these same two events take place at different times, $t'_A \neq t'_B$, to an observer in the O' frame (in motion with respect to the O frame).

11. In a spacetime diagram, depict two events with a temporal order of $t_A > t_B$ in the O frame such that they have a reversed order, $t'_A < t'_B$, in the O' frame. What is the condition (Exercise 3.14) that Δx, Δt, and v (the relative speed between the O and O' frames) must satisfy in order to have this reversal of temporal order? Explain why event A cannot possibly cause event B.

Equivalence of Gravitation and Inertia

- After a review of the Newtonian theory of gravitation in terms of its potential function, we take the first step in the study of general relativity (GR) with the introduction of the equivalence principle (EP).

- The weak EP, the equality of the gravitational and inertial masses, was extended by Einstein to the strong EP, the equivalence between inertia and gravitation for all interactions. This implies the existence of a local inertial frame (the frame in free fall) at every spacetime point. In a sufficiently small region, a local inertial observer will sense no gravitational effects.

- The equivalence of acceleration and gravity means that GR (physics laws valid in all coordinate systems, including accelerating frames) must be a theory of gravitation.

- The strong EP has physical consequences on time: gravitational redshift, time dilation, and the gravitational retardation of light speed leading to bending of light rays.

- Motivated by the EP, Einstein proposed a curved spacetime description of the gravitational field.

Soon after completing his formulation of special relativity (SR) in 1905, Einstein started working on a relativistic theory of gravitation. In this chapter, we cover mostly the period 1907–1911, when Einstein relied heavily on the equivalence principle (EP) to extract some general relativity (GR) results. Not until the end of 1915 did he work out fully the ideas of GR. By studying the consequences of the EP, he concluded that the proper language of GR is Riemannian geometry. In this, Einstein was helped by his ETH classmate and later colleague Marcel Grossmann (1878–1936). The mathematics of curved space will be introduced in Chapter 5. The curved spacetime representation of gravitational fields immediately suggests the geodesic equation as the GR equation of motion. In Chapter 6, we then begin the more difficult subject of the GR field equation and its solution.

A College Course on Relativity and Cosmology. First Edition. Ta-Pei Cheng.
© Ta-Pei Cheng 2015. Published in 2015 by Oxford University Press.

4.1 Seeking a relativistic theory of gravitation

Before discussing GR, Einstein's field theory of gravitation, we review Newton's theory, whose field equation and equation of motion can be expressed in terms of the gravitational potential.

4.1.1 Newtonian potential: a summary

Newton formulated his theory of gravitation using a force that acts instantaneously between distant objects (action-at-a-distance force):

$$\mathbf{F}(\mathbf{r}) = -G_{\mathrm{N}} \frac{mM}{r^2} \hat{\mathbf{r}}, \tag{4.1}$$

where G_{N} is Newton's constant, M the point-source mass, m the test mass, and $\mathbf{r}$ the displacement of m from M.

Just as in electrostatics, where electric field is force per unit charge, $\mathbf{F}(\mathbf{r}) = q\mathbf{E}(\mathbf{r})$, we can cast Newton's gravitational force in the form

$$\mathbf{F} = m\mathbf{g}. \tag{4.2}$$

This defines the gravitational field $\mathbf{g}(\mathbf{r})$ as the gravitational force per unit mass. In terms of this field, Newton's law of gravitation for a point source of mass M reads

$$\mathbf{g}(\mathbf{r}) = -G_{\mathrm{N}} \frac{M}{r^2} \hat{\mathbf{r}}. \tag{4.3}$$

Just as Coulomb's law is equivalent to Gauss's law for the electric field, this field (4.3) can be expressed for an arbitrary mass distribution as Gauss's law for the gravitational field:

$$\oint_S \mathbf{g} \cdot d\mathbf{A} = -4\pi\, G_{\mathrm{N}} M. \tag{4.4}$$

The area integral on the left-hand side is the gravitational field flux out of a closed surface S, and M on the right-hand side is the total mass enclosed inside S. This integral representation of Gauss's law (4.4) can be converted into a differential equation. We first turn both sides into volume integrals. On the left-hand side, we use the divergence theorem (the area integral of the field flux equals the volume integral of the divergence of the field), while on the right-hand side, we express the mass in terms of the mass density function ρ:

$$\int \nabla \cdot \mathbf{g}\, dV = -4\pi\, G_{\mathrm{N}} \int \rho\, dV. \tag{4.5}$$

Since this relation holds for any volume, the integrands on both sides must be equal:

$$\nabla \cdot \mathbf{g} = -4\pi\, G_{\mathrm{N}} \rho. \tag{4.6}$$

This is Newton's field equation in differential form. We define the gravitational potential[1] $\Phi(\mathbf{r})$ through the field with $\mathbf{g} \equiv -\boldsymbol{\nabla}\Phi$, so the field equation (4.6) becomes

$$\nabla^2\Phi = 4\pi G_\mathrm{N}\rho. \tag{4.7}$$

To obtain the gravitational equation of motion, we insert (4.2) into Newton's second law, $\mathbf{F} = m\mathbf{a}$, canceling mass to get

$$\frac{d^2\mathbf{r}}{dt^2} = \mathbf{g}. \tag{4.8}$$

Thus the gravitational motion of a particle is totally independent of any of its properties (mass, charge, etc.). The acceleration can be expressed in terms of the gravitational potential as

$$\frac{d^2\mathbf{r}}{dt^2} = -\boldsymbol{\nabla}\Phi. \tag{4.9}$$

We note that the Newtonian field theory of gravitation, as embodied in (4.7) and (4.9), is not compatible with SR, because space and time coordinates are not treated on equal footings. Newtonian theory is a *static* field theory. Stated in another way, these equations are comparable to Coulomb's law in electrostatics. They do not account for the effects of motion (i.e., magnetism). This incompleteness just reflects the underlying assumption of an action-at-a-distance force, which implies an infinite speed of signal transmission, incompatible with the basic postulate of relativity.

4.1.2 Einstein's motivation for general relativity

Einstein's theory of gravitation has a unique history. It was not prompted by any empirical failure of Newton's theory, but resulted from pure thought, the theoretical speculation of one person drawing the consequences of fundamental principles. It sprang fully formed from Einstein's mind. As an old physicists' saying goes, "Einstein just stared at his own navel and came up with GR!"

From Einstein's published papers, one can infer a few interconnected motivations:

1. **Seeking a relativistic theory of gravitation** The Newtonian theory is not compatible with special relativity, as it invokes an action-at-a-distance force, implying infinitely fast signal transmission. Furthermore, inertial frames of reference, which are fundamental to SR, lose their privileged status in the presence of gravity.

2. **"Space is not a thing"** This is how Einstein phrased his conviction that physics laws should not depend on reference frames, which express the relationships among physical processes in the world but do not have independent existence.

3. **Why are inertial and gravitational masses equal?** Einstein strove for a deeper understanding of this empirical fact.

[1] Recall the familiar example of the potential $\Phi = -G_\mathrm{N}M/r$ for a spherically symmetric source with total mass M.

Einstein generalized Newton's field equation (4.7) and the equation of motion (4.9) to develop a relativistic theory valid in all coordinate systems.

4.2 The equivalence principle: from Galileo to Einstein

This section presents several properties of gravitation. They all follow from what Einstein called the principle of the equivalence of gravity and inertia. The final formulation of Einstein's theory of gravitation, the general theory of relativity, automatically and precisely includes this EP. Historically, it motivated a series of discoveries that ultimately led Einstein to the geometric theory of gravity, which models the gravitational field as warped spacetime.

4.2.1   Inertial mass vs. gravitational mass

One of the distinctive features of gravity is that its equation of motion (4.9) is totally independent of the test particle's properties. This comes about because of the cancellation of the mass factors in mg and ma. Actually, these two masses correspond to very different concepts:

- The inertial mass m_I in Newton's second law,

$$\mathbf{F} = m_I \mathbf{a}, \tag{4.10}$$

enters into the description of the response of a particle to all forces.

- The gravitational mass m_G in Newton's law of gravitation,

$$\mathbf{F} = m_G \mathbf{g}, \tag{4.11}$$

reflects the response of a particle to a particular force: gravity. The gravitational mass m_G may be viewed (in analogy to electromagnetic theory) as the gravitational charge placed in a given gravitational field $\mathbf{g}$.

Now consider two objects, A and B, composed of different materials: one of copper and the other of wood. When they are let go in a given gravitational field $\mathbf{g}$ (e.g., dropped from the Leaning Tower of Pisa), they will, according to (4.10) and (4.11), obey the equations of motion:

$$(\mathbf{a})_A = \left(\frac{m_G}{m_I}\right)_A \mathbf{g}, \quad (\mathbf{a})_B = \left(\frac{m_G}{m_I}\right)_B \mathbf{g}. \tag{4.12}$$

Part of Galileo's great legacy is his experimental observation that all bodies fall with the same acceleration—that free fall is universal. The equality $(\mathbf{a})_A = (\mathbf{a})_B$ then leads to

$$\left(\frac{m_G}{m_I}\right)_A = \left(\frac{m_G}{m_I}\right)_B. \tag{4.13}$$

Because this mass ratio is universal for all substances, it can be set, by appropriate choice of units, equal to unity. We can simply say

$$m_I = m_G. \tag{4.14}$$

Even at the level of atomic physics, matter is made up of protons, neutrons, and electrons (all having different interactions) bound together with different binding energies. It is difficult to find an a priori reason to expect such a relation as (4.13). As we shall see, this is the empirical foundation underlying the geometric formulation of GR, the relativistic theory of gravity.

There is no record that Galileo ever measured the acceleration of freely falling objects (dropped from the Tower of Pisa or elsewhere)[2]. But he did measure the acceleration of objects sliding down an inclined plane; this slowed-down motion made measurements feasible. Newton achieved the same end by using a pendulum to measure the ratio m_I/m_G for different objects.

[2] Interestingly Galileo provided a theoretical argument, a thought experiment, in the first chapter of his *Discourse and Mathematical Demonstration of Two New Sciences*, in support of the idea that all substances should fall with the same acceleration. Consider any falling object: without this universality of free fall, the tendency of different components of the object to fall differently would give rise to internal stress and could cause certain objects to spontaneously disintegrate. The nonobservation of this phenomenon could then be taken as evidence for equal accelerations.

Exercise 4.1 Physical examples of m_I/m_G dependence

(a) For the frictionless inclined plane (with angle θ) in Fig. 4.1(a), find the acceleration's dependence on the ratio m_I/m_G. Thus a violation of the equivalence principle would show up as a material dependence in the time required for a material block to slide down the plane. (b) For the simple pendulum (with string length L) in Fig. 4.1(b), find the oscillation period's dependence on the ratio m_I/m_G.

4.2.2   Einstein: "my happiest thought"

In the course of writing a review article on SR in 1907, Einstein came upon what he later termed, "my happiest thought." He recalled the fundamental experimental result of Galileo that all objects fall with the same acceleration. Since all bodies accelerate the same way, an observer in a freely falling laboratory will not be able to detect any gravitational effect (on a point particle) in this frame. That is to say, gravity is transformed away in reference frames in free fall.

Principle of equivalence stated   Imagine an astronaut in a freely falling spaceship. Because all objects fall with the same acceleration, a released object will not fall with respect to the spaceship. Thus, from the viewpoint of the astronaut, gravity is absent; everything becomes weightless. To Einstein, this vanishing of the gravitational effect was so significant that he elevated it (in order to focus on it) to a physical principle, the equivalence principle:

$$\left(\begin{array}{c} \text{Physics in a frame freely falling in a gravitational field} \\ \text{is equivalent to} \\ \text{physics in an inertial frame without gravity} \end{array} \right).$$

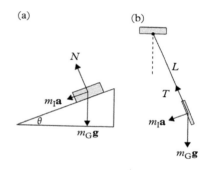

Figure 4.1 *Both the gravitational mass and inertial mass enter in these phenomena: (a) a sliding object on an inclined plane, where N is the normal force, and (b) oscillations of a pendulum, where T is the tension force in the string.*

Namely, within a freely falling frame, where the acceleration exactly cancels the uniform gravitational field, no sign of either acceleration or gravitation can be found by any physical means. Correspondingly,

$$\left(\begin{array}{c} \text{Physics in a nonaccelerating frame with a gravitational field } \mathbf{g} \\ \text{is equivalent to} \\ \text{physics in a frame without gravity but accelerating with } \mathbf{a} = -\mathbf{g} \end{array} \right).$$

Absence of gravity in an inertial frame   According to the EP, accelerating frames of reference can be treated in exactly the same way as frames with gravity. This suggests a new definition of an inertial frame, without reference to any external environment such as fixed stars, as a frame in which there is no gravity. Our freely falling spaceship can thereby be deemed an inertial frame of reference; our astronaut is now an inertial observer. Einstein realized the unique position of gravitation in the theory of relativity. Namely, he understood that the question was not how to incorporate gravity into SR, but rather how to use gravitation as a means to broaden the principle of relativity from inertial frames to all coordinate systems, including accelerating frames.

From EP to gravity as the structure of spacetime   If we confine ourselves to the physics of mechanics, the EP is just a restatement of $m_I = m_G$. But once promoted to a principle, it allowed Einstein to extend this equivalence between inertia and gravitation to all physics—not just to mechanics, but also to electromagnetism, etc. This generalized version is sometimes called the strong equivalence principle. Thus the weak EP is just the statement of $m_I = m_G$, while the strong EP is the principle of equivalence applied to all physics.[3] Henceforth, we shall still call the strong equivalence principle EP for short. Because the motion of a test body in a gravitational field is independent of the properties of the body, Einstein came up with the idea that the effect of gravity on the body can be attributed directly to some spacetime feature, and that gravity is nothing but the structure of a warped spacetime.

[3] Or, the EP says that one can form an inertial frame at any point in spacetime in which matter satisfies the laws of SR. The strong EP implies the validity of this equivalence principle for all laws of nature, not just mechanics.

[4] Referenced in, e.g., (French 1979, p 131).

Exercise 4.2 Two EP brain-teasers

Even in mechanics, in some instances, the (weak) EP can be very useful in helping us to obtain physics results with simple analysis. Here are two notable examples: (a) Use the EP to explain the observation (see Fig. 4.2a) that a helium balloon leans forward in a (forward-) accelerating vehicle. (b) On his 76th birthday, Einstein received a gift from his Princeton neighbor Eric Rogers.[4] It was a toy composed of a ball attached by a spring to the inside of a bowl (a toilet plunger), which was just the right size to hold the ball. The upright bowl was fastened to a broomstick as in Fig 4.2(b). What is the surefire way, suggested by the EP, to pop the ball back into the bowl each time?

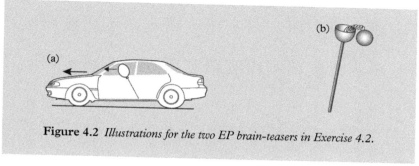

Figure 4.2 *Illustrations for the two EP brain-teasers in Exercise 4.2.*

4.3   EP leads to gravitational time dilation and light deflection

The strong EP implies that gravity can bend a light ray, shift the frequency of an electromagnetic wave, and cause clocks to run slow. Ultimately, these results suggested to Einstein that the proper framework to describe the relativistic effects of gravity is a curved spacetime. These gravitational time dilation phenomena will be interpreted as reflecting the warping of spacetime in the time direction.

4.3.1   Gravitational redshift and time dilation

To deduce the SR effects of relative motion, we often compare observations made in different coordinate frames. Similarly, one can obtain the effects of gravity on certain physical phenomena using the following general procedure:

1. One first considers the description by an observer inside a spaceship in free fall. According to the EP, there is no gravitational effect in this inertial frame, and SR applies.

2. One then considers the same situation from the viewpoint of an observer watching the spaceship from outside: there is a gravitational field in which the first (freely falling) observer is seen to be accelerating.

3. The combined effects of acceleration and gravity, as seen by the second observer, must then reproduce the SR description as recorded by the inertial observer in free fall. Physics should be independent of coordinate choices.

Bending of a light ray—a qualitative account

Let us first study the effect of gravity on a light ray traveling (horizontally) across a spaceship that is falling in a constant (vertical) gravitational field **g**. The EP informs us that from the viewpoint of the astronaut in the spaceship, there is no

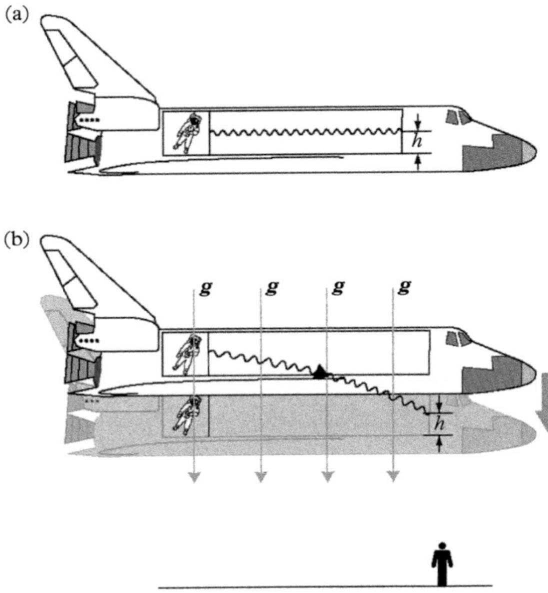

Figure 4.3 *According to the EP, a light ray will fall in a gravitational field. (a) To the astronaut in the freely falling spaceship (an inertial observer), the light trajectory is straight. (b) To an observer outside the spaceship, the astronaut is accelerating (falling) in a gravitational field. The light ray will be bent so that it reaches the opposite side at a height $y = gt^2/2$ below the initial point; it falls with the spaceship.*

detectable effect associated either with gravity or with acceleration: the light travels straight across the spaceship from one side to the other; it is emitted at a height h and received at the same height h on the opposite side of the spaceship as in Fig. 4.3(a). But to an observer outside, the spaceship is accelerating (falling) in a gravitational field **g**. To this outside observer, the light ray bends as it traverses the falling spaceship as in Fig. 4.3(b). Thus the EP predicts that gravity bends light.

We do not ordinarily see light curving; for the gravitational field and distance scale with which we are familiar, this bending effect is unobservably small. Consider a lab with a width of 300 m. A light ray travels across the lab in 1 μs. During this interval, the light drops (is bent by) an extremely small amount: $y = gt^2/2 \simeq 5 \times 10^{-12}\,\mathrm{m} = 0.05\,\text{Å}$. This suggests that one needs to seek such effects where large masses and great distances are involved—as in astronomical settings.

Gravitational redshift

Above, we discussed the effect of a gravitational field on a light ray whose trajectory is transverse to the field direction. Now let us consider the situation when the field is parallel (or antiparallel) to the ray's path as in Fig. 4.4.

Here we have a receiver placed at a distance h directly above the emitter in a downward-pointing gravitational field **g**. Just as in the transverse case considered above, we first describe the situation from the viewpoint of the astronaut (in free fall), Fig. 4.4(a). The EP informs us that the spaceship in free fall is an inertial

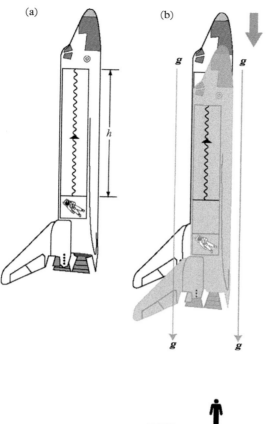

(a)

(b)

h

g g

g g

Figure 4.4 *The EP implies a gravitational redshift. Light is sent upward and received at a height h. (a) To an inertial observer in the freely falling spaceship, there is no frequency shift. (b) To an observer outside the spaceship, this astronaut is accelerating in a gravitational field. In this frame, a gravitational redshift cancels the Doppler blueshift, so both observers agree (as they must) that the frequency does not change.*

frame, with no physical effects associated with gravity or acceleration. The astronaut should not detect any frequency shift; the received light frequency ω_{rec} is the same as the emitted frequency ω_{em}:

$$(\Delta\omega)_{\text{ff}} = (\omega_{\text{rec}} - \omega_{\text{em}})_{\text{ff}} = 0, \qquad (4.15)$$

where the subscript ff reminds us that these are the values as seen by an observer in free fall.

From the viewpoint of the observer outside, the spaceship is accelerating (falling) in a gravitational field as in Fig. 4.4(b). Assume that this spaceship starts to fall at the moment of light emission. It takes a finite amount of time $\Delta t = h/c$ for the light signal to reach the receiver on the ceiling (to lowest order in gh/c^2, we can neglect the change in the receiver's position during the light's travel). During that time, the falling receiver accelerates to a downward velocity $\Delta u = g\Delta t$. The familiar Doppler formula in the low-velocity approximation (3.54) would lead us to expect a frequency shift of

$$\left(\frac{\Delta\omega}{\omega}\right)_{\text{Doppler}} = \frac{\Delta u}{c}. \tag{4.16}$$

Since the receiver above is moving toward the emitter below, the light waves must be compressed; this shift must be toward the blue:

$$(\Delta\omega)_{\text{Doppler}} = (\omega_{\text{rec}} - \omega_{\text{em}})_{\text{Doppler}} > 0. \tag{4.17}$$

But, as stated in (4.15), the inertial observer (in free fall) sees no such shift; the received frequency does not deviate from the emitted frequency. This physical result must hold for both observers, so the blueshift in (4.17) must somehow be canceled. To the observer outside the spaceship, gravity is also present. We can recover the nullshift result if the light is redshifted by gravity by just the right amount to cancel the Doppler blueshift of (4.16):

$$\left(\frac{\Delta\omega}{\omega}\right)_{\text{gravity}} = -\frac{\Delta u}{c}. \tag{4.18}$$

We now express the relative velocity on the right-hand side in terms of the gravitational potential difference $\Delta\Phi$ between the two locations:

$$\Delta u = g\Delta t = \frac{gh}{c} = \frac{\Delta\Phi}{c}. \tag{4.19}$$

By combining (4.18) and (4.19), we obtain the gravitational frequency shift:

$$\frac{\Delta\omega}{\omega} = -\frac{\Delta\Phi}{c^2}; \tag{4.20}$$

namely,[5]

$$\frac{\omega_1 - \omega_2}{\omega_2} = -\frac{\Phi_1 - \Phi_2}{c^2}. \tag{4.21}$$

[5] Whether the denominator is ω_1 or ω_2, the difference is of higher order and can be ignored in these leading-order formulae.

Light emitted at a lower gravitational potential (Φ_2) will be received at a higher gravitational potential ($\Phi_1 > \Phi_2$) with a lower frequency ($\omega_1 < \omega_2$); that is, it is redshifted, even though the emitter and the receiver are not in relative motion. Likewise, light emitted at a higher potential appears blueshifted to a receiver at a lower potential.

The Pound–Rebka experiment In principle, this gravitational redshift can be tested by a careful examination of the spectral emission lines from an astronomical object (hence from a deep gravitational potential well). However, such an effect can easily be masked by the standard Doppler shifts due to the thermal motion of the emitting atoms. Consequently, conclusive data did not exist in the first few decades after Einstein's paper. It was not until 1960 that the gravitational redshift was first convincingly verified in a series of terrestrial experiments in which Robert Pound (1919–2010) and his collaborators measured the tiny frequency shift of radiation traveling up $h = 22.5$ m, the height of an elevator shaft in the building housing the Harvard Physics Department:

$$\left|\frac{\Delta\omega}{\omega}\right| = \left|\frac{gh}{c^2}\right| = O(10^{-15}). \tag{4.22}$$

Normally, it is not possible to fix the frequency of an emitter or absorber to a very high accuracy, because of the energy shift due to thermal recoils of the atoms. However, owing to the **Mössbauer effect**,[6] the emission line width in a rigid crystal is as narrow as possible—limited only by the quantum mechanical uncertainty principle, $\Delta t \Delta E \geq \hbar$, where Δt is the lifetime of the unstable (excited) state. Thus a long-lived state would have a particularly small energy/frequency spread. The Harvard experimenters (Pound and Rebka 1960) worked with an excited isotope of iron, $^{57}Fe^*$, produced through the nuclear beta decay of cobalt-57. It transitions to the ground state by emitting a gamma ray: $^{57}Fe^* \rightarrow {}^{57}Fe + \gamma$. The gamma ray emitted this way at the bottom of the elevator shaft, after climbing the 22.5 m, could no longer be resonantly absorbed by a sheet of iron in the ground state placed at the top of the shaft. To prove that the radiation had been redshifted by just the right $O(10^{-15})$ amount, Pound and Rebka introduced an (ordinary) Doppler blueshift by moving the detector slowly toward the emitter at just the right speed to compensate for the gravitational redshift. Thus, the radiation could again be absorbed. What was the speed with which they had to move the receiver? From (4.20) and (4.16), we have

$$\frac{gh}{c^2}\bigg|_{\text{gravity}} = \left|\frac{\Delta\omega}{\omega}\right|_{\text{Doppler}} = \frac{u}{c}, \tag{4.23}$$

with

$$u = \frac{gh}{c} = \frac{9.8 \text{ m/s}^2 \times 22.5 \text{ m}}{3 \times 10^8 \text{ m/s}} = 7.35 \times 10^{-7} \text{ m/s}. \tag{4.24}$$

This is such a small speed that it would take $h/u = c/g = O(3 \times 10^7 \text{ s}) \simeq 1$ year to cover the same elevator shaft height. Of course this velocity is just the one attained by an object freely falling for the time interval $O(10^{-7} \text{ s})$ that it takes the light to climb the elevator shaft. This is the compensating effect we invoked in our derivation of the gravitational redshift at the beginning of this section.

Gravitational time dilation

From our understanding of the Doppler effect, this gravitational frequency shift looks absurd. How can an observer, **stationary** with respect to the emitter, receive

---

[6] The Mössbauer effect: Atomic recoil can reduce the energy of an emitted photon. Since the emitting atom is surrounded by other atoms in thermal motion, this recoil is uncontrollable. (We can picture the atom as part of a vibrating lattice.) As a result, the photon energies in different emission events can vary considerably, resulting in a significant spread of their frequencies. This rules out high-precision measurements of the atomic frequency for purposes such as testing the gravitational redshift. But, in 1958, Rudolf Mössbauer (1929–2011) made a breakthrough when he pointed out, and verified by observation, that crystals with high Debye–Einstein temperature (i.e., having a rigid crystalline structure) could pick up the recoil by the entire crystal. Namely, in such a situation, the emitting atom has a huge effective mass. Consequently, the atom carries away no recoil energy; the photon can pick up all the energy lost by the emitting atom, and the frequency of the emitted radiation is as precise as it can be.

a different number of wave crests per unit time than the emitted rate? Here is Einstein's radical yet simple answer: while the number of wave crests does not change, the time unit itself changes in the presence of gravity. Clocks at different gravitational potentials run at different rates; there is a *gravitational time dilation* effect.

Frequency is proportional to the inverse of the local proper time rate $\omega \sim 1/d\tau$; the gravitational frequency shift formula (4.21) can be converted to a time dilation formula:

$$\frac{d\tau_1 - d\tau_2}{d\tau_2} = \frac{\Phi_1 - \Phi_2}{c^2}. \tag{4.25}$$

Namely, a clock at higher gravitational potential ($\Phi_1 > \Phi_2$) will run faster ($d\tau_1 > d\tau_2$); a lower clock runs slow. The fast/slow descriptions reflect the larger/smaller elapsed time intervals. All observers agree on this, since $d\tau_1$ and $d\tau_2$ are scalars. Another derivation of (4.25) will be presented in Box 4.1.

Contrast this gravitational time dilation with the distinct SR effect of (2.22): $dt = \gamma \, d\tau$. If two observers are in relative motion, each perceives the other's clock to run slow by a factor of γ. Obviously, one must be careful to understand which clocks rate is being described and by whom. Recall the twin paradox discussed in Box 2.3. The time dilation of Al's clock in motion means its time $d\tau_A$ is measured by a stationary Bill to run slow, $dt_B > d\tau_A$ as $dt_B = \gamma_A \, d\tau_A$; when the twins meet, Bill has aged 50 years compared with Al's 30. Consider a comparable case in which Bill lives on the top floor of a high-rise building and Al at the bottom ($\Phi_B > \Phi_A$). The twins will again age differently, now because of gravitational time dilation. Bill's clock runs faster; as the years pass by, he will be older than the low-level-dwelling Al. In Section 4.3.2, orbiting satellites provide yet another concrete example illustrating these different types of time dilation.

Box 4.1 Gravitational time dilation—another derivation

As discussed in our SR chapters, the standard method to compare clock readings is through the exchange of light signals. Our way of arriving at the gravitational time dilation result (4.25) followed this standard method—an exchange of light signals and a comparison of light frequencies—leading to the gravitational redshift and then time dilation results.

Here we present another derivation of (4.25) that will display its relation to (and its compatibility with) the familiar SR effect of time dilation due to relative motion. Two clocks are located at different gravitational potential points, Clock-1 at Φ_1 and Clock-2 at Φ_2. Let Clock-3 fall freely in this gravitational field (see Fig. 4.5); when it passes Clock-1, it has the speed u_1 and when it passes Clock-2, the speed u_2. At the instant when Clock-3 passes Clock-1, both clocks are at the *same* gravitation potential Φ_1. Thus a comparison of the clocks' rates involves only the SR effect due to their relative motion of speed u_1. A similar comparison can be carried out when Clock-3 passes Clock-2 at Φ_2 with

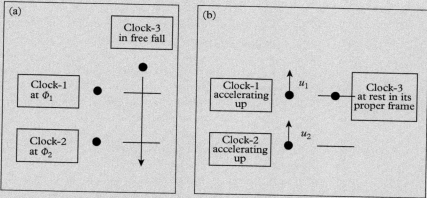

Figure 4.5 *Comparing clock rates at different gravitational potential points: (a) Clock-3 in free fall passing Clock-1 at Φ_1 with speed u_1 and Clock-2 at Φ_2 with speed u_2. (b) Clock-1 is seen moving with speed $-u_1$ in the inertial (free-fall and gravity-free) frame of Clock-3 when the latter passes Clock-1. A picture similar to (b) can be drawn when Clock-3 passes Clock-2.*

speed u_2. One can relate the proper clock rates $d\tau_1$ and $d\tau_2$ to the coordinate time rate in Clock-3's freely falling frame of reference. By the SR time dilation formula at the respective Φ_1 and Φ_2 points,

$$dt_3^{(1)} = \gamma_1\, d\tau_1, \quad \text{and} \quad dt_3^{(2)} = \gamma_2\, d\tau_2, \tag{4.26}$$

with $\gamma_{1,2} = (1 - u_{1,2}^2/c^2)^{-1/2}$. Because Clock-3 is in free fall, gravity is absent in its inertial reference frame; time dt_3 passes at a constant rate as the other clocks accelerate by: $dt_3^{(1)} = dt_3^{(2)}$. The time dilation result (4.25) can then be derived by connecting the two equations in (4.26):

$$\frac{d\tau_1}{d\tau_2} = \frac{\gamma_2}{\gamma_1} = \left(\frac{1 - u_1^2/c^2}{1 - u_2^2/c^2} \right)^{1/2}$$

$$\simeq 1 - \frac{1}{2} \frac{u_1^2 - u_2^2}{c^2} = 1 + \frac{\Phi_1 - \Phi_2}{c^2}, \tag{4.27}$$

where, to reach the second line, we have dropped terms $O(u^4/c^4)$ in the power series expansions of the denominator and of the square root. At the last equality, we have used the low-velocity version (consistent with our presentation) of the energy conservation relation for the freely falling Clock-3. The change in kinetic energy must equal minus the change in potential energy: $\frac{1}{2} m \Delta u^2 = -m \Delta \Phi$.

This derivation of (4.25) shows that gravitational time dilation is entirely compatible with the previously known SR time dilation effect—just as we showed its compatibility with the Doppler frequency shift (4.16) in our first derivation of gravitational redshift (4.20).

4.3.2   Relativity and the operation of GPS

The Global Positioning System (GPS) is capable of fixing locations on earth by exchange of electromagnetic signals between the ground and a network of orbiting satellites. More than 24 satellites are distributed more-or-less uniformly among orbits around the globe (e.g., four satellites on each of six equally spaced orbital planes), so that for any point on earth, $\mathbf{r}$, there are at least four satellites at $\mathbf{r}_i$ ($i = 1, 2, 3, 4$) above its horizon (Fig. 4.6). Radar signals encode each satellite's position and time at transmission ($\mathbf{r}_i, t_i$). The GPS receiver on the ground then calculates the four unknowns of its position and time of reception ($\mathbf{r}, t$) by solving the four simultaneous equations $|\mathbf{r}_i - \mathbf{r}| = c|t_i - t|$. Since distance is obtained from timing measurements, $\Delta r = c\Delta t$, one needs an extremely accurate knowledge of the transmission time, as a difference of one nanosecond (10^{-9} s) translates into a distance of about one foot, 10^{-9} s $\times$ 3×10^8 m/s $= 0.3$ m.

To achieve such accuracy, each satellite carries an atomic clock. Still, all the time measurements must account for the fact that the clocks on the satellites are moving at high speeds and are at different gravitational potentials with respect to the ground location. How large are the resulting SR and gravitational time dilation effects? In the next paragraph, we carry out an order-of-magnitude calculation to show that both effects are significant, and furthermore that the gravitational effect is opposite to, and about six times bigger than, the effect due to relative motion.

Satellites high up in space and moving with high speed   To determine the relativistic effects, we need to know the orbital radius of the satellite (r_S) and how fast it is moving (v_S). Each of the satellites is set to have a period of 12 hours, so that a satellite passes overhead of a given observer on earth twice each day. We will find (r_S, v_S) by solving the coupled equations representing the orbital period (12 hours) and the gravitational equation of motion with the centripetal acceleration $a = v^2/r$:

$$\frac{2\pi r_S}{v_S} = 12 \text{ h} = 12 \times 3600 \text{ s},$$

$$\frac{G_N M_\oplus}{r_S^2} = \frac{F}{m} = a = \frac{v_S^2}{r_S}, \tag{4.28}$$

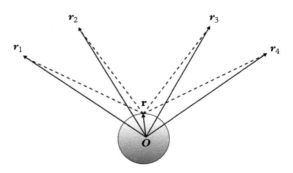

Figure 4.6 *Light signals are exchanged between a location* $\mathbf{r}$ *on earth and four GPS satellites located at* $\mathbf{r}_i$ ($i = 1,2,3,4$).

where $M_\oplus$ is the mass of the earth. Solving these two equations, we obtain the satellite's orbital radius in terms of the earth's radius $r_\oplus$,

$$r_S = 2.7 \times 10^7 \text{m} = 4.2 \, r_\oplus, \tag{4.29}$$

and the tangential speed $v_S = 3.87$ km/s. Such a velocity gives a relativistic beta factor of $\beta_S = v_S/c = 1.3 \times 10^{-5}$ and a Lorentz gamma factor of

$$\gamma_S = (1 - \beta_S^2)^{-1/2} = 1 + 0.83 \times 10^{-10}. \tag{4.30}$$

Time dilation due to relative motion From this, we can calculate the SR time dilation correction, as both the satellite clock and the earthbound clock are in motion. Let t_0 be the time recorded by a clock at rest with respect to the coordinate origin (the center of the earth). The time dilation of the moving satellite clock is $dt_0 = \gamma_S \, d\tau_S$ and that of the clock on earth, $dt_0 = \gamma_\oplus \, d\tau_\oplus$. Thus the relative dilation effect due to the motion of the satellite and the ground clocks around the earth is given by

$$\left(\frac{d\tau_\oplus}{d\tau_S}\right)_{\text{mo}} = \frac{\gamma_S}{\gamma_\oplus} = 1 + \frac{v_S^2 - v_\oplus^2}{2c^2}. \tag{4.31}$$

Since the ground clock speed $v_\oplus$ depends on its location on earth and is small,[7] we will drop the $v_\oplus^2$ term. The fractional time difference due to relative motion is, according to (4.30),

$$\mathcal{F}_{\text{mo}} \equiv \left(\frac{d\tau_\oplus - d\tau_S}{d\tau_S}\right)_{\text{mo}} = \left(\frac{d\tau_\oplus}{d\tau_S}\right)_{\text{mo}} - 1$$
$$= \frac{1}{2}\beta_S^2 = 0.83 \times 10^{-10}. \tag{4.32}$$

[7] This is the case even for a receiver located on the equator with $v_\oplus = 2\pi R_\oplus/24$ h. Comparing it with $v_S = 2\pi r_S/12$ h, we have $v_S/v_\oplus = 2r_S/R_\oplus = 8.4$; thus the correction $(v_\oplus/v_S)^2$ is only about 1%.

With $d\tau_S < d\tau_\oplus$, we say that the satellite clock runs slower than the clock on the ground. This is so because the satellite time is dilated more than the ground clock, reflecting their relative speed $\gamma_S > \gamma_\oplus$.

Time dilation due to relative heights The gravitational time dilation formula (4.25) yields a fractional time difference of

$$\mathcal{F}_{\text{grav}} \equiv \left(\frac{d\tau_\oplus - d\tau_S}{d\tau_S}\right)_{\text{grav}} = \frac{\Phi_\oplus - \Phi_S}{c^2}, \tag{4.33}$$

where $\Phi_S > \Phi_\oplus$ as the satellite is located at a potential that is less negative,

$$\Phi_\oplus = -\frac{G_N M_\oplus}{r_\oplus} \quad \text{and} \quad \Phi_S = -\frac{G_N M_\oplus}{r_S} = \frac{\Phi_\oplus}{4.2}. \tag{4.34}$$

In the last term, we have inserted the result from (4.29). Given that $\Phi_\oplus = -gr_\oplus$, where $g = 9.8\,\text{m/s}^2$ is the gravitational acceleration at the earth's surface, one immediately obtains

$$\mathcal{F}_{\text{grav}} = \left(\frac{d\tau_\oplus}{d\tau_S}\right)_{\text{grav}} - 1 \tag{4.35}$$

$$= -\frac{gr_\oplus}{c^2}\left(1 - \frac{1}{4.2}\right) = -5.3 \times 10^{-10}.$$

With the elapsed satellite time interval being larger, $d\tau_S > d\tau_\oplus$, we say that the satellite clock runs faster than the clock on the ground, in contrast to the effect due to relative motion.

Full relativistic correction   We see that this gravitational correction is about six times as great as that due to SR time dilation (4.32) and in the opposite direction. When we combine the leading-order corrections due to the gravitational potential difference (4.33) and to the relative motion (4.31), the total relativistic correction may be written as

$$\left(\frac{d\tau_\oplus}{d\tau_S}\right)_{\text{rel}} = \left(\frac{d\tau_\oplus}{d\tau_S}\right)_{\text{grav}} \times \left(\frac{d\tau_\oplus}{d\tau_S}\right)_{\text{mo}} \tag{4.36}$$

$$= 1 + \frac{\Phi_\oplus - \Phi_S}{c^2} - \frac{v_\oplus^2 - v_S^2}{2c^2}.$$

As we shall see, this result is contained in the full GR solution to be discussed in Chapter 6 (cf. Example 6.1). Without taking the relativistic correction into consideration, errors will rapidly accumulate in the GPS time measurements. For example, in a period of only one minute, by ignoring time dilation due to relative motion, one would incur an error of $\mathcal{F}_{\text{mo}} \times 60\,\text{s} \simeq 5\,\text{ns}$, while the error from ignoring the gravitational time dilation would be $\mathcal{F}_{\text{grav}} \times 60\,\text{s} \simeq -30\,\text{ns}$; taken together, one gets an error of about 25 ns. As discussed in the introductory paragraph, one nanosecond corresponds to a distance of about a foot. Without taking relativistic effects into account, there is no way that the GPS system could pinpoint locations within a few feet or so in accuracy.

Exercise 4.3   A GPS calculation

We expect that the gravitational time dilation could be reduced if the satellite traveled in a lower orbit. Figure out how low the satellite's orbital radius r_S must be so that the time dilation effects due to gravity and due to relative motion cancel each other? What will be its period? Just as we did above, you may neglect $v_\oplus$. (a) From the requirement that $\mathcal{F}_{\text{mo}} + \mathcal{F}_{\text{grav}} = 0$, you should be able to deduce the relation of velocity v_S to orbital radius r_S:

$$v_S^2 = 2gr_\oplus\left(1 - \frac{r_\oplus}{r_S}\right). \tag{4.37}$$

Note that for this problem, it is entirely adequate to approximate the gamma factor by $\gamma \equiv (1-\beta^2)^{-1/2} \simeq 1 + \frac{1}{2}\beta^2$. (b) From the centripetal acceleration equation, together with (4.37), one can then find the solution:

$$v_S^2 = \frac{2}{3} g r_\oplus \qquad \text{and} \qquad r_S = \frac{3}{2} r_\oplus. \tag{4.38}$$

(c) What will be the resultant orbital period (in contrast to the real GPS system's 12-hour period)? From these numbers, you should be able to conclude that such a system of low-flying satellites may have some practical problems.

4.3.3 The EP calculation of light deflection

The EP implies that clocks run at different rates at locations where the gravitational potentials are different. Such effects will lead to different speed measurements—even the speed of light can be measured to have different values! We are familiar with light speeds in different media being characterized by varying indices of refraction. Gravitational time dilation implies that even in vacuum there is an effective index of refraction when a gravitational potential is present. Since potential usually varies in space (i.e., its gradient, the gravitational field, is usually nonzero), this index is generally a position-dependent function.

Gravity-induced index of refraction in free space At a given position r with gravitational potential $\Phi(r)$, a determination of light speed with respect to the local proper time $d\tau$ and local proper length $d\rho$ gives

$$\frac{d\rho}{d\tau} = c. \tag{4.39}$$

This speed c is a universal constant. On the other hand, the light speed will deviate from c according to the elapsed time and distance displacement measured by the clock and ruler at a different position. In fact, a common choice of coordinate time and distance is that given by the clock and ruler located far away from the gravitational source. Equation (4.25) then gives us the relation between the local time ($d\tau = d\tau_1$) measured at r and the coordinate time ($dt = d\tau_2$) measured where $\Phi_2 = 0$:

$$d\tau = \left(1 + \frac{\Phi(r)}{c^2}\right) dt. \tag{4.40}$$

What about the gravitational effect on length measurement? The deflection of light by a gravitational source was first predicted by Einstein in 1911. The calculation was based on EP-implied gravitational time dilation alone. This was before Einstein had developed the idea of gravity as structure of spacetime. Since most of the first strong-EP effects to be discussed were those of gravity on time measurement, Einstein did not discuss in his paper the influence of gravity on length measurement. Thus, in effect, he set $d\rho = dr$, and this is what we do here as well. In this way, to a remote observer, light speed is reduced by gravity[8] (Φ being negative):

$$[c(r)]_{\mathrm{EP}} \equiv \frac{dr}{dt} = \frac{1 + \Phi(r)/c^2}{d\tau}\, d\rho = \left(1 + \frac{\Phi(r)}{c^2}\right) c, \tag{4.41}$$

which varies from position to position as the gravitational potential varies. For the faraway observer, the effect of the gravitational field can be viewed as introducing an index of refraction in space:

$$\frac{1}{[n(r)]_{\mathrm{EP}}} \equiv \frac{[c(r)]_{\mathrm{EP}}}{c} = \left(1 + \frac{\Phi(r)}{c^2}\right). \tag{4.42}$$

Let us reemphasize some key concepts behind this position-dependent speed of light. We are not suggesting that the deviation of $c(r)$ from the constant c means that the speed of light measured by a local observer has changed, or that the velocity of light is no longer a universal constant in the presence of gravitational fields. Rather, it reflects the physics that clocks at different gravitational potentials run at different rates. For an observer located far from the gravitational source (whose proper time is conveniently taken to be the coordinate time), the velocity of light appears to slow down. A dramatic example is offered by black holes (to be discussed in Chapter 7). Because of infinite gravitational time dilation, it would take an infinite amount of coordinate time for a light signal to leave a black hole (thus, to the remote observer, no light can escape from a black hole), even though to a local observer, his proper time seems to flow normally.

Bending of light ray calculated using Huygens' construction   We can use this position-dependent index of refraction to calculate the bending of a light ray by a transverse gravitational field via the Huygens' construction. Consider a plane light wave propagating in the $+x$ direction. At each time interval Δt, the wavefront advances a distance $c\Delta t$; see Fig. 4.7(a). The existence of a transverse gravitational field (in the y direction) means a nonvanishing derivative of the gravitational potential, $d\Phi/dy \neq 0$. A change of the gravitational potential implies a change in $c(r)$, which leads to tilting of the wavefronts. We can calculate the angle of the bending of the light ray by using the diagram in Fig. 4.7(b):

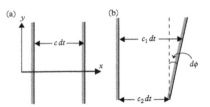

Figure 4.7 *Wavefronts of light trajectories: (a) Wavefronts of a straight-moving trajectory in the absence of gravity. (b) Tilting of wavefronts in a medium with an index of refraction varying in the vertical y direction so that $c_1 > c_2$. The bending of the resultant trajectory is signified by the small angular deflection $d\phi$.*

$$d\phi \simeq \frac{(c_1 - c_2)\, dt}{dy} \simeq \frac{d[c(r)](dx/c)}{dy}. \tag{4.43}$$

Working in the limit of weak gravity with small $\Phi(r)/c^2$ (or equivalently $n \simeq 1$), we can relate $d[c(r)]$ to a change in the index of refraction via (4.42):

$$d\left[\frac{c(r)}{c}\right]_{\mathrm{EP}} = d\left[\frac{1}{n(r)}\right]_{\mathrm{EP}} = \frac{d\Phi(r)}{c^2}. \tag{4.44}$$

Thus, integrating (4.43) over the entire path, we obtain the total deflection angle:

$$[\delta\phi]_{\rm EP} = \int [d\phi]_{\rm EP} = \frac{1}{c^2} \int_{-\infty}^{\infty} \frac{\partial\Phi}{\partial y}\, dx = \frac{1}{c^2} \int_{-\infty}^{\infty} (\boldsymbol{\nabla}\Phi \cdot \hat{\mathbf{y}})\, dx. \tag{4.45}$$

The integrand is the gravitational acceleration perpendicular to the light path. We shall apply this formula to the case of a spherical source with $\Phi = -G_{\rm N}M/r$ and $\boldsymbol{\nabla}\Phi = G_{\rm N}M\hat{\mathbf{r}}/r^2$. Although the gravitational field will no longer be a simple uniform field in the $\hat{\mathbf{y}}$ direction, our approximate result can still be used, because the bending takes place mostly in the small region of $r \simeq r_{\rm min}$. (See Fig. 4.8.) We have

$$[\delta\phi]_{\rm EP} = \frac{G_{\rm N}M}{c^2} \int_{-\infty}^{\infty} \frac{\hat{\mathbf{r}}\cdot\hat{\mathbf{y}}}{r^2}\, dx = \frac{G_{\rm N}M}{c^2} \int_{-\infty}^{\infty} \frac{y}{r^3}\, dx, \tag{4.46}$$

where we have used $\hat{\mathbf{r}}\cdot\hat{\mathbf{y}} = \cos\theta = y/r$. An inspection of Fig. 4.8 also shows that, for small deflection, we can approximate $y \simeq r_{\rm min}$; hence

$$r = (x^2 + y^2)^{1/2} \simeq (x^2 + r_{\rm min}^2)^{1/2}, \tag{4.47}$$

leading to

$$[\delta\phi]_{\rm EP} = \frac{G_{\rm N}M}{c^2} \int_{-\infty}^{\infty} \frac{r_{\rm min}}{\left(x^2 + r_{\rm min}^2\right)^{3/2}}\, dx = \frac{2G_{\rm N}M}{c^2 r_{\rm min}}. \tag{4.48}$$

It should be noted again: this deflection result follows from an EP calculation which is based on gravitational time dilation alone. As we shall discuss in Section 6.4.2, this is half of the GR result $[\delta\varphi]_{\rm GR} = 2[\delta\varphi]_{\rm EP}$, because the curved spacetime description of GR implies in addition a gravitational length contraction.

4.3.4 Energetics of light transmission in a gravitational field

Because light gravitates (i.e., it bends and redshifts in a gravitational field), it is tempting to imagine that a photon has a (gravitational) mass. This may well lead to some erroneous conclusions regarding the energetics of light traveling in a gravitational field. Box 4.2 addresses this problem.

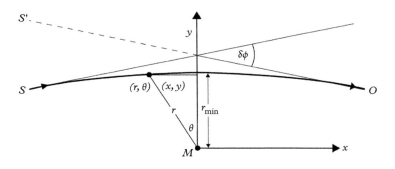

Figure 4.8 *Angle of deflection $\delta\phi$ of light by a mass M. A point on the light trajectory (solid curve) can be labeled either as (x, y) or as (r, θ). The source at S would appear to the observer at O to be located at a shifted position S'.*

[9] For instance, the (Pound and Rebka 1960) paper has the title, "Apparent weight of photons."

[9a] Equation (4.49) is quoted in small-angle approximation of a general result that can be found in textbooks on mechanics. See, e.g., Eq. (4.37) in (Kibble 1985).

[10] We have used the fact that the energy of a light ray is proportional to its frequency. For most of us, the quantum relation $E = \hbar\omega$ comes immediately to mind, but this proportionality also holds in classical electromagnetism, where the field can be pictured as a collection of harmonic oscillators; see, e.g., (Cheng 2013, Section 3.1).

Box 4.2 Energy considerations for gravitating light

Erroneous energy considerations

When considering the propagation of a light ray in a gravitational field, one might argue as follows: from the viewpoint of relativity, there is no fundamental difference between mass and energy, $E = m_I c^2$. The equivalence $m_I = m_G$ means that any energy also has a nonzero gravitational mass $m_G = E/c^2$, and hence will gravitate. The gravitational redshift formula (4.20) can be derived by regarding such a light pulse as losing kinetic energy when climbing out of a gravitational potential well.[9] Applying similar reasoning to the problem of gravitational light deflection, one can derive the result (4.48) by using the Newtonian mechanics formula[9a] for a moving mass with velocity u being gravitationally deflected by a spherically symmetric mass M (just as in Fig. 4.8),

$$[\delta\phi]_{\text{Newton}} = \frac{2G_N M}{u^2 r_{\min}}. \qquad (4.49)$$

For the case of a photon with $u = c$, this would just reproduce the EP result shown in (4.48). Thus it so happens that the EP and Newtonian results coincide. As stated earlier, the predicted deflection is half of the correct GR prediction. Nevertheless, such an approach to understanding the effect of gravity on a light ray is conceptually incorrect, because

- A photon has no mass, so it cannot be described as a nonrelativistic massive object having a gravitational potential energy.

- This approach makes no connection to the underlying physics of gravitational time dilation.

The correct energy consideration

The energetics of gravitational redshift should be properly considered as follows. Light is emitted and received through atomic transitions between two atomic energy levels of a given atom:[10] $E_1 - E_2 = \hbar\omega$. We can treat the emitting and receiving atoms as nonrelativistic massive objects. Thus, when sitting at a higher gravitational potential, the receiver atom has more energy than the emitter atom:

$$E_{\text{rec}} = E_{\text{em}} + mgh. \qquad (4.50)$$

We can replace the mass by E/c^2, so that, to leading order, $E_{\text{rec}} = (1 + gh/c^2)E_{\text{em}}$. There is a multiplicative energy shift of the atomic levels. This implies that all the energy levels (and their differences) of the receiving atom are blueshifted (increased energy, increased frequency) with respect to those of the emitter atom by

$$(E_1 - E_2)_{\text{rec}} = \left(1 + \frac{gh}{c^2}\right)(E_1 - E_2)_{\text{em}};\qquad(4.51)$$

hence there is a fractional shift of atomic energy

$$\left(\frac{\Delta E}{E}\right)_{\text{atom}} = \frac{gh}{c^2} = \frac{\Delta\Phi}{c^2}.\qquad(4.52)$$

On the other hand, the traveling light pulse, neither gaining nor losing energy along its trajectory, has the *same* energy as the emitting atom. But it will be seen by the blueshifted receiver atom as redshifted:

$$\left(\frac{\Delta E}{E}\right)_{\gamma} = -\frac{\Delta\Phi}{c^2} = \frac{\Delta\omega}{\omega},\qquad(4.53)$$

which is the previously obtained result (4.20). This approach is conceptually correct, as

- Atoms can be treated as nonrelativistic objects having gravitational potential energy *mgh*.
- This derivation is entirely consistent with gravitational time dilation. The gravitational frequency shift does not result from any change in photon properties. It comes about because the standards of frequency (i.e., time) are different at different locations. In fact, this present approach gives us a physical picture of why clocks must run at different rates at different gravitational potentials. An atom is the most basic form of a clock, whose time rates are determined by transition frequencies. The fact that atoms have different gravitational potential energies (hence different energy levels) naturally give rise to different transitional frequencies, and hence different clock rates.

The various results called "Newtonian"

We should also clarify the often-encountered practice of calling results such as (4.49) Newtonian. By this it is meant that the result can be derived in the pre-Einsteinian-relativity framework, in which particles can take on any speed we wish them to have. Consequently, it is entirely correct to use the Newtonian mechanics formula for a light particle that happens to propagate with speed *c*. However, one should be aware of the difference between this Newtonian (pre-relativistic) framework and the proper Newtonian limit of relativistic physics (which we shall specify in Section 5.3.3) of nonrelativistic velocity and a static weak gravitational field. In this contemporary sense, (4.48) is not a result valid in the Newtonian limit (cf. Section 6.4.2).

Einstein's inference of a curved spacetime

[11] In particular, that the gravitational equation of motion (4.9) is totally independent of any property of the test particle suggested to Einstein that the gravitational field, unlike other force fields, is related to some fundamental feature of spacetime.

Aside from the principle of relativity, the EP is the most important physical principle underlying Einstein's formulation of a geometric theory of gravity. Not only did it allow accelerating frames to be treated on equal footing with inertial frames, thus giving early glimpses of GR phenomenology, but also the study of the EP physics of time change led Einstein to propose that gravity represents the structure of a curved spacetime.[11] We shall explain this connection after learning in the following chapters some mathematics of curved spaces.

Review questions

1. Write out, in terms of the gravitational potential $\Phi(x)$, the field equation and the equation of motion for Newton's theory of gravitation. What is the distinctive feature of this equation of motion (as opposed to that for other forces)?

2. What is inertial mass? What is gravitational mass? Give the simplest experimental evidence that their ratio is a universal constant (i.e., independent of the material composition of the object).

3. What is the equivalence principle? What is the weak EP? The strong EP?

4. Give a qualitative argument that the EP implies gravitational bending of a light ray.

5. Provide two derivations of the formula for gravitational frequency shift: $\Delta\omega/\omega = -\Delta\Phi/c^2$. (a) Use the idea that gravity can be transformed away by taking a reference frame in free fall. (b) Use the idea that atomic energy levels will be shifted in a gravitational field.

6. Derive the gravitational time dilation formula, $\Delta\tau/\tau = \Delta\Phi/c^2$, in two ways: (a) from the gravitational frequency shift formula; (b) from the consideration of three identically constructed clocks—two stationary at potential points Φ_1 and Φ_2, and the third in free fall passing by the first two.

7. GPS requires very precise time reading of clocks on the satellites that send electromagnetic signals to fix locations within a few meters on earth. What relativistic effects must be taken into account in order for this arrangement to work?

8. The presence of a gravitational field implies the presence of an effective index of refraction in free space. Does this mean that the speed of light is not absolute? What is the physical consequence of this index of refraction. When viewed from later developments, why was Einstein's 1911 calculation not complete?

9. Find the deviation from c when the light speed measured by an observer far from the gravitational source, when only gravitational time dilation is taken into account.

General Relativity as a Geometric Theory of Gravity

- By focusing on the equivalence principle (EP), Einstein discovered that the gravitational field can be modeled as a curved spacetime. The mathematics of such a curved manifold is that of Riemannian geometry. We present in Section 5.1 some of its basic elements: Gaussian coordinates and the metric tensor.

- We use the calculus of variations to deduce the geodesic equation for curves of minimum length (i.e., extreme invariant spacetime interval). We also discuss how this equation describes a straight line.

- We next present a geometric description of the EP physics of gravitational time dilation. In this geometric theory, the metric $g_{\mu\nu}(x)$ plays the role of a relativistic gravitational potential.

- The identification of curved spacetime as the gravitational field naturally suggests that spacetime's geodesic equation is the GR equation of motion, which reduces to the Newtonian equation of motion in the limit of a nonrelativistic moving particle ($v \ll c$) in a static and weak gravitational field. This also clarifies how Newton's theory is extended by Einstein's general-relativistic theory of gravitation.

The road from the EP to GR can be viewed as follows. The equivalence of an accelerated frame to one with gravity means that we cannot say for sure that gravity causes a particle's acceleration. We could just as easily attribute the acceleration to some property of the reference frame itself. Thus Einstein proposed a geometric theory modeling gravity as a warping of spacetime. From this point of view, there is no gravitational force. Particles move freely through spacetime with gravity; however, the motion of such particles following straight paths in a warped spacetime may be nontrivial. Gravitational phenomena thereby reflect the curvature of spacetime.

In this chapter, we mainly study the GR equation of motion, the geodesic equation, which describes the motion of a test particle in a curved spacetime.

A College Course on Relativity and Cosmology. First Edition. Ta-Pei Cheng.
© Ta-Pei Cheng 2015. Published in 2015 by Oxford University Press.

In Chapter 6, we will take up the GR field equation, the Einstein equation, which describes how a mass/energy source gives rise to a curved spacetime. But before presenting the geometric gravitational theory itself, we first provide a short mathematical introduction to some of the basic elements of a metric description of a curved space.

5.1 Metric description of a curved manifold

Differential geometry is the branch of mathematics that uses calculus techniques to study geometry. Its sub-branch Riemannian geometry concerns in particular the non-Euclidean n-dimensional spaces that can be described by distance measurements. Since most of us can only visualize, and only have any familiarity with, curved surfaces of two dimensions, we shall often use this simpler case, which was pioneered by Carl Friedrich Gauss (1777–1855), to illustrate the more general theory. The extension of Gauss's theory to higher dimensions was first[1] studied by his student, Bernhard Riemann (1826–1866).

[1] Non-Euclidean geometry was independently discovered by János Bolyai (1802–1860) and Nikolai Lobachevsky (1792–1856).

5.1.1 Gaussian coordinates and the metric tensor

Gaussian coordinates

Many of us intuitively think of a curved surface as a 2D surface embedded in 3D Euclidean space, for example a spherical surface (of radius R) described in 3D Cartesian coordinates (X, Y, Z) by

$$X^2 + Y^2 + Z^2 = R^2. \tag{5.1}$$

More generally, the embedding coordinates are subject to a constraint condition, $f(X, Y, Z) = 0$. This is an **extrinsic geometric description**; the space of interest (here the curved surface) is described using entities outside the space. We are most interested in an **intrinsic geometric description**, a characterization of the physical space without such external reference. Specifically, we would like to describe a 2D surface solely through measurements made by an inhabitant who never leaves that surface. Gauss introduced a generalized parametrization whose coordinates (x^1, x^2) are free to range over their respective domains without constraint:

$$X = X(x^1, x^2), \quad Y = Y(x^1, x^2), \quad Z = Z(x^1, x^2). \tag{5.2}$$

The **Gaussian coordinates** (x^1 and x^2, the number of which corresponds to the dimensionality of the embedded space) make the embedding coordinates (X, Y, and Z) superfluous; hence the geometric description can be purely intrinsic.

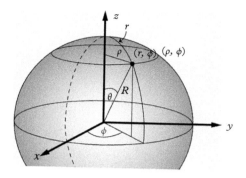

For the space of a 2-sphere of radius R, described extrinsically by (5.1) and illustrated in Fig. 5.1, using a 3D Euclidean embedding space, we provide two examples of Gaussian coordinate systems:

1. **The polar coordinate system:** To set up a Gaussian coordinate system $(x^1, x^2) = (\theta, \phi)$ to label points in this 2D surface, first pick a point on the surface (the north pole) and a longitudinal great circle through the pole (the prime meridian). The radial coordinate r of a point is the arclength between the point and the north pole. Instead of r, one can equivalently use the coordinate $\theta = r/R$ (the polar angle[2] or colatitude). The azimuthal angle ϕ (i.e., the longitude) is measured against the prime meridian. The coordinate domains are $0 \leq \theta \leq \pi$ and $0 \leq \phi < 2\pi$. In this case, (5.2) is specified by

$$X = R \sin\theta \cos\phi, \quad Y = R \sin\theta \sin\phi, \quad Z = R\cos\theta. \tag{5.3}$$

One may easily verify that this parametrization and the following one, (5.4), are consistent with the extrinsic description, (5.1).

2. **The cylindrical coordinate system:** We can choose another set of Gaussian coordinates to label points in the 2D surface by defining a different radial coordinate $\rho = R\sin\theta$, with a domain $0 \leq \rho \leq R$. Thus $(x^1, x^2) = (\rho, \phi)$, so that (5.2) is specified by[3]

$$X = \rho \cos\phi, \quad Y = \rho \sin\phi, \quad Z = \pm\sqrt{R^2 - \rho^2}. \tag{5.4}$$

From now on, we will no longer use extrinsic coordinates such as (X, Y, Z). By coordinates, we shall always mean Gaussian coordinates such as (x^1, x^2) that label points on a 2D space. Since one can choose from any number of coordinate systems, and geometric relations should be independent of such choices, a proper formulation of geometry must be invariant under general coordinate transformations.

[2] It may be helpful to visualize the coordinate θ of a point as the angle formed (at the center of the sphere in the embedding space) between the polar (z) axis and the radial (R) axis as displayed in Fig. 5.1. Of course, the entire point of the intrinsic description is to discard the embedding space. The surface itself has no center. When we extend this method to higher dimensions or pseudo-Euclidean spaces, such mental/visual crutches will serve less well in any case.

[3] If the spherical surface is embedded in a 3D Euclidean space, ρ is interpreted as the perpendicular distance to the Z axis as shown in Fig. 5.1. Perhaps the term cylindrical coordinates becomes more understandable if, instead of ρ, we use Z directly: $(x^1, x^2) = (Z, \phi)$, where $\rho^2 = R^2 - Z^2$. While $Z = R\cos\theta$, with a domain of $-R \leqslant Z \leqslant R$, covers all latitudes, ρ covers only half of the sphere.

Exercise 5.1   Coordinate choice

Clearly the ideal choice of coordinate system often depends on the task at hand. Consider the calculation in the space of a 2D plane of the circumference of a circle of radius R ($2\pi R$ of course). It is easy in polar coordinates (r, θ), but rather complicated in Cartesian coordinates (x, y). Carry out the calculations in both coordinate systems.

It must be emphasized that the coordinates $\{x^a\}$ generally do not form a vector space (elements of which can be added and multiplied by scalars, etc.). They are labels of points in the curved space and are devoid of any physical significance in their own right. Cartesian coordinates in flat space are the exception. We learn in our first physics course that the displacement between two distant points is a vector. Indeed, we have already applied rotation and boost transformations to coordinate displacements. We cannot do this in curved spaces! However, we may treat infinitesimal displacements as vectors, because they reside in a flat space, as we will discuss in the context of the flatness theorem (Section 5.1.3).

The metric tensor

The basic idea of Riemannian geometry is that the geometry (angles, lengths, and shapes) of a space can be described by length measurements. To illustrate this for the case of a 2D spherical surface (of radius R), one first sets up a Gaussian coordinate system (e.g., polar or cylindrical coordinates) to label points on the globe, then measures the infinitesimal distances between neighboring points (Fig. 5.2). These length measurements are summarized in an entity called the metric g_{ab} (whose indices range over the coordinates). It relates length measurements to differentials in the chosen Gaussian coordinates at any given point in the space by (3.19), which may be written as a matrix product:

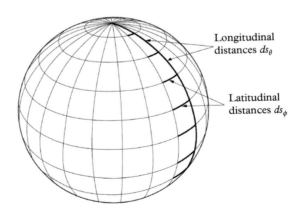

Longitudinal distances ds_θ

Latitudinal distances ds_ϕ

Figure 5.2 *Using distance measurements along links of constant longitude (ds_θ) and latitude (ds_ϕ) to specify the shape of the spherical surface.*

$$ds^2 = g_{ab}\, dx^a\, dx^b \tag{5.5}$$

$$= \left(dx^1\ dx^2 \right) \begin{pmatrix} g_{11} & g_{12} \\ g_{21} & g_{22} \end{pmatrix} \begin{pmatrix} dx^1 \\ dx^2 \end{pmatrix}$$

$$= g_{11}(dx^1)^2 + 2g_{12}\, dx^1\, dx^2 + g_{22}(dx^2)^2.$$

Note that the metric is always a symmetric matrix ($g_{12} = g_{21}$), because $dx^a\, dx^b = dx^b\, dx^a$. Also recall that the metric is directly related to the basis vectors of the space: $g_{ab} = \mathbf{e}_a \cdot \mathbf{e}_b$, cf. (3.8). ds^2 is formally interpreted as the infinitesimal squared length.[4]

The metric is an intrinsic geometric quantity Once the coordinate system has been chosen, the elements of the metric at a point $x = (x^1, x^2)$ can then be determined by infinitesimal distance measurements between x and nearby points. If we choose to measure the length ds_1 along the dx^1 direction (i.e., $dx^2 = 0$), then (5.5) reduces to $(ds_1)^2 = g_{11}(dx^1)^2$. Similarly, $(ds_2)^2 = g_{22}(dx^2)^2$. Let the measured length between x and the nearby point $(x^1 + dx^1, x^2 + dx^2)$ be ds_{12}. In this way, without invoking any extrinsic embedding space, the metric elements (g_{11}, g_{22}, g_{12}) at x can all be expressed in terms of measured lengths (ds_1, ds_2, ds_{12}) and coordinate differentials (dx^1, dx^2):

$$g_{11} = \frac{ds_1{}^2}{(dx^1)^2}, \quad g_{22} = \frac{ds_2{}^2}{(dx^2)^2},$$

$$g_{12} = \frac{ds_{12}^2 - ds_1^2 - ds_2^2}{2\, dx^1\, dx^2} = -\frac{ds_1\, ds_2 \cos\alpha}{dx^1\, dx^2}. \tag{5.6}$$

To reach the last equality, we have used the law of cosines $ds_{12}^2 = ds_1^2 + ds_2^2 - 2\, ds_1\, ds_2 \cos\alpha$, where α is the angle subtended by the axes. We expect that the off-diagonal metric elements should describe the deviation of the basis vectors from orthogonality ($g_{12} = \hat{\mathbf{e}}_1 \cdot \hat{\mathbf{e}}_2 \sim \cos\alpha$); thus, if the coordinates are orthogonal ($\alpha = \pi/2$), the metric matrix must be diagonal ($g_{12} = 0$). We emphasize again that the coordinates $\{x^a\}$ themselves do not measure distance. Only through the metric as in (5.5) are they connected to distance measurements.

As we have already noted prior to this subsection, because an infinitesimally small area on a curved surface can be treated as a flat plane (per the flatness theorem), flat-space geometric relations such as the law of cosines and the Pythagorean theorem apply.

The metrics for a 2-sphere Here we will work out a concrete example by writing down the metric for a 2-sphere in the polar coordinate system. One finds that the latitudinal distances $ds_\phi = R\sin\theta\, d\phi$ (subtended by some fixed $d\phi$ between two points having the same radial distance/latitude, $ds_\theta = 0$) become ever smaller as one approaches the poles ($\theta \to 0, \pi$). Meanwhile, the longitudinal distances ds_θ (subtended by $d\theta$ between two points having the same longitude $d\phi = 0$) have the same value, $ds_\theta = R\, d\theta$, over the whole range of θ and ϕ. (See Fig. 5.2.)

[4] Equation (5.5) is understood to mean $ds^2 = \sum_{a,b} g_{ab}\, dx^a\, dx^b$; i.e., the Einstein summation convention has been employed. For Cartesian coordinates in a Euclidean space, it is simply the Pythagorean theorem, $ds^2 = dx^2 + dy^2 + \cdots$. Recall the result worked out in Exercise 3.1 that a contraction between symmetric and antisymmetric tensors vanishes. Thus, if the metric had an antisymmetric part $g_{ab}^A = -g_{ba}^A$, it would not contribute to the length, because $g_{ab}^A\, dx^a\, dx^b = 0$.

Such distance measurements completely describe this spherical surface. These distance measurements can be compactly expressed in terms of the metric tensor elements. Because we have chosen orthogonal coordinates, $g_{\theta\phi} = \mathbf{e}_\theta \cdot \mathbf{e}_\phi = 0$, the infinitesimal length between two nearby points with coordinate displacement $(d\theta, d\phi)$ can be expressed using the Pythagorean theorem as

$$[ds^2]_{(\theta,\phi)} = (ds_\theta)^2 + (ds_\phi)^2$$
$$= R^2 \, d\theta^2 + R^2 \sin^2\theta \, d\phi^2. \tag{5.7}$$

Matching terms of (5.7) and (5.5) yields the metric tensor for this (θ, ϕ) coordinate system:

$$g_{ab}^{(\theta,\phi)} = R^2 \begin{pmatrix} 1 & 0 \\ 0 & \sin^2\theta \end{pmatrix}. \tag{5.8}$$

Exercise 5.2 Cylindrical coordinate metric

Find the metric tensor for the cylindrical coordinates (ρ, ϕ) on a 2-sphere.

Suggestion: From Fig. 5.1, note that the radial coordinate is related to the polar angle by $\rho = R \sin\theta$; then show that

$$g_{ab}^{(\rho,\phi)} = \begin{pmatrix} R^2/(R^2 - \rho^2) & 0 \\ 0 & \rho^2 \end{pmatrix}. \tag{5.9}$$

Coordinate transformation in a curved space

A key difference between curved and flat spaces is that curved space must necessarily have position-dependent coordinate bases, while a flat space can have constant (Cartesian) coordinates. As a consequence, the metric and coordinate transformation matrices are position-dependent in any curved space.[5]

Because a flat space can still have curvilinear coordinates (such as polar coordinates), the main features of coordinate transformation can be illustrated in a flat space with a curvilinear coordinate system. Take the simplest case of a 2D plane. Consider the transformation from Cartesian coordinates $\{x^a\} = (x, y)$ to polar coordinates $\{x^a\} \to \{x'^a\} = (r, \theta)$. A coordinate transformation can in general be written as a matrix of partial derivatives:

[5] The contrast between flat and curved spaces will also be discussed in Section 6.1, as well as in the introductory paragraph of Chapter 11. For examples of a transformation as a matrix of partial derivatives, see Exercise 2.4 and Box 5.2.

$$dx^a = [\Lambda]^a{}_b dx'^b \quad \text{with} \quad [\Lambda]^a{}_b = \frac{\partial x^a}{\partial x'^b}; \tag{5.10}$$

this expression follows from an application of the chain rule of differentiation. In this 2D example, taking derivatives of the relations $x = r \cos\theta$ and $y = r \sin\theta$ leads to

$$\begin{pmatrix} dx \\ dy \end{pmatrix} = \begin{pmatrix} \cos\theta & -r\sin\theta \\ \sin\theta & r\cos\theta \end{pmatrix} \begin{pmatrix} dr \\ d\theta \end{pmatrix}. \tag{5.11}$$

The elements of the transformation matrix are derivatives, for example,

$$[\Lambda]^1_{\ 1} = \cos\theta = \frac{\partial x}{\partial r} = \frac{\partial x^1}{\partial x'^1} \quad \text{and} \quad [\Lambda]^1_{\ 2} = \frac{\partial x^1}{\partial x'^2}. \tag{5.12}$$

We note that, unlike the rotation matrix (1.5) or the Lorentz transformation (2.12), the transformation matrix elements $[\Lambda]^a_{\ b}$ are position-dependent—here they depend on $\{x'^a\} = (r, \theta)$.

Exercise 5.3 Transformation in curved space

Find the coordinate transformation matrix $[\Lambda]$ (showing its coordinate dependence) that changes the polar coordinates (θ, ϕ) to the cylindrical coordinates (ρ, ϕ) on a 2-sphere:

$$\begin{pmatrix} d\rho \\ d\phi \end{pmatrix} = [\Lambda] \begin{pmatrix} d\theta \\ d\phi \end{pmatrix}. \tag{5.13}$$

5.1.2 The geodesic equation

As previously stated, in Riemannian geometry, the spatial geometry is determined by (infinitesimal) length measurements, which are codified in the metric tensor. Namely, once the (Gaussian) coordinate system has been chosen, the metric elements $g_{ab}(x)$ relate the coordinate differentials to the measured lengths at all x. In this way, the geometry of the space can be determined from the metric.[6] Here we shall work out such an example: how to find, from the metric, the equation that describes the shortest curve between two fixed points in a space.

Any curve in a space can be written in the form[7] $x^a(\tau)$, where τ is some parameter (which might be, but need not be, proper time) having some range, for example $[\tau_i, \tau_f]$. We are interested in finding, for a given space, the shortest curve, called the **geodesic**, that connects initial $x^a(\tau_i)$ and final $x^a(\tau_f)$ positions. (A discussion of the geodesic as a straight line can be found in Box 5.2.) The squared length (invariant interval) ds^2 of an infinitesimal segment of any curve is given by (5.5). We integrate $ds = \sqrt{|ds^2|}$ (the length of a spacelike segment or the proper time of a timelike one) along this curve, changing to Greek indices as is conventional for 4D spacetime (also denoting $\dot{x}^\mu = dx^\mu/d\tau$):

$$s = \int ds = \int \sqrt{|g_{\mu\nu}dx^\mu dx^\nu|} = \int L(x, \dot{x})\, d\tau, \tag{5.14}$$

[6] A note of caution: while the metric determines the geometry, geometry may not fix the metric. For example, a metric with constant elements describes a space with zero curvature, but a space with no curvature does not necessarily imply a constant metric. More specifically, a flat plane can be described by polar coordinates, whose metric is position-dependent.

[7] Namely, as the parameter τ varies $(\tau_1, \tau_2, \tau_3, \ldots)$, one obtains a continuous set of coordinates $x^a(\tau_1), x^a(\tau_2), x^a(\tau_3), \ldots$. An example of such a parametric description of a curve would be the trajectory of a particle parameterized by its time.

where

$$L(x, \dot{x}) = \sqrt{|g_{\mu\nu}\dot{x}^{\mu}\dot{x}^{\nu}|}.$$

(5.15)

To determine the shortest line in the curved space, we impose the extremization condition[8] for variation of the path with endpoints fixed:

$$\delta s = \delta \int L(x, \dot{x}) \, d\tau = 0,$$

(5.16)

which the calculus of variations can translate (cf. Box 5.1) into a partial differential equation—the Euler–Lagrange equation:

$$\frac{d}{d\tau}\frac{\partial L}{\partial \dot{x}^{\mu}} - \frac{\partial L}{\partial x^{\mu}} = 0.$$

(5.17)

[8] In contrast to the extremization of a function $f(x)$, in which a single variable x is varied resulting in the condition $df/dx = 0$, here one varies an entire function $x(\tau)$ (all possible curves running from $x(\tau_i)$ to $x(\tau_f)$), resulting in the condition (5.16), which yields the Euler–Lagrange equation (5.17).

Box 5.1 Euler–Lagrange equation from action extremization

In physics, one often uses the calculus of variations to formulate equations of motion and field equations from the least-action principle. Namely, these equations are derived as the Euler–Lagrange equations from the extremization of some action integral:

$$S = \int_{\tau_1}^{\tau_2} L(x, \dot{x}) \, d\tau.$$

(5.18)

The integrand, called the Lagrangian, is the difference between the kinetic energy and the potential energy for a classical system parameterized by time:

$$L(x, \dot{x}) = T(\dot{x}) - V(x).$$

(5.19)

The principle of least action states that the action is extremal with respect to the variation of the trajectory $x^{\mu}(\tau)$ with its endpoints fixed at initial and final positions $x^{\mu}(\tau_1)$ and $x^{\mu}(\tau_2)$. This extremization requirement can be translated into a partial differential equation as follows. The variation of the Lagrangian is

$$\delta L(x, \dot{x}) = \frac{\partial L}{\partial x^{\mu}} \delta x^{\mu} + \frac{\partial L}{\partial \dot{x}^{\nu}} \delta \dot{x}^{\nu}.$$

(5.20)

Thus the condition for extremization[9] of the action integral becomes

$$0 = \delta S = \delta \int_{\tau_1}^{\tau_2} L(x, \dot{x}) \, d\tau = \int_{\tau_1}^{\tau_2} \left(\frac{\partial L}{\partial x^{\mu}} \delta x^{\mu} + \frac{\partial L}{\partial \dot{x}^{\nu}} \frac{d}{d\tau} \delta x^{\nu} \right) d\tau$$

$$= \int_{\tau_1}^{\tau_2} \left(\frac{\partial L}{\partial x^{\mu}} - \frac{d}{d\tau}\frac{\partial L}{\partial \dot{x}^{\mu}} \right) \delta x^{\mu} \, d\tau.$$

(5.21)

[9] Extremization means minimization or maximization. This allows us to avoid the square roots and absolute values in (5.14) and (5.15) by integrating ds^2 (rather ds) as our action, cf. (5.23). For timelike curves such as the trajectories of a particle with mass, the action is negative and maximized (least negative) by the geodesic path. Taylor and Wheeler (2000) gave it the evocative name "the principle of extremal aging."

To reach the last expression, we have performed an integration by parts on the second term:

$$\int_{\tau_1}^{\tau_2} \frac{\partial L}{\partial \dot{x}^\nu} d(\delta x^\nu) = \left[\frac{\partial L}{\partial \dot{x}^\nu} \delta x^\nu (\tau) \right]_{\tau_1}^{\tau_2} - \int_{\tau_1}^{\tau_2} (\delta x^\nu) d\left(\frac{\partial L}{\partial \dot{x}^\nu} \right). \tag{5.22}$$

The first term on the right-hand side can be discarded, because the endpoint positions are fixed: $\delta x^\nu(\tau_1) = \delta x^\nu(\tau_2) = 0$. Since δS must vanish for arbitrary variations $\delta x^\mu(\tau)$, the expression in parentheses on the right-hand side of (5.21) must also vanish. The result is the Euler–Lagrange equation (5.17). For the simplest case of $L = \frac{1}{2}m\dot{x}^2 - V(\mathbf{x})$, the Euler–Lagrange equation is just the familiar $\mathbf{F} = ma$ equation, as it yields $m\ddot{\mathbf{x}} + \nabla V = 0$.

The geodesic determined by the Euler–Lagrange equation As a mathematical exercise, one can show that the *same* Euler–Lagrange equation (5.17) following from (5.15) follows as well from a Lagrangian of the form

$$L(x, \dot{x}) = g_{\mu\nu} \dot{x}^\mu \dot{x}^\nu, \tag{5.23}$$

which, without the square root, is much easier to work with. With L in this form, the derivatives in (5.17) become

$$\frac{\partial L}{\partial \dot{x}^\mu} = 2g_{\mu\nu}\dot{x}^\nu, \quad \frac{\partial L}{\partial x^\mu} = \frac{\partial g_{\lambda\rho}}{\partial x^\mu} \dot{x}^\lambda \dot{x}^\rho, \tag{5.24}$$

where we have used the fact that the metric function $g_{\mu\nu}$ depends on x^μ but not on $\dot{x}^\mu$. Substituting these relations back into (5.17), we obtain the geodesic equation,

$$\frac{d}{d\tau}(g_{\mu\nu}\dot{x}^\nu) - \frac{1}{2}\frac{\partial g_{\lambda\rho}}{\partial x^\mu} \dot{x}^\lambda \dot{x}^\rho = 0, \tag{5.25}$$

which determines the geodesic, the curve that extremizes the invariant interval (the spatial path of the shortest length or the timelike trajectory of maximal proper time) between two points.

Exercise 5.4 Geodesics on simple surfaces

Use the geodesic equation (5.25) to confirm the familiar results that the geodesic is (a) a straight line on a flat plane and (b) a great circle on a spherical surface.

Suggestion: For case (b), working out the full parametrization can be complicated; just check that the great circle given by $\phi = $ constant and $\theta = \alpha + \beta\tau$ solves the relevant geodesic equation.

Casting the geodesic equation into standard form We can rewrite (5.25) in a more symmetric form. Differentiating the first term and noting that the metric's dependence on the curve parameter τ is entirely through $x^{\mu}(\tau)$, we have

$$g_{\mu\nu}\frac{d^2x^{\nu}}{d\tau^2} + \frac{\partial g_{\mu\rho}}{\partial x^{\lambda}}\frac{dx^{\lambda}}{d\tau}\frac{dx^{\rho}}{d\tau} - \frac{1}{2}\frac{\partial g_{\lambda\rho}}{\partial x^{\mu}}\frac{dx^{\lambda}}{d\tau}\frac{dx^{\rho}}{d\tau} = 0. \qquad (5.26)$$

In the second term, we have relabeled the dummy index $\nu \to \rho$; also, its coefficient can be decomposed into a symmetric (with respect to the exchange of λ and ρ indices) and an antisymmetric term:

$$\frac{\partial g_{\mu\rho}}{\partial x^{\lambda}} = \frac{1}{2}\left(\frac{\partial g_{\mu\rho}}{\partial x^{\lambda}} + \frac{\partial g_{\mu\lambda}}{\partial x^{\rho}}\right) + \frac{1}{2}\left(\frac{\partial g_{\mu\rho}}{\partial x^{\lambda}} - \frac{\partial g_{\mu\lambda}}{\partial x^{\rho}}\right). \qquad (5.27)$$

Since the product $(dx^{\lambda}/d\tau)(dx^{\rho}/d\tau)$ in the second term on the left-hand side of (5.26) is symmetric, the antisymmetric part will not survive their contraction (cf. Exercise 3.1). In this way, the geodesic equation (5.25) becomes

$$g_{\mu\nu}\frac{d^2x^{\nu}}{d\tau^2} + \frac{1}{2}\left(\frac{\partial g_{\mu\rho}}{\partial x^{\lambda}} + \frac{\partial g_{\mu\lambda}}{\partial x^{\rho}} - \frac{\partial g_{\lambda\rho}}{\partial x^{\mu}}\right)\frac{dx^{\lambda}}{d\tau}\frac{dx^{\rho}}{d\tau} = 0. \qquad (5.28)$$

We can remove the first metric $g_{\mu\nu}$ factor by contracting the whole equation with the inverse metric $g^{\mu\sigma}$. In this way, the geodesic equation can now be written in its standard form,

$$\frac{d^2x^{\sigma}}{d\tau^2} + \Gamma^{\sigma}_{\lambda\rho}\frac{dx^{\lambda}}{d\tau}\frac{dx^{\rho}}{d\tau} = 0, \qquad (5.29)$$

where

$$\Gamma^{\sigma}_{\lambda\rho} = \frac{1}{2}g^{\sigma\mu}\left(\frac{\partial g_{\mu\rho}}{\partial x^{\lambda}} + \frac{\partial g_{\mu\lambda}}{\partial x^{\rho}} - \frac{\partial g_{\lambda\rho}}{\partial x^{\mu}}\right). \qquad (5.30)$$

The set $\Gamma^{\mu}_{\lambda\rho}$ defined by this particular combination of the first derivatives of the metric tensor are called the Christoffel symbols (also known as the **affine** or **connection coefficients**). Recall that the metric is directly related to the coordinate bases, $g_{\mu\nu} = \mathbf{e}_{\mu}\cdot\mathbf{e}_{\nu}$, as in (3.8). $\Gamma^{\mu}_{\nu\lambda}$ are nonvanishing because the bases, and hence the metric, are position-dependent. They are called *symbols*, because, despite their appearance with indices, $\Gamma^{\mu}_{\nu\lambda}$ are not tensor elements.[10] Namely, they do not transform as tensor components under a coordinate transformation, cf. (3.27). Clearly, (5.29) is applicable for all higher-dimensional spaces (just by varying the range of the indices). Of particular relevance to us, this geodesic equation for 4D spacetime turns out to be the equation of motion in the GR theory of gravitation.

[10] As we shall demonstrate in Chapter 11, in particular (11.84), the geodesic equation (5.29) is a proper tensor equation, even though $\Gamma^{\mu}_{\lambda\rho}$ and the first derivative term have extra terms in their transformation. But these extra terms mutually cancel.

Box 5.2 Geodesic as a straight line

In the above, we have introduced the geodesic as the shortest line between two fixed endpoints in a curved space by a variational calculation. Here we will provide another view of (5.29) as the equation describing a straight line in coordinate systems with position-dependent bases.

A curved space must necessarily have position-dependent coordinates. In a flat space, it is possible to have constant coordinate bases (Cartesian coordinates), but we can still have curvilinear coordinates such as polar coordinates. Thus this second interpretation of the geodesic equation can be illustrated by the case of a straight line in a flat 2D plane. In the Cartesian system $\{x^a\} = (x, y)$, a straight line is the curve $x^a(\tau)$ with the tangent $(dx^a/d\tau)$ unchanged along the curve (see, e.g., Fig. 11.2a):

$$\frac{d^2 x^a}{d\tau^2} = 0. \tag{5.31}$$

We now transform this equation to another system such as polar coordinates $\{x'^a\} = (r, \theta)$. The coordinate derivatives (with respect to the curve parameter) transform in the same way as the coordinate differentials of (5.10):

$$\frac{dx^a}{d\tau} = \frac{\partial x^a}{\partial x'^b} \frac{dx'^b}{d\tau}. \tag{5.32}$$

In this way, the left-hand side of (5.31) can be written as

$$\frac{d}{d\tau}\left(\frac{dx^a}{d\tau}\right) = \frac{d}{d\tau}\left(\frac{\partial x^a}{\partial x'^b} \frac{dx'^b}{d\tau}\right) = \frac{\partial x^a}{\partial x'^b} \frac{d^2 x'^b}{d\tau^2} + \frac{d}{d\tau}\left(\frac{\partial x^a}{\partial x'^b}\right)\frac{dx'^b}{d\tau}. \tag{5.33}$$

The last term contains

$$\frac{d}{d\tau}\left(\frac{\partial x^a}{\partial x'^b}\right) = \frac{\partial^2 x^a}{\partial x'^b \partial x'^c} \frac{dx'^c}{d\tau} \tag{5.34}$$

with the first factor on the right-hand side being the position derivatives of the transformation matrix,

$$\frac{\partial^2 x^a}{\partial x'^b \partial x'^c} = \frac{\partial}{\partial x'^b}\left(\frac{\partial x^a}{\partial x'^c}\right), \tag{5.35}$$

which are nonvanishing because the transformation is position-dependent when the bases are position-dependent. The straight-line equation thus takes the form

$$\frac{\partial x^a}{\partial x'^b} \frac{d^2 x'^b}{d\tau^2} + \frac{\partial^2 x^a}{\partial x'^b \partial x'^c} \frac{dx'^b}{d\tau} \frac{dx'^c}{d\tau} = 0. \tag{5.36}$$

In order to compare this straight-line equation with the geodesic equation (5.29), we multiply it by the *inverse* coordinate transformation $\partial x'^b/\partial x^c$, with the property $(\partial x'^b/\partial x^c)(\partial x^a/\partial x'^b) = \delta_c^a$. After relabeling some indices, we can write (5.36) as

$$\frac{d^2 x'^a}{d\tau^2} + \left[\frac{\partial x'^a}{\partial x^d} \frac{\partial^2 x^d}{\partial x'^b \partial x'^c}\right]\frac{dx'^b}{d\tau} \frac{dx'^c}{d\tau} = 0. \tag{5.37}$$

continued

Box 5.2 *continued*

This is the geodesic equation (5.29) when we identify the square bracket as the Christoffel symbols (5.30): $[\ldots] = \Gamma^a_{bc}$. We will not work out their explicit correspondence, but only remark that both expressions are related to the moving bases—in Γ^a_{bc}, we have the derivatives of the metric, and in $[\ldots]$, we have derivatives of the coordinate transformation. Although our discussion has been carried out in a flat space, the key ingredient is the position dependence of the coordinates. This property is a necessary feature of a curved space, while it is optional in a flat space. Thus this demonstration that the geodesic equation (5.29) describes a straight line is also valid in a curved space. The proper proof, involving the concepts of covariant differentiation and parallel transport, will be presented in Section 11.1.

5.1.3 Local Euclidean frames and the flatness theorem

A different choice of coordinates leads to a different metric, which is generally position-dependent. What distinguishes a flat space from a curved one is that for a flat space it is possible to find a coordinate system for which the metric is a constant, like Cartesian coordinates in Euclidean space with $[g] = [1]$ or in the Minkowski space of SR with $[g] = \mathrm{diag}(-1, 1, 1, 1) \equiv [\eta]$.

While it is clear that flat and curved spaces are different geometric entities, they are closely related. We are familiar from our experience with curved surfaces that any curved space can be approximated locally by a flat plane. This is the content of the **flatness theorem**.

In a curved spacetime with a general coordinate system x^μ and a metric value $g_{\mu\nu}$ at a given point P, it is always possible to find a coordinate transformation $x^\mu \to \bar{x}^\mu$ and $g_{\mu\nu} \to \bar{g}_{\mu\nu}$ such that the metric is flat at P (which can be taken to be the origin of the transformed coordinates, $P \to 0$), with $\bar{g}_{\mu\nu} = \eta_{\mu\nu}$ and $\partial \bar{g}_{\mu\nu}/\partial \bar{x}^\lambda = 0$; thus

$$\bar{g}_{\mu\nu}(\bar{x}) = \eta_{\mu\nu} + \gamma_{\mu\nu\lambda\rho}(0)\bar{x}^\lambda \bar{x}^\rho + \cdots . \tag{5.38}$$

Namely, the metric in the neighborhood of the origin (P) will differ from $\eta_{\mu\nu}$ only by the second- and higher-order derivatives. This is simply a Taylor series expansion of the metric at the origin; there is the constant $\bar{g}_{\mu\nu}(0)$ plus second-order derivative terms $\gamma_{\mu\nu\lambda\rho}(0)\bar{x}^\lambda \bar{x}^\rho$. That $\bar{g}_{\mu\nu}(0) = \eta_{\mu\nu}$ should not be a surprise. For a metric value at one point, it is always possible to find an orthogonal system so that $\bar{g}_{\mu\nu}(0) = 0$ for $\mu \neq \nu$. The diagonal elements can be scaled to unity so that the new coordinate bases all have unit length. Thus the metric is an identity matrix or whatever is the correct orthogonal flat metric with the appropriate signature. The nontrivial content of (5.38) is the absence of the first derivative.

In short, only in a flat manifold is it possible to have a constant metric for the entire space. However, in a curved space, it is still possible to have local Euclidean frames $\{\bar{x}^\mu\}$. The flatness theorem informs us that the general spacetime metric

$g_{\mu\nu}(x)$ is characterized at a point (P) not so much by the value $g_{\mu\nu}|_P$, since that can always be chosen to be its flat-space value, $\bar{g}_{\mu\nu}|_P = \eta_{\mu\nu}$, or by its first derivative, which can always be chosen to vanish, $\partial\bar{g}_{\mu\nu}/\partial\bar{x}^\lambda|_P = 0$, but by the second derivative of the metric, $\partial^2\bar{g}_{\mu\nu}/\partial x^\lambda\partial x^\rho$, which characterizes the curved space. It is related to the curvature of the space, to be discussed in Chapter 6.

5.2 From the equivalence principle to a metric theory of gravity

How did Einstein get the idea for a geometric theory of gravitation? What does one mean by a geometric theory?

A geometric physics theory

By a geometric theory or a geometric description of any physical phenomenon, we mean a theory that attributes the results of physical measurements directly to the underlying geometry of space and time. This can be illustrated by the description of a spherical surface (Fig. 5.2) that we discussed earlier in this chapter. The length measurements on the surface of a globe are different in different directions: the east/west distances between pairs of points separated by the same azimuthal angle $\Delta\phi$ are smaller for pairs farther from the equator, while the north/south lengths for a given $\Delta\theta$ are all the same. We could, in principle, interpret such results in two equivalent ways:

1. Without considering that the 2D space is curved, we could say that the physics (i.e., dynamics) is such that the measuring ruler changes its scale at different points or when pointing in different directions—much in the same manner as the Lorentz–FitzGerald length contraction of SR was originally interpreted.

2. Alternatively, we could use a standard ruler with a fixed scale, and attribute the varying length measurements to the underlying geometry of a curved spherical surface per Fig. 5.2. This geometry is expressed mathematically by a position-dependent metric tensor $g_{ab}(x) \neq \delta_{ab}$.

EP physics and a warped spacetime

In Chapter 4, we deduced several physical consequences from the empirical principle of gravity–inertia equivalence. In a geometric theory, these gravitational phenomena are attributed to the underlying curved spacetime, which has a metric as defined in (5.5):

$$ds^2 = g_{\mu\nu}\,dx^\mu\,dx^\nu. \tag{5.39}$$

For SR, we have the geometry of a flat spacetime with a position-independent metric $g_{\mu\nu} = \eta_{\mu\nu} \equiv \text{diag}(-1, 1, 1, 1)$. The study of EP physics led Einstein to propose that gravity represents the structure of a curved spacetime, with $g_{\mu\nu} \neq \eta_{\mu\nu}$, and that gravitational phenomena are just the effects of curved spacetime on a test object. For instance, gravitational time dilation just reflects the bending of spacetime in the time direction.

Gravitational time dilation due to $g_{00} \neq -1$ For gravitational time dilation, instead of working with a complicated scheme of clocks running at different rates, this physical phenomenon can be interpreted geometrically as showing the presence of a nontrivial metric. Namely, we can interpret (4.40) in terms of a nontrivial metric. Recall our discussion relating the metric elements to the defined coordinates and distance measurements, cf. (5.6); the time–time element of the metric can be fixed by

$$g_{00} = \frac{ds_0^2}{(dx^0)^2},   \tag{5.40}$$

where $dx^0 = c\,dt$ is the coordinate time interval and $ds_0^2 = [ds^2]_{dx^i=0}$ is the measured interval in the time direction. The coordinate time is measured by the clock at a location far from the source of gravity ($\Phi = 0$), while $ds_0^2 = -c^2\,d\tau^2$ is directly related to the proper time $\tau(x)$ measured by stationary clocks located at x. Thus (5.40) is simply the relation $d\tau^2 = -g_{00}\,dt^2$, which, by (4.40), implies

$$g_{00} = -\left(1 + \frac{\Phi(x)}{c^2}\right)^2 \simeq -\left(1 + \frac{2\Phi(x)}{c^2}\right).   \tag{5.41}$$

The metric element g_{00} deviates from its flat-spacetime value of $\eta_{00} = -1$ because of the presence of gravity. Thus the geometric interpretation of gravitational time dilation is that gravity warps spacetime (in this case in the time direction), changing the spacetime metric element from a constant $\eta_{00} = -1$ to an x-dependent function $g_{00}(x)$.

5.2.1   Curved spacetime as gravitational field

We provide further arguments for identifying warped spacetime as the gravitational field. Adopting a geometric interpretation of EP physics, we find that the resultant spacetime geometry has characteristics of a warped manifold such as a position-dependent metric and deviations from Euclidean geometric relations. Moreover, at every location, we can always transform gravity away to obtain a locally inertial spacetime, just as one can always find a locally flat region in a curved space (per the flatness theorem).

Position-dependent metric

The metric tensor in a curved space is necessarily position-dependent as in (5.41). In Einstein's geometric theory of gravitation, the metric function completely describes the gravitational field. The metric $g_{\mu\nu}(x)$ plays the role of the relativistic gravitational potential, just as $\Phi(x)$ is the Newtonian gravitational potential.

Non-Euclidean relations

In a curved space, Euclidean relations no longer hold. For example, the ratio of a circle's circumference to its radius may differ from 2π. The EP also implies non-Euclidean relations among geometric measurements. We illustrate this with a simple example. Consider a cylindrical room in high-speed rotation around its axis. The centripetal acceleration of the reference frame, according to the EP, is equivalent to a centrifugal gravitational field. (This is one way to produce artificial gravity.) For such a rotating frame, one finds that, because of SR (longitudinal) length contraction, the circular circumference of the cylinder is no longer equal to 2π times the radius, which does not change because the frame's velocity is transverse to the radial direction (see Fig. 5.3). Thus Euclidean geometry fails in the presence of gravity. We reiterate this connection: the rotating frame, according to the EP, is a frame with gravity; the rotating frame, according to SR length contraction, has a relation between its radius and circumference that is not Euclidean. Hence we say that in the presence of gravity the measuring rods map out a non-Euclidean geometry.

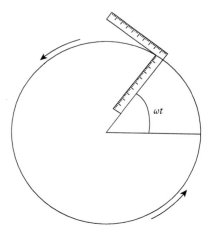

Figure 5.3 *Rotating cylinder with length contraction in the tangential direction but not in the radial direction, resulting in a non-Euclidean relation between circumference and radius.*

Local flat metric and local inertial frame

In a curved space, a small region can always be described approximately as a flat space per the flatness theorem, cf. (5.38). Now, if we identify the curvature of our spacetime as the gravitational field, the corresponding flatness theorem must be satisfied. Indeed, the EP informs us that gravity can always be transformed away locally. In the absence of gravity, spacetime is flat. Thus Einstein put forward this elegant theory that identifies gravity as the structure of spacetime, thereby incorporating the EP in a fundamental way.

A 2D illustration of geometry as gravity The possibility of using a curved space to represent a gravitational field can be illustrated with the following example involving a 2D curved surface. Two masses on a spherical surface start out at the equator and move along two longitudinal great circles (which are geodesics, as shown in Exercise 5.4). As they move along, the distance between them decreases (Fig. 5.4). We can attribute this convergence to some attractive force between them or simply to the curvature of space. We will discuss such tidal effects in more detail in Section 6.2.2.

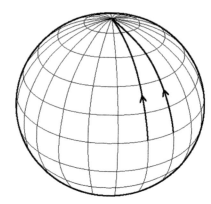

Figure 5.4 *The convergence of two particle trajectories can be explained by either an attractive force or the underlying geometry of a spherical surface.*

Example 5.1 Curved spacetime and gravitational redshift

In Chapter 4, we showed that the strong EP implies a gravitational redshift (in a static gravitational field) of light frequency ω, cf. (4.20),

$$\frac{\Delta\omega}{\omega} = -\frac{\Delta\Phi}{c^2}. \tag{5.42}$$

continued

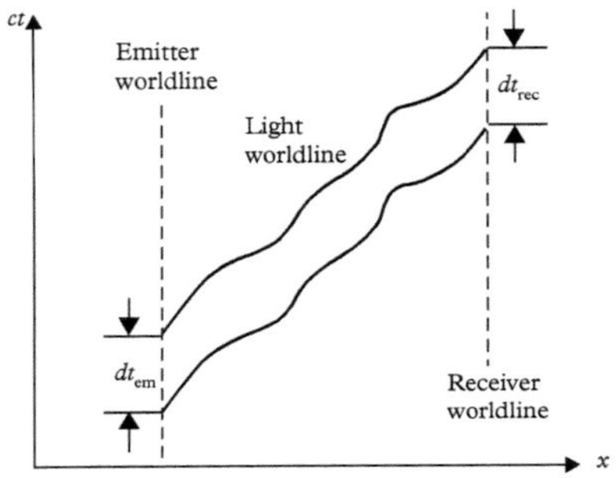

Figure 5.5 *Worldlines for two light wavefronts propagating from emitter to receiver in a static curved spacetime.*

Example 5.1 *continued*

From this result, we heuristically deduced that in the presence of nonzero gravitational potential, the metric must deviate from its flat-space value. Namely, from the gravitational redshift, we deduced a curved spacetime. Now we shall show the converse: that a curved spacetime implies redshift. In this chapter, we have seen that Einstein's theory based on a curved space-time yields the result (5.41) in the Newtonian limit. This can be stated as a relation between the proper time τ and the coordinate time t:

$$d\tau = \sqrt{-g_{00}}\, dt, \qquad \text{with} \qquad g_{00} = -\left(1 + 2\frac{\Phi}{c^2}\right). \qquad (5.43)$$

Here we wish to show how the gravitational frequency shift result (5.42) emerges in this curved-spacetime description. In Fig. 5.5, the two curvy lines are the lightlike worldlines of two wavefronts emitted at an interval dt_{em} apart. They curve because the spacetime is warped by gravity. (In flat spacetime, they would be two straight 45° lines.) Because the gravitational field is static (hence spacetime curvature is time-independent), this dt_{em} time interval between the two wavefronts is maintained throughout the trip from emission to reception:

$$dt_{\text{em}} = dt_{\text{rec}}. \qquad (5.44)$$

On the other hand, because the frequency is inversely proportional to the proper time interval $\omega = 1/d\tau$, we can use (5.43) and (5.44) to derive the redshift:

$$\frac{\omega_{\text{rec}}}{\omega_{\text{em}}} = \frac{d\tau_{\text{em}}}{d\tau_{\text{rec}}} = \frac{\sqrt{-(g_{00})_{\text{em}}}\, dt_{\text{em}}}{\sqrt{-(g_{00})_{\text{rec}}}\, dt_{\text{rec}}} = \left(\frac{1 + 2(\Phi_{\text{em}}/c^2)}{1 + 2(\Phi_{\text{rec}}/c^2)}\right)^{1/2}$$

$$= 1 + \frac{\Phi_{\text{em}} - \Phi_{\text{rec}}}{c^2} + O\left(\Phi^2/c^4\right), \qquad (5.45)$$

which is the claimed result of (5.42):

$$\frac{\omega_{\text{rec}} - \omega_{\text{em}}}{\omega_{\text{em}}} = -\frac{\Phi_{\text{rec}} - \Phi_{\text{em}}}{c^2}. \qquad (5.46)$$

5.2.2   GR as a field theory of gravitation

Recall that a field-theoretical description of the interaction between a source and a test particle involves two steps:

| Source particle | $\xrightarrow[\text{Field equation}]{}$ | Field | $\xrightarrow[\text{Equation of motion}]{}$ | Test particle |

Instead of acting directly on the test particle through some instantaneous action-at-a-distance force, the source particle creates a field everywhere, which then acts on the test particle locally. The first step is governed by the field equation, which, given the source distribution, dictates the field everywhere. In the case of electromagnetism, this is Maxwell's equations. In the second step, the equation of motion determines a test particle's motion from the field at its location. The electromagnetic equation of motion follows directly from the Lorentz force law.

Based on his study of EP phenomenology, Einstein made the conceptual leap (a logical deduction, but a startling leap nevertheless) that curved spacetime is the gravitational field. A source mass gives rise to a curved spacetime, which in turn influences the motion of a test mass:

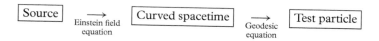

While spacetime in SR, as in all pre-GR physics, is fixed, it is dynamic in the general theory of relativity and is determined by the matter/energy distribution. GR fulfills Einstein's conviction that "space is not a thing." Spacetime is merely an expression of the ever-changing relations among physical processes. Thus the metric,[11] which describes the geometry, is ever-changing. Furthermore, the laws of physics should not depend on reference frames. Physics equations should be tensor equations under general coordinate transformations. This principle of general covariance is a key feature of GR.

Next we shall study the GR equation of motion, the geodesic equation, which describes the motion of a test particle in a curved spacetime. The more difficult topic of the GR field equation, the Einstein equation, is deferred to Chapter 6, after we have briefly discussed the Riemann curvature tensor.

5.3 Geodesic equation as the GR equation of motion

In GR, the mass/energy source determines the metric function through the field equation. Namely, the metric $g_{\mu\nu}(x)$ is the solution of the GR field equation. From $g_{\mu\nu}(x)$, one can write down the geodesic equation, which fixes the trajectory of the test particle. In this approach, gravity is regarded as the structure of spacetime rather than as a force (which would bring about acceleration through Newton's second law). That is, a test body will move freely in such a curved spacetime; it is natural to expect[12] it to follow the shortest and straightest possible trajectory, the geodesic curve. Thus the particle's coordinate acceleration comes from the geodesic equation (5.29) rather than Newton's second law.

[11] It is important to note that the gravitational field is not a scalar, nor is it a four-component vector, but rather a symmetric tensor $g_{\mu\nu} = g_{\nu\mu}$ with ten independent components; in contrast, the antisymmetric electromagnetic field tensor $F_{\mu\nu} = -F_{\nu\mu}$ has six components.

[12] The correctness of this heuristic choice will be justified by a formal derivation of the geodesic equation in Section 11.3.1.

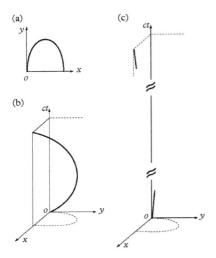

Figure 5.6 *(a) Particle trajectory in the (x, y) plane. (b) Particle worldline with projection onto the (x, y) plane as shown in (a), plotted with a greatly compressed time axis. (c) Spacetime diagram with the time axis restored to its proper scale.*

[13] One can picture the spacetime being curved by the (gravitational) change of the time intervals when moving away from the origin.

Box 5.3 The geodesic is the worldline of a test particle

It may appear somewhat surprising to hear that a test particle will follow a "straight line" in the presence of a gravitational field. After all, our experience is just the opposite: when we throw an object in the earth's gravitational field, it follows a parabolic trajectory. Was Einstein saying that the parabolic trajectory is actually straight? All such paradoxes result from confusing 4D spacetime with ordinary 3D space. The GR equation of motion tells us that a test particle will follow a geodesic line in spacetime (whose invariant interval has a negative part from the time coordinate) rather than a geodesic line in the 3D space (which minimizes ordinary length). A geodesic in spacetime (a worldline) generally does not imply a straight trajectory in its spatial subspace. A simple illustration using a spacetime diagram should make this clear.

Consider an object thrown to a height of 10 m over a distance of 10 m. Its spatial trajectory is displayed in Fig. 5.6(a). When we represent the corresponding worldline in the spacetime diagram, we must also plot the time axis ct; see Fig. 5.6(b). The object takes 1.4 s to reach the highest point and another 1.4 s to come down. But a 2.8 s time interval will be represented by almost one million kilometers of ct in the spacetime diagram (more than the round-trip distance to the moon), leading to a very nearly straight worldline as depicted in Fig. 5.6(c).

The main point here is not so much the straightness of the worldline, which reflects the practically flat spacetime in the very weak terrestrial gravitational field (recall that $\Phi_\oplus / c^2 \simeq 10^{-10}$). Rather, the point is that one must not confuse the trajectory in regular 3D space with the geodesic curve in spacetime.[13]

The interval extremized by the geodesic in spacetime is not simply the spatial length (cf. Box 5.3). In fact, the invariant interval of a particle's worldline is its proper time. We shall demonstrate that this geodesic equation is the relativistic generalization of the Newtonian equation of motion (4.9). To do so, we must define more precisely the sense in which Einstein's theory is an extension of Newtonian gravity; it is much more than an extension to higher-speed motion.

5.3.1   The Newtonian limit

To support our claim that the geodesic equation is the GR equation of motion, we shall now show that the geodesic equation (5.29) does reduce to the Newtonian equation of motion (4.9) in the **Newtonian limit** of

a test particle moving with nonrelativistic velocity $v \ll c$,
in a weak and static gravitational field.

We now take such a limit of the GR equation of motion (5.29):

- Nonrelativistic speed, $dx^i/dt \ll c$: This inequality, $dx^i \ll c\, dt$, implies that

$$\frac{dx^i}{d\tau} \ll c\frac{dt}{d\tau} = \frac{dx^0}{d\tau}. \qquad (5.47)$$

Keeping only the dominant term $(dx^0/d\tau)(dx^0/d\tau)$ in the double sum over indices λ and ρ in the geodesic equation (5.29), we have

$$\frac{d^2 x^\mu}{d\tau^2} + \Gamma^\mu_{00}\frac{dx^0}{d\tau}\frac{dx^0}{d\tau} = 0. \qquad (5.48)$$

- Static field, $\partial g_{\mu\nu}/\partial x^0 = 0$: Because all time derivatives vanish, the Christoffel symbols of (5.30) take a simpler form:

$$\Gamma^\mu_{00} = -\frac{1}{2}g^{\mu\nu}\frac{\partial g_{00}}{\partial x^\nu}. \qquad (5.49)$$

- Weak field, $h_{\mu\nu} \ll 1$: We assume that the metric is not too different from the flat-spacetime metric $\eta_{\mu\nu} = \text{diag}(-1, 1, 1, 1)$:

$$g_{\mu\nu} = \eta_{\mu\nu} + h_{\mu\nu}, \qquad (5.50)$$

where $h_{\mu\nu}(x)$ is a small correction field. $\eta_{\mu\nu}$ is constant, so $\partial g_{\mu\nu}/\partial x^\lambda = \partial h_{\mu\nu}/\partial x^\lambda$. The Christoffel symbols, being derivatives of the metric, are of order $h_{\mu\nu}$. To leading order, (5.49) is

$$\Gamma^\mu_{00} = -\frac{1}{2}\eta^{\mu\nu}\frac{\partial h_{00}}{\partial x^\nu},$$

which, because $\eta_{\nu\mu}$ is diagonal, has (for a static h_{00}) the following components:

$$\Gamma^0_{00} = \frac{1}{2}\frac{\partial h_{00}}{\partial x^0} = 0 \quad \text{and} \quad \Gamma^i_{00} = -\frac{1}{2}\frac{\partial h_{00}}{\partial x^i}. \qquad (5.51)$$

We can now evaluate (5.48) by using (5.51): the $\mu = 0$ equation leads to

$$\frac{dx^0}{d\tau} = \text{constant}, \qquad (5.52)$$

and the three $\mu = i$ equations are

$$\frac{d^2 x^i}{d\tau^2} + \Gamma^i_{00}\frac{dx^0}{d\tau}\frac{dx^0}{d\tau} = \left(\frac{d^2 x^i}{c^2\, dt^2} + \Gamma^i_{00}\right)\left(\frac{dx^0}{d\tau}\right)^2 = 0, \qquad (5.53)$$

where we have used the chain rule $dx^i/d\tau = (dx^i/dx^0)(dx^0/d\tau)$ and the condition (5.52) to conclude that $d^2x^i/d\tau^2 = (d^2x^i/dx^{0\,2})(dx^0/d\tau)^2$. Equation (5.53), together with (5.51), implies

$$\frac{d^2x^i}{c^2\,dt^2} - \frac{1}{2}\frac{\partial h_{00}}{\partial x^i} = 0, \tag{5.54}$$

which can be compared with the Newtonian equation of motion (4.9). Thus $h_{00} = -2\Phi/c^2$, and, using the definition (5.50), we recover (5.41), first obtained heuristically in the previous discussion.

We can indeed regard the metric tensor as the relativistic generalization of the gravitational potential. The expression (5.50) also provides us with a criterion to characterize a field as weak:

$$\left[\,|h_{00}| \ll |\eta_{00}|\,\right] \quad \Rightarrow \quad \left[\frac{|\Phi|}{c^2} \ll 1\right]. \tag{5.55}$$

Consider the gravitational potential at the earth's surface. It is equal in magnitude to the gravitational acceleration times the earth's radius, $|\Phi_\oplus| = g \times R_\oplus = O(10^7\,\mathrm{m^2/s^2})$, or $|\Phi_\oplus|/c^2 = O(10^{-10})$. Thus any gravitational field less than ten billion g's (acting over distances comparable to the earth's radius) may be considered weak.

Review questions

1. What is an intrinsic geometric description (vs. an extrinsic description)? Describe the intrinsic geometric operations that fix the metric elements.

2. What is the relation of the geodesic equation to the length-extremization condition?

3. What is the fundamental difference between coordinate transformations in a curved space and those in flat space (e.g., Lorentz transformations in flat Minkowski space)?

4. What is a local Euclidean frame of reference? What is the flatness theorem?

5. What does one mean by a "geometric theory of physics"? Use distance measurements on the surface of a globe to illustrate your answer.

6. How can the phenomenon of gravitational time dilation be described in geometric terms? Use this to argue that the spacetime metric can be regarded as the relativistic gravitational potential.

7. Use the simple example of a rotating cylinder to illustrate how EP physics can imply a non-Euclidean geometric relation.

8. What significant conclusion did Einstein draw from the analogy between the facts that a curved space is locally flat and that gravity can be transformed away locally?

9. Give the heuristic argument that the GR equation of motion is the geodesic equation.

10. What is the Newtonian limit? In this limit, what relation can one infer between the Newtonian gravitational potential and a certain metric tensor component of the spacetime. Use this relation to derive the gravitational redshift.

Einstein Equation and its Spherical Solution

6

- At every spacetime point, one can construct a free-fall frame in which gravity is transformed away. However, in a finite-sized region, one can detect the residual tidal forces that stem from the second derivatives of the gravitational potential. The curvature of spacetime characterizes these tidal effects.

- The GR field equation, a set of coupled partial differential equations, directly relates spacetime's curvature to the mass/energy distribution. Its solution (the metric) determines the geometry of spacetime.

- We briefly introduce the concept of curvature in differential geometry.

- The nonrelativistic theory of tidal forces yields the Newtonian deviation equation, whose relativistic analog is the equation of geodesic deviation, which relates spacetime curvature to the relative motion of nearby particles.

- We explore the symmetries and contractions of the Riemann curvature in search of a rank-2 symmetric curvature tensor appropriate for the GR field equation.

- A spherically symmetric metric has two unknown scalar functions: g_{00} and g_{rr}. The Schwarzschild solution to the GR field equation yields $g_{00} = -g_{rr}^{-1} = -(1 - r^*/r)$, with $r^* = 2G_N M/c^2$. An embedding diagram allows us to visualize such a warped space.

- GPS time corrections and the precession of Mercury's perihelion are worked out as successful applications of GR.

- Gravitational deflection of light is calculated, showing comparable contributions from gravitational time dilation and from gravitational length contraction. This basic GR effect is the basis of the gravitational lensing technique, an important tool of modern cosmology.

In Chapter 5, we introduced Einstein's geometric theory of gravitation. Its equation of motion is identified with the geodesic equation. Here we present the GR field equation, Einstein's equation, whose key element is spacetime curvature. We introduce in Section 6.1 the geometry of curvature, which can be physically

A College Course on Relativity and Cosmology. First Edition. Ta-Pei Cheng.
© Ta-Pei Cheng 2015. Published in 2015 by Oxford University Press.

interpreted as relativistic tidal gravity. The Newtonian theory of tidal forces is discussed in Section 6.2. After finding the GR field equation, we present its most important solution: the Schwarzschild exterior solution, which describes the spacetime outside a spherical source. Some notable examples of the geodesics in such a geometry are worked out.

The curvature of spacetime and Einstein's equation

With the identification of curved spacetime as the gravitational field and of its metric as the relativistic generalization of Newton's gravitational potential,

$$[\Phi(x)] \quad \longrightarrow \quad [g_{\mu\nu}(x)], \tag{6.1}$$

the GR generalization of Newton's equation of motion (4.9) is naturally identified with the geodesic equation (5.29):

$$\left[\frac{d^2\mathbf{r}}{dt^2} + \nabla\Phi = 0\right] \longrightarrow \left[\frac{d^2x^\sigma}{d\tau^2} + \Gamma^\sigma_{\lambda\rho}\frac{dx^\lambda}{d\tau}\frac{dx^\rho}{d\tau} = 0\right]. \tag{6.2}$$

Now our task is to find the GR generalization of Newton's field equation (4.7):

$$\left[\nabla^2\Phi = 4\pi G_{\mathrm{N}}\rho\right] \longrightarrow \quad [?]. \tag{6.3}$$

Let us lay out a strategy to find this GR field equation—the Einstein equation.

Seeking the GR field equation In Section 3.2.4, we learned that the mass density ρ is proportional to T_{00}, a component of the symmetric relativistic energy–momentum tensor. Thus, the right-hand side of the Einstein equation should be proportional to $T_{\mu\nu} = T_{\nu\mu}$. The GR field equation must be covariant and hence a tensor equation; the left-hand side (which incorporates the field, i.e., $g_{\mu\nu}$) must have the same tensor structure:

$$[\hat{O}g]_{\mu\nu} = \kappa\, T_{\mu\nu}, \tag{6.4}$$

where κ is some proportionality constant presumably related to G_{N}. Namely, the left-hand side should likewise be a symmetric tensor of rank 2, $[\hat{O}g]_{\mu\nu} = [\hat{O}g]_{\nu\mu}$, resulting from some differential operator $\hat{O}$ acting on the metric $[g]$. Since we expect $[\hat{O}g]_{00}$ to have a Newtonian limit proportional to $\nabla^2\Phi$, $[\hat{O}]$ must be a second-derivative operator. Besides the $\partial^2 g$ terms, we also expect it to contain nonlinear terms of the $(\partial g)^2$ type. This follows from the fact that energy, just like mass, is a source of gravity, and gravitational fields themselves hold energy. Just as electromagnetic fields have an energy density quadratic in the fields $\sim (\mathbf{E}^2 + \mathbf{B}^2)$, the gravitational field energy density must be quadratic in the gravitational field strength ∂g. In terms of Christoffel symbols $\Gamma \sim \partial g$ (cf. (5.30)), we anticipate $[\hat{O}g]$ to contain not only $\partial\Gamma$ but also Γ^2 terms.

Thus, seeking the GR field equation reduces to finding such a tensor $[\hat{O}g]$:

$[\hat{O}g]$ must be
a symmetric rank-2 tensor, composed of terms like $(\partial^2 g), (\partial g)^2 \sim \partial\Gamma, \Gamma^2.$

(6.5)

In our discussion of the flatness theorem in Section 5.1.3, we identified the second derivative of the metric with the curvature of spacetime. We shall elaborate this geometric view of curvature in Section 6.1. We will then present in Section 6.2 another aspect of curvature—as relativistic tidal forces. This will be followed by a discussion of Einstein's equation itself in Section 6.3.

6.1 Curvature: a short introduction

Only a flat space can have a metric tensor whose elements are all constant. However, one cannot determine that a space is curved just because its metric elements are coordinate-dependent. Consider, for instance, a 2D flat surface. Its metric elements in Cartesian coordinates are all constant, $[g^{(x,y)}] = [\mathbb{I}]$, but they are not so in polar coordinates (r, ϕ),

$$[g^{(r,\phi)}]_{ab} = \begin{pmatrix} 1 & 0 \\ 0 & r^2 \end{pmatrix}.$$ (6.6)

Thus, to conclude that a space is curved, one has to ensure that there does not exist *any* coordinate system in which the metric is constant.

Gaussian curvature of a 2D curved surface

To characterize 2D flat and curved surfaces, Gauss discovered a unique invariant called the curvature K. It is composed of the second derivative of the metric tensor $\partial^2 g$ and the nonlinear term $(\partial g)^2$ in such a way, independent of the coordinate choice, $K = 0$ for flat surfaces and $K \neq 0$ for curved surfaces. Since this curvature K is expressed entirely in terms of the metric and its derivatives, it is an intrinsic geometric property. With no loss of generality, we shall quote Gauss's result for a diagonalized metric $g_{ab} = \mathrm{diag}(g_{11}, g_{22})$:

$$K = \frac{1}{2g_{11}g_{22}} \left\{ -\frac{\partial^2 g_{11}}{(\partial x^2)^2} - \frac{\partial^2 g_{22}}{(\partial x^1)^2} \right.$$

$$\left. + \frac{1}{2g_{11}} \left[\frac{\partial g_{11}}{\partial x^1} \frac{\partial g_{22}}{\partial x^1} + \left(\frac{\partial g_{11}}{\partial x^2}\right)^2 \right] + \frac{1}{2g_{22}} \left[\frac{\partial g_{11}}{\partial x^2} \frac{\partial g_{22}}{\partial x^2} + \left(\frac{\partial g_{22}}{\partial x^1}\right)^2 \right] \right\}.$$ (6.7)

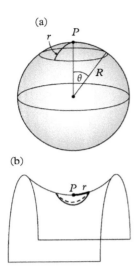

Figure 6.1 *A circle with radius r centered on a point P on a curved surface: (a) on a spherical surface with curvature $K = 1/R^2$; (b) on the middle portion of a saddle-shaped surface, which has negative curvature $K = -1/R^2$.*

[1] One can think of this 3D embedding space as Minkowski spacetime with one temporal and two spatial coordinates. The usual spacetime diagrams are attempts to represent the pseudo-Euclidean geometry in a 3D Euclidean space. In such a diagram, a pseudosphere is portrayed as a hyperbolic bowl of points displaced from the origin by a fixed proper time (a negative interval $x^2 + y^2 - c^2 t^2 = -c^2 \tau^2$). Such a bowl in Euclidean space would have positive curvature—for example, try to draw or envision in 3D spacetime a 2-sphere, the set of points displaced from the origin by a fixed positive (spacelike) interval. The hyperbolic tube appears to have negative curvature. As demonstrated by our discussion in Box 5.3, our Euclidean brains need lots of practice in order to grasp something as counterintuitive as pseudo-Euclidean geometry.

We will not present its derivation (see however Sidenote 11.7), but only indicate that for spaces of constant curvature (or for any infinitesimal surface),

$$K = \frac{k}{R^2}, \tag{6.8}$$

where R is the radius of curvature (which in the case of a spherical surface is simply the radius of the sphere) and k is the **curvature signature**: $k = 0$ for a flat surface; $k = +1$ for a spherical surface (called a **2-sphere** in geometry); and $k = -1$ for a hyperbolic surface, a **2-pseudosphere**.

Exercise 6.1 Gaussian curvature is coordinate-independent

Check the coordinate independence of the curvature (a) for a flat plane in Cartesian and polar coordinates and (b) for a spherical surface with radius R in polar coordinates and cylindrical coordinates, by plugging their respective metrics (6.6), (5.8), and (5.9) into the curvature formula (6.7).

Pseudosphere   We have already displayed the embedding of a spherical surface in a 3D Euclidean space (with metric $g_{ij} = \delta_{ij} = \text{diag}(1, 1, 1)$) as in (5.1); cf. Fig. 5.1. Unlike the cases of the plane ($k = 0$) or the sphere ($k = +1$), there is no simple way to visualize the whole ($k = -1$) pseudosphere, because its natural embedding is not into a 3D Euclidean space but into a flat 3D space with a pseudo-Euclidean metric $\eta_{ij} = \text{diag}(1, 1, -1)$. In such a space[1] with coordinates (X, Y, Z), the embedding condition is $X^2 + Y^2 - Z^2 = -R^2$. While we cannot draw the whole pseudosphere in an ordinary 3D Euclidean space, the central portion of a saddle surface does have a negative curvature; see Fig. 6.1(b).

Exercise 6.2 Pseudosphere metric from embedding coordinates

(i) *Recover the result (5.8) for the metric g_{ab} of a 2D $k = +1$ surface in the Gaussian coordinates $(x^{1,2} = \theta, \phi)$ by way of its 3D embedding coordinates $X^i(\theta, \phi)$ as shown in (5.3). From the invariant interval in the 3D embedding space, extended to the Gaussian coordinate differentials by the chain rule of differentiation, we have*

$$ds^2 = \delta_{ij} \, dX^i \, dX^j = \delta_{ij} \frac{\partial X^i}{\partial x^a} \frac{\partial X^j}{\partial x^b} \, dx^a \, dx^b. \tag{6.9}$$

By comparing this with (5.5), you can identify the metric for the 2D space:

$$g_{ab} = \delta_{ij} \frac{\partial X^i}{\partial x^a} \frac{\partial X^j}{\partial x^b}, \tag{6.10}$$

which can be viewed as the transformation of the metric from Cartesian to polar coordinates.

(ii) *For a k = −1 2D pseudosphere, its hyperbolic Gaussian coordinates $(x^{1,2} = \psi, \phi)$ can be related to the 3D embedding coordinates, in analogy with (5.3), by*

$$X^1 = R \sinh \psi \cos \phi, \quad X^2 = R \sinh \psi \sin \phi, \quad X^3 = R \cosh \psi. \quad (6.11)$$

This embedding space has the metric $\eta_{ij} = \mathrm{diag}(1, 1, -1)$. Follow the above steps to deduce the 2D space metric

$$g_{ab}^{(\psi, \phi)} = R^2 \begin{pmatrix} 1 & 0 \\ 0 & \sinh^2 \psi \end{pmatrix}. \quad (6.12)$$

(iii) *Show that this metric leads, via (6.7), to a negative Gaussian curvature $K = -1/R^2$.*

(iv) *Show that while the circumference of a circle with radius r on a spherical surface is $2\pi R \sin(r/R)$ which is smaller than $2\pi r$ as shown in Fig. 6.1(a), on a hypersphere it is $2\pi R \sinh(r/R)$, and hence greater than $2\pi r$.*

(v) *A k = +1 2D sphere has no boundary but a finite area: $A_2 = \int ds_\theta \, ds_\phi = 4\pi R^2$. Similarly demonstrate that a k = −1 2D pseudosphere has no boundary but an area that is infinite: $\tilde{A}_2 = \int ds_\psi \, ds_\phi = \infty$.*

In cosmology, we shall encounter 4D spacetimes with 3D spatial subspaces having constant curvature. Besides the 3D flat space, it is also possible to have 3D spaces with positive and negative curvature: 3-spheres and 3-pseudospheres. When they are embedded in 4D space with metric $g_{\mu\nu} = \mathrm{diag}(\pm 1, 1, 1, 1)$ and coordinates (W, X, Y, Z), these 3D spaces obey the constraint conditions

$$\pm W^2 + X^2 + Y^2 + Z^2 = \pm R^2. \quad (6.13)$$

Box 6.1 Metric tensors for spaces with constant curvature

Let us first collect the metric expressions in polar coordinates for the three 2D surfaces with constant curvature $K = k/R^2$. For the flat plane ($k = 0$), we have (6.6); for the sphere ($k = +1$), we have (5.8); and for the pseudosphere ($k = -1$), we have (6.12). In terms of the dimensionless radial coordinate $\chi \equiv r/R$ (previously we had $\chi = \theta$ and $\chi = \psi$), they all take a similar form:

continued

Box 6.1 *continued*

$$[ds^2]^{(k)}_{2D,\chi} = \begin{cases} R^2(d\chi^2 + \sin^2\chi\, d\phi^2) & \text{for } k = +1, \\ R^2(d\chi^2 + \chi^2\, d\phi^2) & \text{for } k = 0, \\ R^2(d\chi^2 + \sinh^2\chi\, d\phi^2) & \text{for } k = -1. \end{cases} \quad (6.14)$$

This similarity among the metrics suggests an even more compact expression that combines these three equations into one:[2]

$$[ds^2]^{(k)}_{2D,\chi} = R^2[d\chi^2 + k^{-1}(\sin^2\sqrt{k}\chi)d\phi^2]. \quad (6.15)$$

3D spaces with constant curvature

We are particularly interested in 3D spaces of constant curvature, which will be relevant in our study of cosmology. The 2D spaces with constant curvature have metrics displayed in (6.14) with polar coordinates. We now make a heuristic argument for the generalization of the 2D results (6.15) to the corresponding 3D spaces.[3] We start with the observation that the 2D $k = 0$ metric for a flat plane in polar coordinates (r, ϕ) is $[ds^2]^{(k=0)}_{2D} = dr^2 + r^2\, d\phi^2$. For the $k = 0$ flat 3D space, the usual spherical coordinate system (r, θ, ϕ) involves an additional (polar) angle coordinate θ. We have the familiar metric $[ds^2]^{(k=0)}_{3D} = dr^2 + r^2\, d\Omega^2$, where $d\Omega^2 = d\theta^2 + \sin^2\theta\, d\phi^2$ is the differential solid angle. We now suggest that even for the curved spaces $k \neq 0$, we can similarly obtain the 3D metrics from their 2D counterparts by the simple substitution $d\phi^2 \rightarrow d\Omega^2$. We can thereby infer from (6.15) the metric for the $(k = 0, \pm 1)$ 3D spaces in spherical polar coordinates (χ, θ, ϕ):

$$[ds^2]^{(k)}_{3D,\chi} = R^2[d\chi^2 + k^{-1}(\sin^2\sqrt{k}\chi)\, d\Omega^2]. \quad (6.16)$$

Extending in a similar way the result shown in (5.9), we can write down the metric for the three 3D spaces of constant curvature in cylindrical coordinates (ρ, θ, ϕ) using the dimensionless radial coordinate $\xi \equiv \rho/R$:

$$[ds^2]^{(k)}_{3D,\xi} = R^2\left(\frac{d\xi^2}{1 - k\xi^2} + \xi^2\, d\Omega^2\right). \quad (6.17)$$

[2] Recall from Box 3.1 (in particular Sidenote 3) that $\sin i\chi = i\sinh\chi$. For $k = 0$, (6.15) is understood to mean the limit as $k \rightarrow 0$, so that $\sin^2\sqrt{k}\chi \rightarrow k\chi^2$.

[3] A rigorous derivation of these results would involve the mathematics of symmetric spaces, Killing vectors, and isometry. Our heuristic argument can also be supported by solving the Einstein equation for a homogeneous and isotropic 3D space; see (Cheng 2010, Section 14.4.1).

Curvature measures the deviation from Euclidean relations   On a flat surface, the familiar Euclidean geometrical relations hold. For example, the circumference of a circle with radius r is $S = 2\pi r$. The curvature measures how curved a surface is because it is directly proportional to the violation of Euclidean relations. Figure 6.1 shows two circles of radius r drawn on surfaces with nonvanishing curvature. The circular circumference S differs from its flat-surface value $2\pi r$ by an amount proportional to the Gaussian curvature K:

$$\lim_{r \to 0} \frac{2\pi r - S}{r^3} = \frac{\pi}{3}K. \qquad (6.18)$$

On a positively curved surface, the circumference is smaller than (or, on a negatively curved surface, larger than) that on a flat surface. Another simple example is that on a curved surface the **angular excess** ϵ of a polygon (the difference between the sum of the interior angles and the corresponding Euclidean sum) is proportional to its area σ, with the proportionality constant being the curvature K:

$$\epsilon = K\sigma. \qquad (6.19)$$

Figure 6.2 illustrates this relation[4] for the case of a triangle on a sphere. The illustration also demonstrates that this angular excess can be measured by the directional change of a vector that is parallel-transported around this polygon. Such a 2D relation can be extended to higher-dimensional spaces, suggesting the way to define the curvature tensor in a general-dimensional space; see Section 11.2.1.

Riemann curvature tensor

Extending Gauss's discovery of the curvature (a single component) in 2D space, his pupil Riemann (with further contributions by Christoffel) established the existence of a rank-4 tensor, the **Riemann curvature tensor**, in an n-dimensional space:

$$R^{\mu}{}_{\lambda\alpha\beta} = \partial_\alpha \Gamma^{\mu}_{\lambda\beta} - \partial_\beta \Gamma^{\mu}_{\lambda\alpha} + \Gamma^{\mu}_{\nu\alpha}\Gamma^{\nu}_{\lambda\beta} - \Gamma^{\mu}_{\nu\beta}\Gamma^{\nu}_{\lambda\alpha}. \qquad (6.20)$$

The Christoffel symbols Γ are first derivatives, as shown in (5.30), so this curvature $R = \partial\Gamma + \Gamma\Gamma$ is a nonlinear second-derivative function of the metric, with terms like $\partial^2 g$ and $(\partial g)^2$. It measures, independently of coordinate choice, the deviation from flat-space relations; thus $R^{\mu}{}_{\lambda\alpha\beta} = 0$ for a flat spacetime.[5] As we shall see, a linear combination of the contractions of this curvature tensor enters directly into the GR field equation, the Einstein equation.

There are many (mutually consistent) ways to derive the expression for the curvature tensor displayed in (6.20). One simple method is to generalize the 2D relation (6.19) to higher dimensions by calculating the change in a vector A parallel-transported around an infinitesimal parallelogram spanned by two infinitesimal vectors a^μ and b^ν. As the relative change of the vector yields the angular excess ($dA/A = \epsilon$ for small changes), the higher-dimensional generalization of (6.19) should then be

$$dA^\mu = -R^{\mu}{}_{\nu\lambda\rho}A^\nu a^\lambda b^\rho. \qquad (6.21)$$

Namely, the vectorial change is proportional to the vector A^ν itself and to the area of a parallelogram spanned by two (infinitesimal) vectors a^λ and b^ρ.

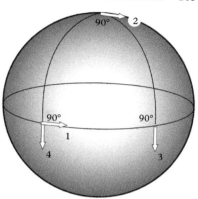

Figure 6.2 *A triangle with all interior angles of 90° on a spherical surface (radius R) has an angular excess $\epsilon = \pi/2$. This satisfies the relation $\epsilon = K\sigma$, with curvature $K = 1/R^2$ and the triangular area $\sigma = \pi R^2/2$ equal to one-eighth of the spherical area. The angular excess can be measured by the parallel transport of a vector around this triangle (vector-1, clockwise to vector-2, to -3, and finally back to the starting point as vector-4), which rotates the vector by 90° (the angular difference between vector-4 and vector-1).*

[4] A somewhat less specific example would be a triangle with two right angles and a third unspecified angle θ; thus $\epsilon = \theta$. Such a triangle has an area θR^2, satisfying the $\epsilon = K\sigma$ relation. For a proof using a general spherical triangle with arbitrary interior angles, see (Cheng 2010, p. 92 in Section 5.3.2). Such a relation for triangles can be extended to polygons, as any polygon can be decomposed into triangles.

[5] It should be noted that in cosmology we shall often refer to a flat universe that has a nontrivial spacetime $R^{\mu}{}_{\lambda\alpha\beta} \neq 0$ but has a 3D subspace with a flat geometry (having a curvature signature $k = 0$).

The coefficient of proportionality $R^{\mu}_{\ \nu\lambda\rho}$ is a quantity with four indices, antisymmetric in λ and ρ because the area $a^{\lambda}b^{\rho}$ should be antisymmetric. We shall take this to be the definition of the curvature tensor of an n-dimensional space. For an explicit derivation, see Section 11.2.1.

6.2 Tidal gravity and spacetime curvature

What is the physical significance of the curvature? Curvature involves second derivatives of the metric. Since the metric tensor can be regarded as the relativistic gravitational potential, its first derivatives are gravitational accelerations, so its second derivatives must then be relative accelerations (per unit separation) between neighboring particles. Namely, the second derivatives give rise to relativistic tidal forces.

6.2.1 Tidal forces—a qualitative discussion

Here we give an elementary and qualitative discussion of the (nonrelativistic) tidal force and its possible geometric interpretation. This will be followed by a more mathematical presentation in terms of the Newtonian deviation equation, which relates the relative motion due to tidal forces between two neighboring particles to the second derivatives of the gravitational potential. This in turn suggests the GR generalization, the equation of geodesic deviation, which relates the relative spacetime motion of two particles to the underlying spacetime curvature.

The EP states that in a freely falling reference frame, the physics is the same as that in an inertial frame with no gravity; SR applies, and the metric is the Minkowski metric $\eta_{\mu\nu}$. As shown in the flatness theorem (Section 5.1.3), one can only approximate $g_{\mu\nu}$ by $\eta_{\mu\nu}$ locally, that is, in an appropriately small region. Gravitational effects can always be detected in a finite-sized free-fall frame, as the gravitational field in reality is never strictly uniform; the second derivatives of the metric come into play (as the first derivatives vanish in a free-fall frame).

Consider the lunar gravitational attraction exerted on the earth. While the earth is in free fall toward the moon (and vice versa), there is still a detectable lunar gravitational effect on earth. Different points on earth feel slightly different gravitational pulls by the moon, as depicted in Fig. 6.3(a). The center-of-mass (CM) force causes the earth and everything on it to fall toward the moon, so this CM gravitational effect is canceled out in the freely falling terrestrial frame. But subtracting out this CM force leaves residual (relative) forces on various parts of the earth, as shown in Fig. 6.3(b). They are a stretching in the longitudinal direction and a compression in the transverse directions. This is just the familiar tidal gravity.[6] Namely, in the free-fall frame, the CM gravitational effect is transformed away, but there are still remnant tidal forces. They reflect the **differences** of the gravitational effects on neighboring points, and are thus proportional to the derivatives of the gravitational field (and hence to the second derivatives of

[6] The ocean is pulled away in opposite directions, giving rise to two tidal bulges. This explains why, as the earth rotates, there are two high tides in a day. This of course is a simplified description, as there are other effects (e.g., the solar tidal forces) that must be taken into account.

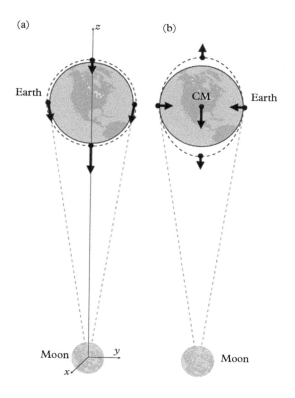

Figure 6.3 *Variations of the gravitational field as tidal forces. (a) Lunar gravitational forces on four representative points on earth. (b) After taking out the center-of-mass (CM) motion, the relative forces on earth are the tidal forces of longitudinal stretching and transverse compression.*

the potential Φ). Since tidal forces cannot be coordinate-transformed away, they should be regarded as the essence of gravitation.

6.2.2 Deviation equations and tidal gravity

Here we provide a quantitative description of gravitational tidal forces in the Newtonian framework, which suggests an analogous GR approach in which the curvature tensor plays the role of the second derivatives of the potential in producing tidal gravity.

Newtonian deviation equation for tidal forces

The tidal effect concerns the relative motion of particles in a nonuniform gravitational field. Let us consider two particles: one has trajectory $\mathbf{x}(t)$ and the other has $\mathbf{x}(t) + \mathbf{s}(t)$. That is, the locations of these two particles measured at the same time have a coordinate separation $\mathbf{s}(t)$. The respective equations of motion ($i = 1, 2, 3$) obeyed by these two particles, according to (4.9), are

$$\frac{d^2x^i}{dt^2} = -\frac{\partial \Phi(x)}{\partial x^i} \quad \text{and} \quad \frac{d^2x^i}{dt^2} + \frac{d^2s^i}{dt^2} = -\frac{\partial \Phi(x+s)}{\partial x^i}. \qquad (6.22)$$

Consider the case in which the separation distance $s^i(t)$ is small, so we can approximate the gravitational potential $\Phi(x + s)$ by a Taylor expansion

$$\Phi(x + s) = \Phi(x) + \frac{\partial \Phi}{\partial x^j} s^j + \cdots. \tag{6.23}$$

From the difference of the two equations in (6.22), we obtain the **Newtonian deviation equation** that describes the separation between two nearby particle trajectories in a gravitational field:

$$\frac{d^2 s^i}{dt^2} = -\left(\frac{\partial^2 \Phi}{\partial x^i \partial x^j}\right) s^j. \tag{6.24}$$

Thus the relative acceleration per unit separation, $(d^2 s^i/dt^2)/s^j$, is given by the tidal force matrix, a rank-2 tensor having the second derivatives of the gravitational potential as its elements.

The spherically symmetric field As an illustrative application of (6.24), let us discuss the case of a spherical gravitational source of mass M (e.g., the gravity due to the moon on earth as shown in Fig. 6.3), $\Phi(x) = -G_N M/r$, where the radial distance is related to the rectangular coordinates by $r = (x^2 + y^2 + z^2)^{1/2}$. Since $\partial r/\partial x^i = x^i/r$ and $\partial \Phi/\partial x^i = G_N M x^i/r^3$, the tidal force tensor is

$$\frac{\partial^2 \Phi}{\partial x^i \partial x^j} = \frac{G_N M}{r^3}\left(\delta_{ij} - \frac{3 x^i x^j}{r^2}\right). \tag{6.25}$$

Consider a reference particle located along the z axis: $x^i = (0, 0, r)$. The Newtonian deviation equation (6.24) for the displacement of a second particle from the reference, with the tidal force tensor given by (6.25), becomes[7]

[7] Since we have taken $x^i = (0, 0, r)$, the term $x^i x^j$ on the right-hand side can be worked out with $x^1 = x^2 = 0$ and $x^3 = r$.

$$\frac{d^2}{dt^2}\begin{pmatrix} s_x \\ s_y \\ s_z \end{pmatrix} = \frac{-G_N M}{r^3}\begin{pmatrix} 1 & 0 & 0 \\ 0 & 1 & 0 \\ 0 & 0 & -2 \end{pmatrix}\begin{pmatrix} s_x \\ s_y \\ s_z \end{pmatrix}. \tag{6.26}$$

We see that there is an attractive tidal force between the two particles in the transverse directions, $d^2 s_{x,y}/dt^2 = -G_N M s_{x,y}/r^3$, which leads to compression, while a tidal repulsion, $d^2 s_z/dt^2 = +2 G_N M s_z/r^3$, leads to stretching in the longitudinal (i.e., radial) direction.[8] This verifies the tidal effects illustrated in Fig. 6.3(b).

[8] One notable feature of the tidal force matrix is that it is traceless. This follows from Newton's equation $\delta_{ij}\partial^2 \Phi/\partial x^i \partial x^j = \nabla^2 \Phi = 0$ in the space exterior to the source.

GR equation of geodesic deviation and the curvature tensor

We shall follow a similar approach in GR. The two equations of motion (6.22) will be replaced by the corresponding GR equations of motion (the geodesic equations). A Taylor expansion of their difference yields the **equation of geodesic** deviation,[9] a generalization of (6.24) replacing the tidal force tensor by a contraction of the Riemann curvature tensor, given by (6.20), with two factors of the relative 4-velocity:

$$\left[\frac{\partial^2 \Phi}{\partial x^i \partial x^j}\right] \quad \longrightarrow \quad \left[R^\mu_{\ \alpha\nu\beta}\dot{x}^\alpha \dot{x}^\beta\right]. \tag{6.27}$$

Thus the relative motion of nearby particles, like that of the converging particles in Fig. 5.4, can be attributed to curvature. Such considerations led Einstein to give gravity a direct geometric interpretation by identifying tidal gravity with the curvature of spacetime.

6.3 The GR field equation

We now discuss the contraction of the rank-4 Riemann curvature tensor (6.20) with the aim of finding, as outlined in (6.5), the appropriate rank-2 tensor for the GR field equation.

6.3.1 Einstein curvature tensor

There are six ways to contract a rank-4 tensor to produce a rank-2 tensor. However, because of the symmetries of the Riemann tensor, all of its contractions either vanish or are trivially related to one another.

Symmetries and contractions of the curvature tensor

We first note that the Riemann curvature tensor with all lower indices,

$$R_{\mu\nu\alpha\beta} = g_{\mu\lambda}R^\lambda_{\ \nu\alpha\beta}, \tag{6.28}$$

has the following symmetry features:

- It is *antisymmetric* with respect to the interchange of its first and second indices and of its third and fourth indices:

$$R_{\mu\nu\alpha\beta} = -R_{\nu\mu\alpha\beta}; \tag{6.29}$$
$$R_{\mu\nu\alpha\beta} = -R_{\mu\nu\beta\alpha}. \tag{6.30}$$

- It is *symmetric* with respect to the interchange of the pair made up of its first and second indices with the pair of third and fourth indices.

$$R_{\mu\nu\alpha\beta} = +R_{\alpha\beta\mu\nu}. \tag{6.31}$$

- It also has the *cyclic* symmetry[10]

$$R_{\mu\nu\alpha\beta} + R_{\mu\beta\nu\alpha} + R_{\mu\alpha\beta\nu} = 0. \tag{6.32}$$

Since the symmetry properties are tensor relations, they are not changed by coordinate transformations. Once they have been proved in one frame, we can

[9] See Section 11.2.2, in particular Eq. (11.67). To carry out this derivation one would have to use the concept of covariant derivative, which will be introduced in Chapter 11.

[10] Symmetries reduce the number of independent elements. Using (6.29)–(6.32), the reader is invited to verify the combinatorial result showing that out of the total of n^4 components of the Riemann tensor in an n-dimensional space, the number of independent components is $N_{(n)} = n^2(n^2 - 1)/12$. Thus $N_{(1)} = 0$; it is not possible for a one-dimensional inhabitant to see any curvature. $N_{(2)} = 1$; this is just the Gaussian curvature K for a curved surface. $N_{(4)} = 20$; there are 20 independent components in the curvature tensor for a 4D curved spacetime. Cf. (Cheng 2010, Section 13.3.2 and Problem 13.9).

then claim their validity in all frames. An obvious choice is the locally inertial frame with $\Gamma = 0$, $\partial\Gamma \neq 0$), where the curvature (6.20) takes a simpler form, $R_{\mu\nu\alpha\beta} = g_{\mu\lambda}(\partial_\alpha\Gamma^\lambda_{\nu\beta} - \partial_\beta\Gamma^\lambda_{\nu\alpha})$. In this way, we can use the fact that $g_{\mu\nu} = g_{\nu\mu}$ and $\Gamma^\mu_{\nu\lambda} = \Gamma^\mu_{\lambda\nu}$ to check easily the validity of the symmetry properties shown in (6.29)–(6.32).

Contractions of the curvature tensor

We contract the Riemann tensor, reducing its rank so it can be used in the GR field equation. Because of the symmetry properties shown above, these contractions are basically unique.

Ricci tensor $R_{\mu\nu}$ This is the Riemann curvature tensor with its first and third indices contracted:

$$R_{\mu\nu} \equiv g^{\alpha\beta} R_{\alpha\mu\beta\nu} = R^\beta_{\ \mu\beta\nu}. \tag{6.33}$$

It follows from (6.31) that the Ricci tensor is symmetric:

$$R_{\mu\nu} = R_{\nu\mu}. \tag{6.34}$$

If we had contracted a different pair of indices (or tried all six possibilities), the result would still be $\sigma R_{\mu\nu}$, with $\sigma = +1, 0,$ or -1. Thus the rank-2 curvature tensor is essentially unique.

Ricci scalar R Contracting the Ricci tensor, we obtain the Ricci scalar field,

$$R \equiv g^{\alpha\beta} R_{\alpha\beta} = R^\beta_{\ \beta}. \tag{6.35}$$

Energy/momentum conservation and the Einstein tensor

These contractions show that there is indeed a ready-made symmetric rank-2 curvature tensor, the Ricci tensor $R_{\mu\nu}$, for the left-hand side of (6.4). Making this choice $[\hat{O}g]_{\mu\nu} = R_{\mu\nu}$, Einstein found that the resultant field equation did not reduce to (4.7) in the Newtonian limit. After a long struggle, he finally found (at the end of 1915) the correct tensor $[\hat{O}g]_{\mu\nu} = G_{\mu\nu}$, now known as the Einstein tensor,

$$G_{\mu\nu} \equiv R_{\mu\nu} - \frac{1}{2} R g_{\mu\nu}. \tag{6.36}$$

He realized that besides the Ricci tensor $R_{\mu\nu}$, there is another rank-2 symmetric curvature tensor, $R g_{\mu\nu}$, the product of the Ricci scalar with the metric tensor itself, that satisfies the criteria for the geometry side of the field equation. In fact, any linear combination of the form $[\hat{O}g]_{\mu\nu} = R_{\mu\nu} + a R g_{\mu\nu}$ with a constant coefficient a will satisfy the search criteria (6.5). What is the correct value of a? To constrain the choice, Einstein invoked the conservation law of energy/momentum in the field system. This led him[11] to the combination (6.36).

[11] The calculation that Einstein undertook was rather complicated. After he published his GR papers in 1916, several eminent mathematicians, including David Hilbert and Felix Klein, were involved in this study. In particular, Emmy Noether (1882–1935) discovered her famous theorem connecting symmetries to conservation laws. This intense search lasting several years did bring about a greatly expedited proof showing that the Bianchi identity could lead directly to the Einstein tensor (6.36); cf. Section 11.2.3.

Remark 6.1 *The issue of energy/momentum conservation in GR is a delicate one. These conservation laws follow (via Noether's theorem) from spacetime translational symmetry. But in GR, space and time are dynamical, and a gravitational field's local energy/momentum densities are not well defined. Based on this, some would prefer to say that these conservation laws do not hold in GR. Since there can still be global conservation, we choose not to make such a strong statement.*

6.3.2 Einstein field equation

With the identification of $G_{\mu\nu}$ with the left-hand side of (6.4), Einstein (1916) arrived at his GR field equation:

$$G_{\mu\nu} \equiv R_{\mu\nu} - \frac{1}{2}Rg_{\mu\nu} = \kappa T_{\mu\nu}. \qquad (6.37)$$

The EP informs us that gravity can always be transformed away locally (by going to a reference frame in free fall); the immutable essence of gravity lies in its differentials (tidal forces). Thus the presence of the spacetime curvature in the GR field equation can be understood. This equation can be written in an alternative form. Contracting the two indices in (6.37) leads to $-R = \kappa T$, where T is the trace $g^{\mu\nu}T_{\mu\nu}$. In this way, we can rewrite the field equation (6.37) as

$$R_{\mu\nu} = \kappa \left(T_{\mu\nu} - \frac{1}{2}Tg_{\mu\nu} \right). \qquad (6.38)$$

Because the source distribution $T_{\mu\nu}$, the Ricci tensor $R_{\mu\nu}$, and the metric $g_{\mu\nu}$ are symmetric tensors, each has ten independent elements. Thus this seemingly simple field equation is actually a set of ten coupled nonlinear partial differential equations whose unknowns are the ten components[12] of the metric $g_{\mu\nu}(x)$. It is nonlinear because the solution potential can itself act as the source of gravitational potential.[13] We reiterate the central point of Einstein's theory: the spacetime geometry in GR, in contrast to SR, is not a fixed entity, but is dynamically ever-changing as dictated by the mass/energy distribution.

The Newtonian limit for a general source

One can examine one aspect of the correctness of this proposed GR field equation by checking its Newtonian limit. A straightforward calculation similar to that carried out in Section 5.3.1 shows that (6.38) does reduce to Newton's field equation (4.7) for a particle moving nonrelativistically in a weak and static field.[14] This also gives us the identification

$$\kappa = 8\pi G_{\mathrm{N}}/c^4. \qquad (6.39)$$

The proportionality constant κ, and hence Newton's constant G_{N}, is the conversion factor between energy density on the right-hand side and the geometric quantity on the left-hand side of Einstein's equation (6.37) or (6.38).

[12] There is subtle point here: actually not all ten components of the metric are independent. The principle of GR requires that a solution of the field equation be the same in any choice of coordinates. This means that if $g_{\mu\nu}(x)$ is a solution, so must $g_{\mu\nu}(x')$, where the x'^μ are related to the x^μ by any general coordinate transformations. If there are only six independent metric elements, do the ten field equations overdetermine them? We shall resolve this puzzle when we discuss the Bianchi identity and the energy/momentum conservation constraint of Einstein equations at the end of Section 11.3.2.

[13] In contrast to electrically neutral photons, the quanta of the electromagnetic field, gravitons, the quanta of the gravitational field, carry energy (i.e., gravitational charge) and thus are themselves sources of the gravitational field.

[14] See Exercise 11.6.

(a)   (b)

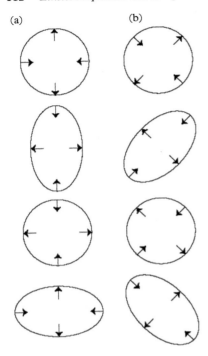

Figure 6.4 *Tidal effects of gravitational waves on a circle of test particles. The two columns represent the effects of waves in two polarization states shown in (6.44): (a) the plus-polarization and (b) the cross-polarization states. Similar to the effects of the transverse forces shown in Fig. 6.3, the circle is alternately stretched and squeezed in directions perpendicular to the wave propagation direction.*

[15] Namely, we will ignore any further gravitational waves generated by the gravitational field itself.

[16] For a derivation, see (Cheng 2010, Section 15.1.2). Just as we have chosen a particular gauge (the Lorentz gauge, $\partial^\mu A_\mu = 0$) to simplify the writing of the electromagnetic wave equation, we have made a particular coordinate choice, the transverse-traceless (TT) gauge, in writing the gravitational wave equation.

In taking the above limit, we have taken the source to be nonrelativistic matter with its energy–momentum tensor completely dominated by the $T_{00} = \rho c^2$ term. Effectively, we have taken the source to be a swarm of noninteracting particles (dust). In many situations, with cosmology being the notable example, we consider the source of gravity to be a plasma having mass density ρ and pressure p. Namely, we need to take $T_{\mu\nu}$ as the energy–momentum tensor for an ideal fluid as in (3.84). A calculation entirely similar to the Newtonian-limit calculation mentioned above leads to

$$\nabla^2 \Phi = 4\pi \, G_N \left(\rho + 3\frac{p}{c^2} \right). \tag{6.40}$$

In this way, Einstein's relativistic theory makes it clear that not only mass but also pressure can be a source of the gravitational field.

6.3.3   Gravitational waves

Newton's theory of gravitation is a static theory; the field due to a source is established instantaneously. Thus the field depends explicitly on the spatial coordinates, but not on time. Einstein's relativistic theory treats space and time on an equal footing. As in Maxwell's theory, the field propagates outward from the source with a finite speed c. Thus, just as one can shake an electric charge to generate electromagnetic waves, one can shake a mass to generate gravitational waves. Since gravity is ordinarily such a weak force, we shall only deal here with the case of a weak gravitational field whose metric field is almost flat:

$$g_{\mu\nu} = \eta_{\mu\nu} + h_{\mu\nu}, \quad \text{with } h_{\mu\nu} \ll O(1). \tag{6.41}$$

This approximation linearizes[15] Einstein's theory. In this limit, gravitational waves may be viewed as small curvature ripples propagating in a background of flat spacetime. Just as the wave equation yields the electromagnetic potential $A_\mu(x)$ due to a current density $j_\mu(x)$,

$$\Box A_\mu = \frac{4\pi}{c} j_\mu, \tag{6.42}$$

the perturbed gravitational field obeys[16]

$$\Box h_{\mu\nu} = -\frac{16\pi \, G_N}{c^4} T_{\mu\nu}. \tag{6.43}$$

Its vacuum solution is a transverse wave having two independent polarization states and traveling at the speed of light. For a plane wave solution (with this particular choice of coordinates), we can display the field's tensor structure:

$$h_{\mu\nu}(z, t) = \begin{pmatrix} 0 & 0 & 0 & 0 \\ 0 & h_+ & h_\times & 0 \\ 0 & h_\times & -h_+ & 0 \\ 0 & 0 & 0 & 0 \end{pmatrix} e^{i\omega(z-ct)/c}. \tag{6.44}$$

Gravitational waves thus have two independent polarizations: the plus (h_+) and the cross $(h_\times)$ states.[17] As we discussed previously, the gravitational field (whether static or time-dependent) affects the relative positions of test bodies. A gravitational wave alternately compresses and elongates space in directions perpendicular to that of its propagation. These tidal effects deform a circle of test particles into an ellipse whose orientation oscillates (as shown in Fig. 6.4) or rotates (by a combination of oscillations with offset phases). Contrast that with an electromagnetic wave, which pushes around a test charge (or a circle of charges in unison), causing it to oscillate back and forth (linear polarization) or in circles (by a combination of linear polarizations with offset phases). A major effort is underway to detect gravitational waves using sensitive interferometers that can measure the minute compressions and elongations of orthogonal lengths that are caused by the passage of such waves (Fig. 6.5).

In the meantime, there is already indirect[18] but convincing evidence for the existence of gravitational waves as predicted by GR. This has come from observations, spanning more than 25 years, of the orbital motion of the relativistic Hulse–Taylor binary pulsar system (PSR 1913+16). The orbiting binary stars (the shaking mass source) generate gravitational waves. Even though the binary is 21 000 light-years away from us, the basic parameters of the system can be deduced by carefully monitoring the radio pulses emitted by the pulsar, which effectively acts as an accurate and stable clock. From this record, we can verify a number of GR effects. GR predicts that gravitational radiation will carry away energy from the system, reducing the orbital period at a calculable rate.[19] The observed orbit rate decrease has been found to be in splendid agreement with the prediction by Einstein's theory (Fig. 6.6).

[17] Because the gravitation metric field is a symmetric tensor of rank 2, we expect its quantum, the graviton, to be a spin-2 particle. While a massive spin-2 particle has $2s + 1 = 5$ spin projection states, the graviton, being massless (as demonstrated by gravity's long-range nature), has only two independent helicity states (projections of spin in the direction of motion). In short, quantum theory leads us to expect exactly what GR gives us: two independent polarizations of gravitational waves.

[18] At the time of this writing, the BI-CEP2 Collaboration (2014) announced that they had obtained more direct evidence of gravitational waves. The B-mode of radiation polarization in the cosmic microwave background (CMB) was observed. This is a signature of gravitational waves generated in the big bang that left their imprint on the particle plasma at the cosmic epoch when the CMB was created. (See further comments in Chapter 10, especially Box 10.3.)

[19] For a more detailed calculation, see Eq. (15.71) in (Cheng 2010).

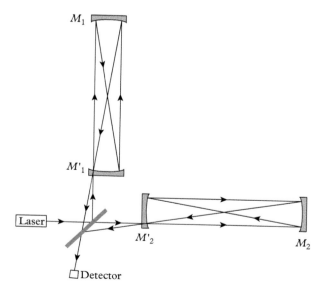

Figure 6.5 *Schematic diagram of a gravitational wave Michelson interferometer. The four mirrors $M_{1,2}$, $M'_{1,2}$ and the beam-splitter mirror are freely suspended. The two arms are optical cavities that increase the optical paths by many factors. A minute length change of the two arms (one expands and the other contracts) will show up as changes in the fringe pattern of the detected light.*

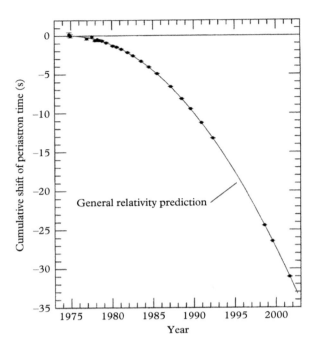

Figure 6.6 *Gravitational radiation damping causes orbital decay of the Hulse–Taylor binary pulsar. Plotted here is the accumulating shift in the epoch of periastron. The parabola is the GR prediction, and observations are depicted by data points. In most cases, the measurement uncertainties are smaller than the line widths. The data gap in the 1990s reflects the downtime when the radio telescope at the Arecibo Observatory was being upgraded. The graph is reproduced with permission from (Weisberg and Taylor 2003). ©2003 Astronomical Society of the Pacific Conference Series.*

6.4 Geodesics in Schwarzschild spacetime

Given the source mass distribution, we can solve the Einstein equation to find the metric $g_{\mu\nu}(x)$ and therefore the spacetime geometry. Here we shall concentrate on the important case of the gravitational field outside a spherical mass. The solution was obtained by Karl Schwarzschild (1873–1916) almost immediately after Einstein finally formulated his field equation in 1915. It is called the Schwarzschild metric. We note that its Newtonian analog is the gravitational potential

$$\Phi(r) = -\frac{G_N M}{r}, \tag{6.45}$$

which is the solution to the field equation $\nabla^2 \Phi(x) = 4\pi G_N \rho(x)$ with a spherically symmetric mass density $\rho(x)$ having a total mass M inside a sphere with radius r.

In this section, we describe the metric and geometry of Schwarzschild spacetime. We apply this to GPS and two other interesting examples of geodesic motion: the deflection of a light by a massive body (e.g., the sun) and the precession of an elliptic orbit (e.g., that of the planet Mercury). For such processes within the solar system, the relativistic corrections are very small. We will study in Chapter 7 the strong gravity of black holes and in Chapters 8–10 cosmology, for which the GR description of gravitation is indispensable.

Box 6.2 Form of a spherically symmetric metric tensor

If a gravitational source is spherically symmetric, the spacetime it generates (as the solution of the Einstein equation) must also have this symmetry. Naturally, we will pick a spherical coordinate system (t, r, θ, ϕ) whose center coincides with that of the spherical source. Here we shall demonstrate explicitly that a spherically symmetric metric tensor has only two unknown scalar functions: g_{00} and g_{rr}.

General considerations of spherical symmetry

The infinitesimal invariant interval $ds^2 = g_{\mu\nu} dx^\mu dx^\nu$ must be quadratic in dr and dt without singling out any particular angular direction.[20] That is, ds^2 must be composed of terms having two powers of dr and/or dt; furthermore, the vectors $\mathbf{r}$ and $d\mathbf{r}$ must appear in the form of dot products so as not to spoil the spherical symmetry:

$$ds^2 = A\,d\mathbf{r} \cdot d\mathbf{r} + B(\mathbf{r} \cdot d\mathbf{r})^2 + C\,dt\,(\mathbf{r} \cdot d\mathbf{r}) + D\,dt^2, \qquad (6.46)$$

where A, B, C, and D are scalar functions of t and $\mathbf{r} \cdot \mathbf{r}$. In a spherical coordinate system with orthonormal basis vectors $\hat{\mathbf{r}}$, $\hat{\boldsymbol{\theta}}$, and $\hat{\boldsymbol{\phi}}$, we have

$$\mathbf{r} = r\hat{\mathbf{r}} \quad \text{and} \quad d\mathbf{r} = dr\,\hat{\mathbf{r}} + r\,d\theta\,\hat{\boldsymbol{\theta}} + r\sin\theta\,d\phi\,\hat{\boldsymbol{\phi}}. \qquad (6.47)$$

Thus

$$\mathbf{r} \cdot \mathbf{r} = r^2, \qquad \mathbf{r} \cdot d\mathbf{r} = r\,dr,$$
$$\text{and} \quad d\mathbf{r} \cdot d\mathbf{r} = dr^2 + r^2(d\theta^2 + \sin^2\theta\,d\phi^2). \qquad (6.48)$$

The invariant interval is now

$$ds^2 = A[dr^2 + r^2(d\theta^2 + \sin^2\theta\,d\phi^2)] + Br^2\,dr^2 + Cr\,dr\,dt + D\,dt^2, \qquad (6.49)$$

or, relabeling the scalar function $A + Br^2 = B'$ as B,

$$ds^2 = A[r^2(d\theta^2 + \sin^2\theta\,d\phi^2)] + B\,dr^2 + Cr\,dr\,dt + D\,dt^2. \qquad (6.50)$$

Simplification by coordinate choices

From our discussion in Chapter 5, we learned that the Gaussian coordinates, as labels of points in the space, can be freely chosen. In the same way, the coordinates in Riemannian geometry have no intrinsic significance until their connection to the physical length ds^2 is specified by the metric function.[21]

[20] Here we adopt the notation $x^\mu = (ct, \mathbf{r})$.

[21] This is sometimes stated as "in Riemannian geometry, coordinates have no metric significance." See in particular (5.5) and the introductory paragraph in Section 5.1.2, where the geodesic equation was first derived.

continued

Box 6.2 *continued*

Thus we are free to choose new coordinates, thereby altering the metric to make it as simple as possible. Of course, in our particular case, the new coordinates should also have spherical symmetry.

1. Pick a new time coordinate to eliminate the cross term $dt\,dr$: we introduce a new coordinate t' such that

$$t' = t + f(r). \tag{6.51}$$

Differentiating, we have $dt' = dt + (df/dr)\,dr$. Squaring both sides and solving for dt^2,

$$dt^2 = dt'^2 - \left(\frac{df}{dr}\right)^2 dr^2 - 2\frac{df}{dr}\,dr\,dt. \tag{6.52}$$

Now the cross term $dt\,dr$ in (6.50) has a coefficient $Cr - 2D(df/dr)$, which can be eliminated[22] by choosing an $f(r)$ that satisfies the differential equation

$$\frac{df}{dr} = +\frac{Cr}{2D}. \tag{6.53}$$

Similarly, the $(df/dr)^2\,dr^2$ term in (6.52) can be absorbed into a new scalar function $B' = B - (df/dr)^2 D$.

2. Scale the radial coordinate to simplify the angular coefficient: We can set the function A in (6.50) to unity by choosing a new radial coordinate,

$$r'^2 = A(r, t)r^2, \tag{6.54}$$

so that the first term on the right-hand side of (6.50) is just $r'^2(d\theta^2 + \sin^2\theta\,d\phi^2)$. The effect of this new radial coordinate on the other terms in ds^2 can again be absorbed by the relabeling of scalar functions. We leave this as an exercise for the reader to complete.

With these new radial and time coordinates (relabeled as r and t), we are left with only two unknown scalar functions in the metric. In this way, the line element interval takes the form

$$ds^2 = g_{00}(r, t)c^2\,dt^2 + g_{rr}(r, t)\,dr^2 + r^2(d\theta^2 + \sin^2\theta\,d\phi^2), \tag{6.55}$$

yielding the metric matrix shown in (6.56).

[22] The absence of any linear factor dt means that the metric is also time-reversal invariant (i.e., changing t to $-t$ does not affect ds^2). In fact, one can see that the metric matrix should be diagonal, because spherical symmetry of the problem implies that the solution should be symmetric under the reflections of each coordinate: $r \to -r$, $\theta \to -\theta$, and $\phi \to -\phi$.

6.4.1 The geometry of a spherically symmetric spacetime

Box 6.2 demonstrated that the metric tensor for spherically symmetric space-time has only two unknown scalar functions. For these particular coordinates[23] ($x^0 = ct, r, \theta, \phi$), the functions are the tensor elements $g_{00}(t, r)$ and $g_{rr}(t, r)$. The metric is diagonal, with $g_{\theta\theta} = r^2$ and $g_{\phi\phi} = r^2 \sin^2 \theta$:

[23] This system is called the Schwarzschild coordinate system. We will see other spherically symmetric coordinate systems in Chapter 7.

$$g_{\mu\nu} = \begin{pmatrix} g_{00} & & & \\ & g_{rr} & & \\ & & r^2 & \\ & & & r^2 \sin^2 \theta \end{pmatrix}. \tag{6.56}$$

The metric functions $g_{00}(t, r)$ and $g_{rr}(t, r)$ can be obtained by solving the Einstein field equation. (cf. the penultimate subsection of Chapter 11). Its solution obtained by Schwarzschild is as follows:

$$g_{00}(t, r) = -\frac{1}{g_{rr}(t, r)} = -1 + \frac{r^*}{r}. \tag{6.57}$$

Thus the invariant interval is

$$ds^2 = -\left(1 - \frac{r^*}{r}\right) c^2 dt^2 + \left(1 - \frac{r^*}{r}\right)^{-1} dr^2 + r^2 d\theta^2 + r^2 \sin^2 \theta \, d\phi^2. \tag{6.58}$$

There is warping in the radial as well as in the time coordinate. For points that are far away from the gravitational source, $g_{\mu\nu}$ approaches the flat-spacetime limit:

$$\lim_{r \to \infty} g_{00}(t, r) \to -1 \qquad \text{and} \qquad \lim_{r \to \infty} g_{rr}(t, r) \to 1. \tag{6.59}$$

Thus the deviation from flat Minkowski spacetime is measured by the ratio r^*/r. The constant length r^* can be fixed by our Newtonian-limit result (5.41):

$$g_{00} = -\left(1 + \frac{2\Phi(x)}{c^2}\right) = -\left(1 - \frac{r^*}{r}\right). \tag{6.60}$$

We then have, from (6.45),

$$r^* = \frac{2G_N M}{c^2}. \tag{6.61}$$

This **Schwarzschild radius** is generally a very small lengthscale. For example, the solar and terrestrial Schwarzschild radii are, respectively,

$$r^*_{\odot} \simeq 3 \text{ km} \qquad \text{and} \qquad r^*_{\oplus} \simeq 9 \text{ mm}. \tag{6.62}$$

For the Schwarzschild exterior solution to apply, the radial coordinate value r must be at least R, the radius of the source mass. Thus the respective maximum solar and terrestrial ratios (i.e., for the minimum r) are

$$\frac{r^*_\odot}{R_\odot} = O(10^{-6}) \quad \text{and} \quad \frac{r^*_\oplus}{R_\oplus} = O(10^{-9}), \tag{6.63}$$

which means that the GR modification r^*/r, which enters into (6.60), is very small. Nevertheless, such tiny corrections have been found to agree with observations whenever high-precision measurements have been performed.

Example 6.1 Curved spacetime and the operation of GPS

As the first application of the Schwarzschild geometry, let us revisit the time dilation corrections required for the proper operation of the GPS system, previously discussed in Section 4.3.2. Recall that the relevant issue is the comparison of the proper time rates recorded by the clock on the satellite ($d\tau_S$) and by the ground clock ($d\tau_\oplus$). Their relation can be efficiently deduced by a consideration of the relevant invariant intervals (i.e., proper time rates): $ds_S^2 = -c^2 d\tau_S^2$ and $ds_\oplus^2 = -c^2 d\tau_\oplus^2$. Since both clocks are in circular orbits (hence $dr = 0$), we have[24]

$$ds_\oplus^2 = -\left(1 + 2\frac{\Phi_\oplus}{c^2}\right)c^2 dt^2 + r^2 d\phi^2 = -c^2 d\tau_\oplus^2, \tag{6.64}$$

where t is the (faraway) coordinate time, and we have also plugged in $r^*/r = -2\Phi/c^2$. Dividing both sides by $-c^2 dt^2$ and taking the square root yields,[25] for the terrestrial time interval,

$$\frac{d\tau_\oplus}{dt} = 1 + \frac{\Phi_\oplus}{c^2} - \frac{v_\oplus^2}{2c^2}, \tag{6.65}$$

where we have also substituted the tangential speed $v = r\, d\phi/dt$. We can carry out similar manipulations on the satellite time $d\tau_S$. In this way, we obtain their ratio

$$\frac{d\tau_\oplus}{d\tau_S} = 1 + \frac{\Phi_\oplus}{c^2} - \frac{\Phi_S}{c^2} - \frac{v_\oplus^2}{2c^2} + \frac{v_S^2}{2c^2}, \tag{6.66}$$

which is exactly the same result we obtained in (4.36). The previous derivation in Chapter 4 may have left the impression that the SR time dilation (due to relative motion) is distinct from GR predictions. But SR correction (as a kinematic result) is naturally included in GR; both effects should be contained in one general formulation.

[24] We choose to define the polar angle $\theta = \pi/2$ (hence $d\theta = 0$ and $\sin\theta = 1$) for each orbit plane.

[25] Throughout our presentation, we often make the approximation of $(1 \pm x)^n \simeq 1 \pm nx$ when $x \ll 1$.

Interpreting the coordinates

Our spherically symmetric metric (6.56) is diagonal. In particular, $g_{i0} = g_{0i} = 0$ (with spatial indices $i = 1, 2, 3$), which means that for a given t, we can discuss the spatial subspace separately. To set up the coordinates, we can first imagine the situation where gravity is absent. For a fixed t, this spherically symmetric coordinate system can be visualized as a series of 2-spheres having different radial coordinate values of r, with their common center at the origin of the spherically symmetric source. Each 2-sphere has a surface area $4\pi r^2$ and can be pictured as a lattice of rigid rods arranged in a grid corresponding to various (θ, ϕ) values, with synchronized clocks attached at each grid point (i.e., each point in this subspace has the same coordinate time).

- In the absence of gravity,[26] the spacetime is flat as in Fig. 6.7(a). The coordinate r is the proper radial distance ρ defined as

$$dr^2 = ds_r^2 \equiv d\rho^2, \qquad (6.67)$$

where ds_r^2 is the invariant interval ds^2 with $dt = d\theta = d\phi = 0$. The coordinate t is the proper time τ for an observer at a fixed location:

$$-c^2 \, dt^2 = ds_t^2 \equiv -c^2 \, d\tau^2, \qquad (6.68)$$

where ds_t^2 is the invariant interval ds^2 with $dr = d\theta = d\phi = 0$. Thus, the coordinates (r, t) can be interpreted as the radial distance and time measured by an observer.

- When gravity is turned on, the spacetime is warped.[27] In the spatial radial direction, the proper radial distance no longer equals the radial coordinate:

$$d\rho = \sqrt{g_{rr}} \, dr. \qquad (6.69)$$

Consequently, the spherical surface area $4\pi r^2$ no longer bears the Euclidean relation with the proper radius ρ. Similarly, the proper time

$$d\tau = \sqrt{-g_{00}} \, dt \qquad (6.70)$$

differs from the coordinate time, because $g_{00} \neq -1$. This gravitational time dilation signifies the warping of the spacetime in the time direction.

An embedding diagram, which may help to visualize warped space, is discussed in Box 6.3.

[26] This also applies for an observer far from the gravitational source, where the spacetime is approximately flat. However, while dr is the measured distance between two radially separated points in the faraway region, r is not the radial distance to the origin, as the spacetime near the origin is curved.

[27] We recall that (6.69) and (6.70) are just the standard expressions for the metric elements in terms of defined coordinates and invariant intervals discussed in (5.6).

Box 6.3 Embedding diagram

A convenient way to visualize warped space is to use an **embedding diagram**. Since it is difficult to visualize the full three-dimensional curved space (embedded in a 4D space), we shall concentrate on the two-dimensional subspace corresponding to a fixed polar angle $\theta = \pi/2$ (and at some given instant of time). That is, we will focus on a 2D space slicing through the middle of the source. In the absence of gravity, this is just a flat plane as depicted in Fig. 6.7(a).

In the presence of gravity, this 2D surface is curved. We would like to visualize the warping of this 2D subspace. A useful technique is to imagine that the curved 2D surface is embedded in a fictitious 3D Euclidean space, as in Fig. 6.7(b). For the particular case of the external Schwarzschild geometry, the line element interval with $dt = d\theta = 0$ is

$$ds^2 = \left(1 - \frac{r^*}{r}\right)^{-1} dr^2 + r^2 \, d\phi^2. \tag{6.71}$$

We now imagine this 2D surface as a subspace of a 3D Euclidean space with a cylindrical polar coordinate system: $x^i = (r, \phi, z)$. We have chosen the polar radial (perpendicular) distance from the cylindrical axis of symmetry (the z axis) of this 3D space to coincide with the Schwarzschild r coordinate, and the azimuthal angle to coincide with the Schwarzschild ϕ angle. Just as a line-curve is represented in 3D space by $x^i(\lambda)$, with a single curve parameter λ, a 2D surface is represented by $x^i(\lambda, \sigma)$, with two surface parameters. In our case, we naturally choose the Schwarzschild coordinates (r, ϕ) as parameters. The spherical symmetry of the Schwarzschild space implies a cylindrical symmetry of the embedding space; there is no ϕ dependence in the metric. Thus, the only nontrivial relation is $z(r)$, which can be worked out as follows. We have the Euclidean interval

$$ds^2 = dz^2 + dr^2 + r^2 \, d\phi^2 = \left[\left(\frac{dz}{dr}\right)^2 + 1\right] dr^2 + r^2 \, d\phi^2. \tag{6.72}$$

A comparison of (6.72) with (6.71) leads to $dz = \pm[r^*/(r - r^*)]^{1/2} \, dr$, which can be integrated to yield $z = \pm 2[r^*(r - r^*)]^{1/2}$, or

$$z^2 = 4r^*(r - r^*), \tag{6.73}$$

which is a sideways parabola in the (r, z) plane for any fixed ϕ. Sweeping the parabola in a circle around the z axis (to include all ϕ) yields the 3D representation of the curved surface as shown in Fig. 6.7(b).

Distances between points on this surface are equal to the proper distances in Schwarzschild space. Note that radial proper lengths are stretched (compared with the coordinate separation) near the throat at $r = r^*$. The surface does not extend to $r < r^*$, where the radial paths have timelike (negative) intervals. It is not quite correct to assume that slices of spacetime with constant t must be entirely spatial. We will see in Chapter 7 that different coordinates can better describe this region around a black hole, an object massive and dense enough to fit within its own Schwarzschild radius r^*.

Finally, note that the upper and lower (positive- and negative-z) halves of Fig. 6.7(b) provide redundant descriptions of Schwarzschild space. The redundant half could be used to describe an extra universe inside the Schwarzschild radius. But even if such a universe existed, it would be causally disconnected from ours.

(a)

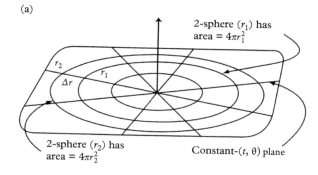

2-sphere (r_1) has
area = $4\pi r_1^2$

r_2

r_1

Δr

2-sphere (r_2) has
area = $4\pi r_2^2$

Constant-(t, θ) plane

(b)

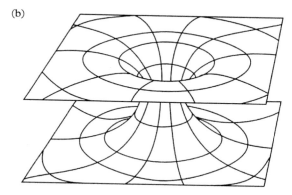

Figure 6.7 *(a) The* $\theta = \pi/2$ *plane* (r, ϕ) *cutting across the spherical source, displayed in a flat coordinate space without gravity. (b) In a fictitious 3D embedding space, the physical 2D subspace of (a) is shown as a curved surface, which truly portrays the proper distance between points in the Schwarzschild space (6.71).*

Spherically symmetric metric is time-independent

We note from (6.57) that the scalar metric functions are time-independent, even though we never assumed a static source. This turns out to be a general result (Birkhoff's theorem): whenever the source is spherically symmetric, the resultant exterior spacetime must be time-independent. We shall not provide a mathematical proof of this theorem, but will note that the same result holds for the Newtonian theory as well—as it should. Recall that the gravitational field outside a spherical source is identical to the gravitational field of a point source having all its mass concentrated at its center. This proof depends only on the spherical symmetry of the setup, irrespective of any possible time dependence. Thus, whether the spherical mass is pulsating or exploding or whatever, the resultant field is the same as long as the spherical symmetry is maintained. The analogous statement in electromagnetism is that there is no monopole radiation (only dipole radiation, quadrupole radiation, etc.)

6.4.2 Curved spacetime and deflection of light

In the time dilation effects on the operation of GPS (Example 6.1) and in the gravitational redshift (Example 5.1), the GR effects are manifestations of the bending of time: $g_{00} \neq -1$. However, GR treats space and time on an equal footing, and involves bending in both the temporal and the spatial directions: $g_{00} \neq -1$ and $g_{rr} \neq 1$. Here we work out the GR result for light deflection, which goes beyond Einstein's 1911 calculation, in which only gravitational time dilation was taken into account.

To an observer far from the source, using the coordinate time and radial distance (t, r), the effective speed of light $c(r)$ can be related[28] to the locally measured universal light speed $c = d\rho/d\tau$ by

$$c(r) \equiv \frac{dr}{dt} = \frac{d\rho}{d\tau} \sqrt{\left| \frac{g_{00}(r)}{g_{rr}(r)} \right|} = c \sqrt{\left| \frac{g_{00}(r)}{g_{rr}(r)} \right|}, \qquad (6.74)$$

[28] An alternative derivation is through a lightlike worldline, $ds^2 = g_{00}c^2 dt^2 + g_{rr} dr^2 = 0$, in a fixed direction, $d\theta = d\phi = 0$. One may wonder about the validity of imposing this fixed angle in the calculation of the light deflection angle $\delta\phi$. Section 4.3.2 showed us that such a calculation requires two steps: first we calculate the speed of light $c(r)$ in terms of gravitational potential; then we relate the differential deflection $d\phi$ to this result for $c(r)$. To calculate $c(r)$, we can work with any propagation direction.

where the proper distance and time (ρ, τ) are related to the coordinate distance and time (r, t) by (6.69) and (6.70). The Schwarzschild solution (6.57) informs us that $g_{rr} = -g_{00}^{-1}$. Thus, alongside the relation between proper and coordinate time (4.40), we also have a relation between the proper and coordinate length intervals:

$$d\rho = \left(1 - \frac{\Phi(r)}{c^2} \right) dr. \qquad (6.75)$$

Clearly, besides gravitational time dilation, we also have a gravitational length contraction effect.[29] The influences of spatial and temporal warping on $c(r)$ in (6.74) are of the same size and in the same direction. The deviation of $c(r)$ from c (thus the change of index of refraction $n(r)$) is **twice** Einstein's 1911 result of taking into account only $g_{00} \neq -1$ as in (4.41):

[29] Recalling the derivation given in Box 4.1 of gravitational time dilation via an arrangement of three identically constructed clocks, one may be tempted to derive gravitational length contraction from the SR length contraction with a similar arrangement involving three identically constructed rods. Such a derivation was criticized in (Rindler 1968) because it inevitably assumes some spacetime geometry, making such a derivation before the introduction of a curved spacetime description of gravity less useful even for pedagogical purposes.

$$[c(r)]_{\mathrm{GR}} \equiv \frac{dr}{dt} = \frac{d\rho}{1 - \Phi(r)/c^2} \frac{1 + \Phi(r)/c^2}{d\tau} \qquad (6.76)$$

$$= \frac{1 + \Phi(r)/c^2}{1 - \Phi(r)/c^2} \frac{d\rho}{d\tau} = \left(1 + \frac{2\Phi(r)}{c^2} \right) c,$$

According to (4.43)–(4.48), the deflection angle $\delta\phi$ is directly proportional to $d[c(r)]$, and thus is twice as large:

$$[\delta\phi]_{\mathrm{GR}} = 2[\delta\phi]_{\mathrm{EP}} = \frac{4G_{\mathrm{N}}M}{c^2 r_{\mathrm{min}}}. \qquad (6.77)$$

The GR prediction verified in the 1919 solar eclipse expeditions   If a light ray is deflected by an angle $\delta\phi$ as shown in Fig. 4.8, the light source at S will appear to the observer at O to be located at S'. Since the deflection is inversely proportional to r_{min}, one can maximize the bending by minimizing r_{min}. For light grazing the surface of the sun, $r_{\mathrm{min}} = R_\odot$ and $M = M_\odot$, the predicted solar deflection of light is $[\delta\phi]_{\mathrm{GR}} = 1.74''$.

The deflection angle (about 1/4000 of the angular width of the sun as seen from earth), predicted by Einstein's equations in 1915, was not easy to detect. One needed a solar eclipse against a background of several bright stars (so that some could be used as reference points). On May 29, 1919, there was an solar eclipse. Two British expeditions were organized by A.S. Eddington (1882–1944) and the Astronomer Royal, F.W. Dyson (1868–1939): one to Sobral in northern Brazil and the other to the island of Principe off the coast of West Africa. The report that Einstein's prediction was successful in these tests created a worldwide sensation—partly for scientific reasons and partly because the world was amazed that so soon after the "Great War" (World War I) the British should finance and conduct an expedition to test a theory proposed by a German citizen.[30]

[30] Eddington was a Quaker conscientious objector. With the backing of Dyson, he was able to continue his research during the war, and, through Willem de Sitter in neutral Holland, obtained Einstein's GR papers in 1916.

(a)

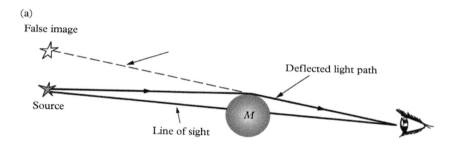

(b)

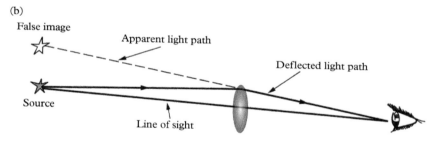

(c)

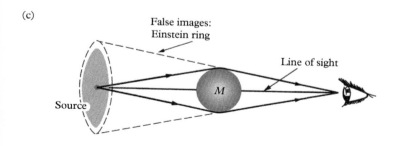

Figure 6.8 *Bending of a light ray as a lensing effect. (a) Light from a distant star is deflected by a lensing mass M lying close to the line of sight from the observer to the source. As a result, the source star appears to be located at a different position. (b) The bending of light in (a) is analogous to that caused by a glass lens. (c) If the line of sight passes directly through the center of a symmetrical lensing mass distribution, the false image appears as a ring, the* **Einstein ring.** *If the alignment is not perfect or the lensing mass is not spherical, the images may be arcs, as seen in Fig. 6.9.*

Figure 6.9 *Gravitational lensing effects due to the galaxy cluster Abell 2218. Nearly all of the bright objects in this picture taken by the Hubble Space Telescope are galaxies in this cluster, which is so massive and so compact that it lenses the light from galaxies that lie behind it into multiple images that appear as long, faint arcs. The image is reproduced with permission from (Fruchter et al. 2001) NASA/ERO - STScI and ST-ECF Team.*

[30a] More spectacularly, the image of a distant bright source (such as a supernova) if lensed by a galactic cluster can appear to us as multiple events separated in time, because the different optical paths it takes to reach our telescope.

Gravitational lensing The gravitational deflection of a light ray discussed above (Fig. 6.8a) has some similarity to the bending of light by a glass lens (Fig. 6.8b). When the source and observer are sufficiently far from the lensing mass, bending from both sides of the mass can produce separate images[30a]—or even a ring image (Fig. 6.8c) if the source, lensing mass, and observer are perfectly aligned. Thus one generally finds stretched arc images of the light source (see, e.g., the photograph in Fig. 6.9). In fact, gravitational lensing has become a powerful tool of modern astrophysics in mapping out astronomical mass distributions. In cosmology, the theoretical predictions are mostly given in terms of mass distributions rather than, say, the distributions of stars and galaxies. Thus lensing results are particularly useful. For an example showing that such distributions can be different, see the discussion of the Bullet Cluster of galaxies in Chapter 8; cf. Fig. 8.4. The key input in the various lensing equations is the deflection result (6.77).

6.4.3 Precession of Mercury's orbit

In this section, we shall discuss the motion of a test mass in Schwarzschild spacetime. A particle under the Newtonian $1/r^2$ gravitational attraction has an elliptical orbit. Here we calculate the GR correction to such a trajectory.

Celestial mechanics based on the Newtonian theory of gravitation has been remarkably successful. However, it was realized around 1850 that there was a discrepancy between the theory and the observed **precession of the perihelion of the planet Mercury**. The pure $1/r^2$ force law of Newton predicts a closed elliptical orbit for a planet, i.e., an orbit with its major axis fixed in space. However, perturbations due to the presence of other planets and astronomical objects lead to a trajectory that is no longer closed. Since the perturbation is small, such a deviation from the closed orbit can be described as an elliptical orbit with a **precessing axis**, or equivalently a precessing perihelion (the point of the orbit closest to the sun); see Fig. 6.10.

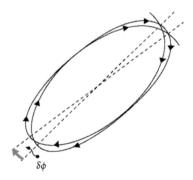

Figure 6.10 *A perturbed $1/r^2$ attraction leads to an open elliptical orbit that may be described as an elliptical orbit with a precessing axis. For planetary motion, this is usually stated as the precession of the point of closest approach to the sun, the perihelion.*

For the case of Mercury, such planetary perturbations could account for most[31] of its perihelion advance of 574″ per century. However, there was still a discrepancy of 43″ left unaccounted for. Following a similar situation involving Uranus that eventually led to the prediction and discovery of the outer planet Neptune in 1846, a new planet, named Vulcan, was predicted to lie inside Mercury's orbit; however, it was never found. This is the perihelion precession problem that Einstein solved by applying his new theory of gravitation. As we shall presently see, GR implies a small correction to the $1/r^2$ force law, which just accounts for the missing 43″ per century advance of Mercury's orbit.

[31] Most of the raw observational value of 5600″ per century is due to the effect of the rotation of our earth-based coordinate system. This leaves the planetary perturbations to account for the remaining 574″, to which Venus contributes 277″, Jupiter 153″, Earth 90″, and Mars and the rest of the planets 10″.

Lagrangian of the geodesic equation

A particle's motion in a curved spacetime such as the Schwarzschild spacetime is determined by the GR equation of motion, the geodesic equation. As discussed in Section 5.1.2, this is the Euler–Lagrange equation

$$\frac{\partial L}{\partial x^\mu} = \frac{d}{d\tau}\frac{\partial L}{\partial \dot{x}^\mu}, \tag{6.78}$$

where the Lagrangian is the particle's 4-velocity squared,

$$L(x,\dot{x}) = g_{\mu\nu}\dot{x}^\mu\dot{x}^\nu. \tag{6.79}$$

As is appropriate for a massive test particle, we have picked its proper time τ to be the curve parameter and have used the notation $\dot{x}^\mu = dx^\mu/d\tau$. The dependence of L on the coordinates x^μ is through the metric tensor $g_{\mu\nu}$.

As in Newtonian mechanics, because angular momentum is conserved under the action of a central force, the trajectory will always remain in the plane spanned by the particle's initial velocity and the radial vector **r** connecting the force center to the test particle. Let us call this the $\theta = \pi/2$ plane. By inserting the Schwarzschild metric, the Lagrangian (6.79) becomes

$$L(x,\dot{x}) = -\left(1-\frac{r^*}{r}\right)c^2\dot{t}^2 + \left(1-\frac{r^*}{r}\right)^{-1}\dot{r}^2 + r^2\dot{\phi}^2. \tag{6.80}$$

In principle, one can obtain the orbit $r(\phi)$ by solving the equation of motion (6.78). Since this equation involves the second-order time derivative, it is simpler if we proceed via the energy balance equation, which involves only the first derivative. The energy balance equation, as will be demonstrated below, is simply the normalization condition of the Lagrangian (i.e., the 4-velocity invariant (6.79)):[32]

$$L(x,\dot{x}) = -c^2. \tag{6.81}$$

[32] Even though this $L = -c^2$ result was first obtained in (3.29) in the context of a flat spacetime, it is still valid for a general metric $g_{\mu\nu}$, as the proper time is always defined by $-c^2\,d\tau^2 \equiv g_{\mu\nu}\,dx^\mu\,dx^\nu$. Dividing both sides by $d\tau^2$ immediately leads to (6.81).

Two constants of motion

Our next step is to express the various terms of (6.80) in terms of constants of motion (i.e., conserved quantities along the particle's path). If the metric $g_{\mu\nu}$ is independent of a particular coordinate, say x^α (for a particular index α), then $\partial L/\partial x^\alpha = 0$. This can be translated, through (6.78), into a conservation statement: $d(\partial L/\partial \dot{x}^\alpha)/d\tau = 0$. In the present case of planetary motion on a fixed plane ($d\theta = 0$), the Lagrangian (6.80) does not depend on t or ϕ; we have two constants of motion. The metric's independence of t leads to the conservation of $\partial L/\partial \dot{x}^0 = \partial(g_{\mu\nu}\dot{x}^\mu \dot{x}^\nu)/\partial \dot{x}^0 \sim g_{00}\dot{x}^0$; we will call this conserved quantity

$$\kappa \equiv -g_{00}c\dot{t} = \left(1 - \frac{r^*}{r}\right)c\dot{t}. \tag{6.82}$$

The independence of ϕ points to the conservation of $\partial L/\partial \dot{\phi} \sim g_{\phi\phi}\dot{\phi}$:

$$l \equiv mg_{\phi\phi}\dot{\phi} = mr^2\dot{\phi}. \tag{6.83}$$

[33] As discussed in Section 3.2.2, the key attribute of energy and momentum is that they are conserved quantities and have the correct flat-spacetime and Newtonian limits; cf. (3.40) and Example 3.1.

Clearly κ is to be identified with the particle's energy[33] (in units of mc), as in the flat-spacetime limit ($r^*/r \to 0$) it reduces directly to the SR expression (3.40) of energy $c\dot{t} = E/mc$; l is the orbital angular momentum $|\mathbf{r} \times \mathbf{p}| = mr^2\dot{\phi}$.

Energy balance equation

Substituting the constants l and κ into (6.80), we can then write the $L = -c^2$ equation (6.81) as

$$-\frac{\kappa^2}{1 - r^*/r} + \frac{\dot{r}^2}{1 - r^*/r} + \frac{l^2}{m^2 r^2} = -c^2. \tag{6.84}$$

After multiplying by $\frac{1}{2}m(1 - r^*/r)$, and using $r^* = 2G_N M/c^2$ on the right-hand side, this may be written as

$$-\frac{m\kappa^2}{2} + \frac{1}{2}m\dot{r}^2 + \left(1 - \frac{r^*}{r}\right)\frac{l^2}{2mr^2} = -\frac{1}{2}mc^2\left(1 - \frac{2G_N M}{c^2 r}\right), \tag{6.85}$$

or

$$\frac{1}{2}m\dot{r}^2 + \left(1 - \frac{r^*}{r}\right)\frac{l^2}{2mr^2} - \frac{G_N mM}{r} = \frac{m\kappa^2}{2} - \frac{1}{2}mc^2. \tag{6.86}$$

[34] In Schwarzschild spacetime, as κ is the particle energy E/mc, $\mathcal{E}$ can be identified as the corresponding kinetic energy: $\mathcal{E} = m(\kappa^2 - c^2)/2 = (E^2 - m^2 c^4)/2mc^2$, because $E = (m^2 c^4 + 2mc^2 \mathcal{E})^{1/2} \simeq mc^2 + \mathcal{E}$ in the Newtonian limit when $\mathcal{E} \ll E$.

Replacing the constant κ by another energy quantity[34]

$$\frac{\mathcal{E}}{m} = \frac{\kappa^2 - c^2}{2} \tag{6.87}$$

simplifies the right-hand side of (6.86):

$$\frac{1}{2}m\dot{r}^2 + \left(1 - \frac{r^*}{r}\right)\frac{l^2}{2mr^2} - \frac{G_N mM}{r} = \mathcal{E}. \tag{6.88}$$

Except for the $(1 - r^*/r)$ factor, this is just the energy balance equation for the nonrelativistic central force problem: (radial and rotational) kinetic energy plus potential energy equals the total (Newtonian) energy $\mathcal{E}$. The extra term,

$$-\frac{r^*}{r}\frac{l^2}{2mr^2} = -\frac{G_N M l^2}{mc^2 r^3},\qquad(6.89)$$

may be regarded here as a small GR correction to the Newtonian potential energy $-G_N mM/r$ by a $1/r^4$ type of force.

The precessing orbit

This relativistic energy equation (6.88) can be cast in the form of an orbit equation. We can solve for $r(\phi)$ using standard perturbation theory (see Box 6.4). With e being the eccentricity of the orbit, $\alpha = l^2/G_N Mm^2 = (1 + e)r_{min}$, and $\epsilon = 3r^*/2\alpha$, the solution is

$$r = \frac{\alpha}{1 + e\cos[(1-\epsilon)\phi]}.\qquad(6.90)$$

Thus the planet returns to its perihelion r_{min} not at $\phi = 2\pi$ but at $\phi = 2\pi/(1-\epsilon) \simeq 2\pi + 3\pi r^*/\alpha$. The perihelion advances (i.e., the whole orbit rotates in the same sense as the planet itself) per revolution by (Fig. 6.10)

$$\delta\phi = \frac{3\pi r^*}{\alpha} = \frac{3\pi r^*}{(1 + e)r_{min}}.\qquad(6.91)$$

With the solar Schwarzschild radius $r^*_\odot = 2.95$ km, Mercury's eccentricity $e = 0.206$, and its perihelion $r_{min} = 4.6 \times 10^7$ km, the numerical value of the advance is

$$\delta\phi = 5.0 \times 10^{-7} \text{ radians per revolution}\qquad(6.92)$$

or 5.0×10^{-7} rad $\times (180°/\pi$ rad$) \times (60'/\text{deg}) \times (60''/\text{min}) = 0.103''$ per revolution. In terms of the advance per century,

$$0.103'' \times \frac{100 \text{ years/century}}{\text{Mercury's period of } 0.241 \text{ years}} = 43'' \text{ per century}.\qquad(6.93)$$

This agrees with the observational evidence.

This calculation explaining the perihelion advance of the planet Mercury from first principles, and the correct prediction for the bending of starlight around the sun, were both obtained by Einstein in an intense two-week period in November, 1915. This gave Einstein great of joy.[34a] This moment of elation was characterized by his biographer Abraham Pais as "by far the strongest emotional experience in Einstein's scientific life, perhaps, in all his life". Afterward, he wrote to Arnold Sommerfeld in a now-famous letter:

[34a] "For a few days, I was besides myself with joyous excitement," Einstein wrote.

This last month I have lived through the most exciting and the most exacting period of my life; and it would be true to say this, it has been the most fruitful. Writing letters has been out of the question. I realized that up until now my field equations of gravitation have been entirely devoid of foundation. When all my confidence in the old theory vanished, I saw clearly that a satisfactory solution could only be reached by linking it with the Riemann variations. The wonderful thing that happened then was that not only did Newton's theory result from it, as a first approximation, but also the perihelion motion of Mercury, as a second approximation. For the deviation of light by the Sun I obtained twice the former amount.

We have already discussed the doubling of the light deflection angle. Recall that for several years Einstein had struggled with the Ricci-tensor-based field equation that could not reproduce the Newtonian limit. The Riemann variation to which he referred above is the linear combination that we now called the Einstein tensor. The mathematics involving the curvature tensor can be found in Chapter 11.

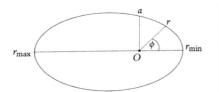

Figure 6.11 *Points on an elliptical orbit around a mass at the origin O are labeled by their coordinates (r, ϕ), with some notable positions at $(r_{min}, 0)$, (r_{max}, π), and $(\alpha, \pi/2)$.*

Box 6.4 The orbit equation and its perturbation solution

We solve the relativistic energy equation (6.88) as a standard central force problem. The relevant kinematic variables are shown in Fig. 6.11.

The orbit equation

To obtain the orbit $r(\phi)$, we first change all the time derivatives into derivatives with respect to the angle ϕ using the angular momentum equation (6.83):

$$d\tau = \frac{mr^2}{l} \, d\phi. \tag{6.94}$$

We then make the change of variable $u \equiv 1/r$ (and thus $u' \equiv du/d\phi = -u^2 \, dr/d\phi = -m\dot{r}/l$). In this way, (6.88) turns into

$$u'^2 + u^2 - \frac{2}{\alpha}u - r^*u^3 = C, \tag{6.95}$$

where $\alpha = l^2/(G_N M m^2)$ and the constant $C = 2m\mathcal{E}/l^2$. This is the equation we need to solve in order to obtain the planet's orbit $r(\phi)$.

Zeroth-order solution

Split the solution $u(\phi)$ into an unperturbed part u_0 and a small (of the order of r^*u^2) correction u_1: $u = u_0 + u_1$, with

$$u_0'^2 + u_0^2 - \frac{2}{\alpha}u_0 = C. \tag{6.96}$$

This unperturbed orbit equation can be solved by differentiating with respect to ϕ and dividing the resulting equation by $2u_0'$:

$$u_0'' + u_0 = \alpha^{-1}, \tag{6.97}$$

which is a simple harmonic oscillator equation in the variable $u_0 - \alpha^{-1}$, with ϕ replacing the time variable and $\omega = 1$ the angular frequency. It has the solution $u_0 - \alpha^{-1} = D \cos\phi$. We choose to write the constant $D \equiv e/\alpha$, so that the unperturbed solution takes on the well-known form of a conic section,

$$r = \frac{\alpha}{1 + e \cos\phi}. \tag{6.98}$$

It is clear (see Fig. 6.11) that $r = \alpha/(1 + e) = r_{\min}$ (perihelion) at $\phi = 0$, and $r = \alpha/(1 - e) = r_{\max}$ (aphelion) at $\phi = \pi$. Geometrically, e is called the **eccentricity** of the orbit. The radial distance at $r(\phi = \pi/2) = \alpha$ can be expressed in terms of perihelion (or aphelion) and eccentricity as

$$\alpha = (1 + e)r_{\min} = (1 - e)r_{\max}. \tag{6.99}$$

Relativistic correction

We now plug the full solution $u = u_0 + u_1$ into the perturbed orbit equation (6.95):

$$(u_0' + u_1')^2 + (u_0 + u_1)^2 - \frac{2}{\alpha}(u_0 + u_1) - r^*(u_0 + u_1)^3 = C, \tag{6.100}$$

and separate out the leading and the next-to-leading terms (with $u_1 = O(r^*u^2) = O(r^*/\alpha^2)$):

$$\left(u_0'^2 + u_0^2 - \frac{2}{\alpha}u_0 - C\right) + \left(2u_0'u_1' + 2u_0u_1 - \frac{2}{\alpha}u_1 - r^*u_0^3\right) = 0, \tag{6.101}$$

where we have dropped the higher-order terms proportional to u_1^2, $u_1'^2$, or u_1r^*. After using (6.96), we can then pick out the first-order equation:

$$2u_0'u_1' + 2u_0u_1 - \frac{2}{\alpha}u_1 = r^*u_0^3, \tag{6.102}$$

where

$$u_0 = \frac{1 + e \cos\phi}{\alpha}, \qquad u_0' = -\frac{e}{\alpha}\sin\phi. \tag{6.103}$$

The equation for u_1 is then

$$-e\sin\phi \frac{du_1}{d\phi} + e\cos\phi\, u_1 = \frac{r^*(1 + e\cos\phi)^3}{2\alpha^2}. \tag{6.104}$$

continued

Box 6.4 *continued*

One can verify that it has the solution

$$u_1 = \frac{r^*}{2\alpha^2}\left[(3 + 2e^2) + \frac{1 + 3e^2}{e}\cos\phi - e^2\cos^2\phi + 3e\phi\sin\phi\right]. \quad (6.105)$$

The first two terms have the form of the zeroth-order solution, $A + B\cos\phi$; thus they represent unobservably small corrections. The third term, whose period (in ϕ) is exactly half that of the unperturbed orbit, is likewise unimportant. We only need to concentrate on the ever-increasing fourth term,[35] whose effects accumulate over many orbits. Plugging this into $u = u_0 + u_1 = 1/r$, we obtain

$$r = \frac{\alpha}{1 + e\cos\phi + \epsilon e\phi\sin\phi}, \quad (6.106)$$

where $\epsilon = 3r^*/2\alpha$ is a small quantity. By approximating $\cos\epsilon\phi \simeq 1$ and $\sin\epsilon\phi \simeq \epsilon\phi$, so that

$$e(\cos\phi + \epsilon\phi\sin\phi) \simeq e\cos(\phi - \epsilon\phi), \quad (6.107)$$

we can put the solution (6.106) in the more transparent form shown in (6.90).

[35] While the detection of the higher-order effects represented by the first three terms would require measurements at impossibly high accuracy, the fourth term, while equally small, is a new effect (a correction to a zero) that is much easier to measure.

In Section 4.3.2, we derived using Huygens' method the gravitational angular deflection $\delta\phi$ of a light ray according to the equivalence principle (accounting for the curvature only in the time direction). In Exercise 6.3, you are asked to obtain $\delta\phi$ by following the more standard procedure of directly applying the geodesic equation—just as we have done in calculating the precession of an elliptical orbit. The steps to solve these two problems are the same. But for the light geodesic $x^\mu(\tau)$, the curve parameter τ cannot be the proper time. Still we can choose to define it so that the light's 4-momentum $p^\mu = \dot{x}^\mu \equiv dx^\mu/d\tau$, which is a null 4-vector. Thus the geodesic Lagrangian (6.79), $L = p_\mu p^\mu = 0$, is the energy balance equation, instead of $L = -c^2$ as in (6.81) for a particle with mass.

Exercise 6.3 Light deflection from solving the geodesic equation

Take the following steps to obtain the result for the bending of light shown in (6.77):

(a) Identify the two constants of motion.

(b) Express the $L = 0$ equation in terms of these constants.

(c) By changing the curve parameter differential $d\tau \rightarrow d\phi$ and changing the radial distance variable to its inverse $u \equiv 1/r$, you should find that the light trajectory obeys[36]

$$u'' + u - \epsilon u^2 = 0, \qquad (6.108)$$

where $u'' = d^2 u/d\phi^2$ and $\epsilon = 3r^*/2$.

(d) Solve (6.108) by perturbation:[37] $u = u_0 + \epsilon u_1$. Suggestion: Parameterize the first-order perturbation solution as $u_1 = \alpha + \beta \cos 2\phi$; then fix the constants α and β.

(e) From this solution of the orbit $r(\phi)$ for the light trajectory, one can deduce the angular deflection $\delta\phi$ result (6.77) by comparing the directions of the initial and final asymptotes ($r = \infty$ at $\phi_i = \pi/2 + \delta\phi/2$ and $\phi_f = -\pi/2 - \delta\phi/2$) as shown in Fig. 6.12(b).

[36] This is the massless particle version of (6.95).

[37] The unperturbed solution u_0 leads to the trajectory as shown in Fig. 6.12(a).

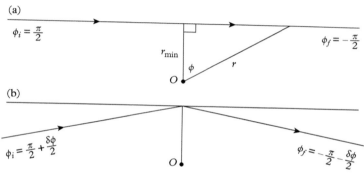

Figure 6.12 (a) Light's geodesic trajectory is straight, $r = r_{min}/\cos\phi$, in the absence of gravity. (b) The light geodesic is bent by an angle $\delta\phi$ by a source mass at O. Thus the trajectory has asymptotes ($r = \infty$) at $\phi_i = \pi/2 + \delta\phi/2$ and $\phi_f = -\pi/2 - \delta\phi/2$ or $\phi_f = 3\pi/2 - \delta\phi/2$.

Review questions

1. A curved surface necessarily has a position-dependent metric. Why is this not a sufficient condition? Illustrate this point with an example of a flat space with a position-dependent metric. Gaussian curvature K provides a good criterion for finding out whether a surface is curved or not. Show that $K = 0$ for your example, and explain why $K = 1/R^2$ for a spherical surface, regardless of what coordinates one uses.

2. What are the three surfaces of constant curvature? Write out the embedding equation of a 3-sphere in a 4D space, as well as the equation of a 3-pseudosphere.

Is the embedding space for the latter a Euclidean space?

3. The Gaussian curvature measures how curved a space is, because it controls the amount of deviation from Euclidean relations. What is angular excess? How is it related to Gaussian curvature? Give a simple example of a polygon on a spherical surface that clearly illustrates this relation. Give another example of such a non-Euclidean relation, showing that the deviation from flatness is proportional to the curvature.

4. What are tidal forces? How are they related to the gravitational potential? Explain how in GR the tidal forces are identified with the curvature of spacetime.

5. Give a qualitative description of the GR field equation. Describe its tensor structure and its various terms: what are its source term and curvature term? What is the solution to this field equation?

6. What is the form of the spacetime metric (when written in terms of the spherical Schwarzschild coordinates) for a spherically symmetric spacetime? Explain very briefly how such a spacetime is curved in space as well as in time.

7. Present a simple proof of Birkhoff's theorem for Newtonian gravity. Explain how it implies that there is no monopole radiation.

8. Write down the metric elements for the Schwarzschild spacetime (i.e., the Schwarzschild solution). If the metric element g_{00} is related to the gravitational potential by $g_{00} = -(1 + 2\Phi/c^2)$ in the Newtonian limit, demonstrate that the Newtonian result $\Phi = -G_N M/r$ is contained in this Schwarzschild solution.

9. A light ray is bent in the presence of a gravitational field. How does the feature $g_{rr} = -g_{00}^{-1}$ in the Schwarzschild metric lead to a bending of the light ray in GR that is twice as much as predicted by Einstein's 1911 EP calculation?

10. Explain qualitatively how GR causes the perihelion of an elliptical orbit to precess.

11. Suppose the metric does not depend on a certain coordinate, say the ϕ angle of the (t, r, θ, ϕ) coordinates. Use the geodesic equation to show that this leads to a constant of motion $\sim g_{\phi\phi}\dot{\phi}$.

Black Holes

- A black hole forms around an object massive enough and dense enough to fit within its event horizon, which is a one-way surface through which particles and light can only traverse inward. Thus an exterior observer cannot receive any signal sent from inside.

- In this chapter, we mostly study the spherically symmetric, nonrotating Schwarzschild black hole. The event horizon is a spherical surface of radius $r = r^* = 2G_N M/c^2$, which is a coordinate singularity of the Schwarzschild metric. The metric elements g_{00} and g_{rr} change signs when crossing from the $r > r^*$ to the $r < r^*$ region, leading to a role reversal between space and time.

- The gravitational energy unleashed when a particle falls into a tightly bound orbit around a black hole can be enormous, more than ten times that released in a nuclear fusion reaction. This powers some of the most energetic phenomena observed in the universe.

- GR-based models show that a rotating star of sufficient final mass ($\gtrsim 3M_\odot$) after it burns out cannot support its own weight, inevitably collapsing into a rotating (Kerr) black hole.

- The physical reality of, and observational evidence for, black holes are briefly discussed.

- There is a mysterious correspondence between the laws of black hole physics and the laws of thermodynamics. In particular, the surface gravity at the event horizon behaves like the temperature of a thermodynamical system; the horizon area behaves like the entropy.

- This correspondence was greatly strengthened by the discovery of Hawking radiation. Quantum fluctuations around the event horizon bring about the thermal emission of particles and light from a black hole. This is allowed because pair-produced particles falling into the black hole can have negative energy; their partners may thereby escape with positive energy without violating energy conservation.

In Chapter 6, we began to describe the Schwarzschild geometry of the spacetime outside a spherical source. In particular, we studied the bending of a light ray by the sun and the precession of the planet Mercury's perihelion.

A College Course on Relativity and Cosmology. First Edition. Ta-Pei Cheng.

© Ta-Pei Cheng 2015. Published in 2015 by Oxford University Press.

For these solar system applications, gravity is relatively weak (and therefore GR corrections are small). Here we study the spacetime structure exterior to any object whose mass is so compressed that its radius is smaller than its Schwarzschild radius $r^* = 2G_N M/c^2$. Such objects have been given (by John Wheeler) the evocative name **black holes**: they are **holes** because radiation and matter can fall into them; they are **black** because nothing, not even light, can escape from them. These structures necessarily involve such strong gravity (i.e., such strongly curved spacetime) that the GR framework is indispensable for their explication.

Even in the context of Newtonian physics, one can consider a gravity so strong that light cannot escape. In the eighteenth century, John Michell (1724–1793) and (independently) Pierre-Simon Laplace (1749–1827), proposed the possibility of a "black star" whose ratio of mass to radius was so large that the required **escape velocity** $v_{esc} = \sqrt{2G_N M/r} = c\sqrt{r^*/r}$ exceeded the light velocity c. Of course, this speculation was based upon the (from our modern perspective) erroneous assumption that light carried a gravitational mass. GR interprets this phenomenon instead in terms of the causal structure of the spacetime outside a strong gravitational source.

Black holes manifest the full power and glory of Einstein's GR. One of its signature features is the equal treatment of space and time; hence spacetime is the natural arena for the description of physical phenomena. GR is the classical field theory of gravitation, in which curved spacetime is the gravitational field. Now, in the case of black holes, gravity is so strong and the spacetime so warped that the roles of space and time are interchanged, leading to many counterintuitive results.

7.1 Schwarzschild black holes

Coordinate singularities The Schwarzschild geometry in Schwarzschild coordinates (ct, r, θ, ϕ) has the metric

$$g_{\mu\nu} = \begin{pmatrix} -\left(1 - \dfrac{r^*}{r}\right) & & & \\ & \left(1 - \dfrac{r^*}{r}\right)^{-1} & & \\ & & r^2 & \\ & & & r^2 \sin^2\theta \end{pmatrix}. \tag{7.1}$$

The metric and its inverse have singularities at $r = 0$ and $r = r^*$, as well as $\theta = 0$ and π. We understand that $\theta = 0$ and π are **coordinate singularities** associated with our choice of the spherical coordinate system. They are not physical, do not show up in physical measurements at $\theta = 0$ and π, and can be removed by a coordinate transformation. However, the $r = 0$ singularity is real. This is not surprising, as the Newtonian gravitational potential ($\Phi \sim 1/r$) for a point mass already has this feature.

What about the $r = r^*$ surface? As we shall demonstrate, it is actually a co-ordinate singularity. We have discussed the Riemann curvature tensor $R_{\mu\nu\lambda\rho}$ (see (6.20) as well as (6.7)), a nonlinear second-derivative function of the metric that is nontrivial only in a curved spacetime. In the case of Schwarzschild geometry, the coordinate-independent product[1] $R_{\mu\nu\lambda\rho}R^{\mu\nu\lambda\rho} = 12r^{*2}/r^6$ is only singular at $r = 0$. This indicates that the singularity at $r = r^*$ must be associated with our choice of the Schwarzschild coordinate system. Namely, it is not physical and can be transformed away in suitable coordinates, for example, the Kruskal–Szekeres coordinates (see Section 7.1.2).

[1] The Ricci scalar is similarly nonsingular, as it is proportional to the trace of energy-momentum tensor $R = -\kappa T$ as discussed in Section 6.3.2.

The event horizon   While physical measurements are not singular at $r = r^*$, that does not mean that this surface is not special. It is an **event horizon**, separating events that can be viewed from afar from those that cannot (no matter how long one waits). That is, the $r = r^*$ surface is the boundary of a region from which it is impossible to send out any signal. It is a boundary of communication, much as earth's horizon is a boundary of our vision. An event horizon is a one-way barrier: any timelike or null worldline can pass through only inward; particles and light rays cannot move outward.

7.1.1   Time measurements around a black hole

We shall begin our discussion of the causal structure of an event horizon with a simple examination of the elapsed time for a particle traveling inward across the $r = r^*$ boundary. While the proper time of the crossing particle is perfectly finite, a faraway observer sees this crossing take an infinite amount of (Schwarzschild coordinate) time. Thus no signal sent from the horizon's surface or its interior can reach such an observer.

The local proper time

We have already mentioned that there is no physical singularity at $r = r^*$. Here we will examine the time measured by an observer traveling across the Schwarzschild surface. The result shows that such a physical measurement is not singular at $r = r^*$.

Let τ be the proper time measured on the surface of a collapsing star (or, alternatively, the time aboard a spaceship traveling radially inward). Recall from Section 6.4.2 that for a particle (with mass) in the Schwarzschild spacetime, we can write a generalized energy balance equation (6.88). This equation can be simplified further for the case of a collapsing star or infalling spaceship starting from rest at $r = \infty$ (so that $\mathcal{E} = 0$), following a radial path along some fixed azimuthal angle ϕ (i.e., $d\phi/d\tau = 0$, so the angular momentum $l = 0$):

$$\frac{1}{2}\dot{r}^2 - \frac{G_N M}{r} = 0; \tag{7.2}$$

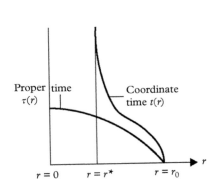

Proper time $\tau(r)$

Coordinate time $t(r)$

$r = 0$ $r = r^*$ $r = r_0$

Figure 7.1 *The contrasting behavior of proper time $\tau(r)$ vs. coordinate time $t(r)$ at the Schwarzschild surface.*

thus

$$\frac{1}{c^2}\left(\frac{dr}{d\tau}\right)^2 = \frac{2G_N M}{c^2 r} = \frac{r^*}{r}; \quad \text{hence} \quad c\,d\tau = \pm\sqrt{\frac{r}{r^*}}\,dr. \tag{7.3}$$

The plus sign corresponds to an exploding star (or an outward-bound spaceship) and the minus sign to a collapsing star (or an inward-bound probe). We pick the minus sign. A straightforward integration yields

$$\tau(r) = \tau_0 - \frac{2r^*}{3c}\left[\left(\frac{r}{r^*}\right)^{3/2} - \left(\frac{r_0}{r^*}\right)^{3/2}\right], \tag{7.4}$$

where τ_0 is the time when the probe is at some reference point r_0.

Thus the proper time $\tau(r)$ is perfectly smooth at the Schwarzschild surface (see Fig. 7.1). An observer on the surface of the collapsing star would not feel anything peculiar when the star passed through the Schwarzschild radius. It would then take a finite amount of proper time to reach the origin, which is a physical singularity.

Exercise 7.1 Travel time from the event horizon to the singular origin

---

(a) How much proper time $\Delta\tau$ (in terms of the Schwarzschild radius r^*) passes for a probe falling from the event horizon to the $r = 0$ singularity? You may assume that the probe fell in radially from rest at infinity as in the discussion above. (b) Evaluate this time interval for the case of a black hole with a mass $3M_\odot$ as well as the case of a supermassive black hole with a mass $10^9 M_\odot$.

The Schwarzschild coordinate time

While the time measured by an observer traveling across the Schwarzschild surface is perfectly finite, this is not the case for an observer far away from the source. Recall that the Schwarzschild coordinate t is the time measured by an observer far away, where the spacetime approaches the flat Minkowski limit. Here we will show that the Schwarzschild coordinate time blows up as the probe approaches the $r = r^*$ surface. To find the coordinate time as a function of the radial coordinate in the $r > r^*$ region, we start with the chain rule: $dt/dr = (dt/d\tau)/(dr/d\tau) = \dot{t}/\dot{r}$. We already have an expression for $\dot{r}$ from (7.3), while $\dot{t}$, according to (6.82), is directly related to the conserved particle energy κ, which is fixed to be c because we are considering a geodesic with zero kinetic energy at infinity, $\mathcal{E} = m(\kappa^2 - c^2)/2 = 0$. In this way, we find $dt/dr = \dot{t}/\dot{r} = -(1 - r^*/r)^{-1}/c(r^*/r)^{1/2}$, so that

$$c\,dt = -\sqrt{\frac{r}{r^*}}\frac{dr}{1 - r^*/r}, \tag{7.5}$$

which shows clearly the singularity at r^*. For $r \simeq r^*$, we can integrate $c\,dt \simeq -r^*\,dr/(r - r^*)$ to display the logarithmic singularity:

$$t - t_0 \simeq -\frac{r^*}{c} \ln \frac{r - r^*}{r_0 - r^*}. \tag{7.6}$$

It takes an infinite amount of coordinate time to reach $r = r^*$. The full function $t(r)$ in the region outside the Schwarzschild surface can be calculated[2] and is displayed in Fig. 7.1.

[2] See, e.g., Problem 8.2 (solved on p. 387) in (Cheng 2010).

Infinite gravitational redshift The above-discussed phenomenon of a distant observer seeing a collapsing star slow to a standstill can also be interpreted as an infinite gravitational time dilation. The relation (6.70) between coordinate and proper time intervals for a stationary observer is given by

$$dt = \frac{d\tau}{\sqrt{-g_{00}}} = \frac{d\tau}{\sqrt{1 - r^*/r}}. \tag{7.7}$$

Clearly, the coordinate time interval dt will blow up as r approaches r^*. In terms of wave peaks, this means that an infinite time interval passes between peaks reaching the faraway receiver. This can be equivalently described as an infinite gravitational redshift. Equation (5.45) showed that the ratio of the received frequency to the emitted frequency is

$$\frac{\omega_{\text{rec}}}{\omega_{\text{em}}} = \sqrt{\frac{(g_{00})_{\text{em}}}{(g_{00})_{\text{rec}}}} = \sqrt{\frac{1 - r^*/r_{\text{em}}}{1 - r^*/r_{\text{rec}}}} = \frac{d\tau_{\text{em}}}{d\tau_{\text{rec}}} \tag{7.8}$$

When $r_{\text{em}} \to r^*$, the received frequency ω_{rec} approaches zero, as the time between received peaks $d\tau_{\text{rec}}$ blows up to infinity. Thus no signal can be transmitted from the black hole.

7.1.2   Causal structure of the Schwarzschild surface

The phenomenon of infinite gravitational redshift implies the impossibility of any signal transmission from the $r < r^*$ region to an outside observer. The Schwarzschild surface is in fact a one-way barrier: while matter and radiation can proceed inward across the horizon, no particle, whether massive or massless, can move outward.

Role change between space and time

To gain a deeper understanding of the Schwarzschild surface as an event horizon, we need to study the causal structure of the geometry exterior to a spherical source. One of the key differences between the $r > r^*$ and $r < r^*$ regions is that the roles of space and time are interchanged, because the metric functions g_{00} and g_{rr} exchange signs at $r = r^*$:

$$r > r^* \quad \text{outside a black hole:} \quad g_{00} < 0, \quad g_{rr} > 0 \quad \text{normal metric;}$$

$$r < r^* \quad \text{inside a black hole:} \quad g_{00} > 0, \quad g_{rr} < 0 \quad \text{flipped metric.} \tag{7.9}$$

Thus, in the $r > r^*$ region, we have the familiar result that the t coordinate is timelike, while the radial r coordinate is spacelike, but inside the Schwarzschild surface, the time axis is actually spacelike and the radial axis timelike. As we discussed in the flat-spacetime diagram, Fig. 3.5, the lightcone structure normally dictates that a timelike or lightlike ($ds^2 \leq 0$) trajectory inevitably moves in the direction of ever-increasing time. You cannot stand still (much less go back) in time, but you can move freely in space. However, (7.9) tells us that inside the $r = r^*$ surface of a black hole, any particle following a timelike or lightlike worldline cannot rest at a fixed radial position (much less move outward) but must proceed toward the $r = 0$ singularity.

Schwarzschild coordinates and their limitation

As we emphasized while discussing the spacetime diagram, the lightcone structure can clarify the causal structure of spacetime. In this context, the tipping of lightcones at the event horizon illustrates the above-discussed role reversal of the Schwarzschild time and space coordinates. For $r > r^*$, the lightcones open in the future direction; for $r < r^*$, they tip toward the singularity at the origin (see Fig. 7.2). We might also like to depict the mechanism by which the event horizon acts as a one-way barrier. Unfortunately, we will see that the coordinate singularity at $r = r^*$ makes Schwarzschild coordinates unsuitable for such an investigation. The faraway observer who measures the coordinate time never sees anything cross the event horizon.

To study lightcones is to study the light geodesics that form them. Recall that in a flat spacetime the radial (i.e., with fixed θ and ϕ) lightlike worldlines, corresponding to the solutions of the condition $ds^2 = -c^2\,dt^2 + dr^2 = 0$, or $c\,dt = \pm dr$, are straight lines of unit slope in the (ct, r) spacetime diagram:

$$ct = \pm r + \text{constant.} \tag{7.10}$$

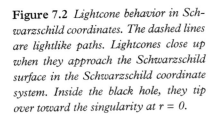

Figure 7.2 *Lightcone behavior in Schwarzschild coordinates. The dashed lines are lightlike paths. Lightcones close up when they approach the Schwarzschild surface in the Schwarzschild coordinate system. Inside the black hole, they tip over toward the singularity at r = 0.*

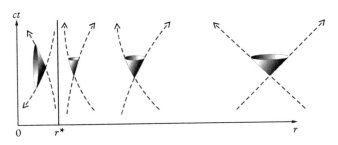

The plus sign is for outgoing (r increasing with t) and the minus sign for incoming light. Timelike worldlines are always contained inside the lightcone; particles with mass must proceed toward the future, cf. Fig. 3.5.

A radial ($d\theta = d\phi = 0$) worldline for a photon in the Schwarzschild coordinates has a null line-element interval:

$$0 = ds^2 = -\left(1 - \frac{r^*}{r}\right)c^2 \, dt^2 + \left(1 - \frac{r^*}{r}\right)^{-1} dr^2$$

$$= -\left(1 - \frac{r^*}{r}\right)\left(c \, dt + \frac{dr}{1 - r^*/r}\right)\left(c \, dt - \frac{dr}{1 - r^*/r}\right). \qquad (7.11)$$

We then have[3]

$$c \, dt = \pm \frac{dr}{1 - r^*/r}, \qquad (7.12)$$

which can be integrated to give

$$ct = \pm(r + r^* \ln |r - r^*| + \text{constant}). \qquad (7.13)$$

Outside the event horizon ($r > r^*$), the plus sign is for the outgoing, and the minus sign for the infalling, lightlike geodesics. In Fig. 7.2, we have drawn several representative lightcones. We note that for the region far from the source where the spacetime becomes flat, the lightcones formed by the null geodesics (7.13) approach their expected form with sides of unit slope; however, as one moves closer to the $r = r^*$ surface, the lightcones close up. Inside the event horizon ($r < r^*$), the causal structure changes as discussed above. The lightcones tip over, opening toward the $r = 0$ singularity rather than toward ever-increasing time.

The fact that the metric becomes singular at $r = r^*$ means that the Schwarzschild coordinates (t, r, θ, ϕ) are not convenient for the description of events near the Schwarzschild surface. One might get the impression from the clammed-up lightcones in Fig. 7.2 that nothing ever crosses the event horizon. In fact, all it means is that a distant observer (whose clock keeps the Schwarzschild coordinate time) never sees it happen. Better-behaved coordinates should better depict such events.

Better-behaved coordinate systems

We now search for coordinates that can describe the Schwarzschild geometry without the $r = r^*$ singularity. In such coordinates, the lightcones should tilt over smoothly.

We start with the (advanced) Eddington–Finkelstein (EF) coordinates, whose time coordinate ($c \, d\bar{t}$) is chosen to equal the distance traveled by an infalling photon ($-dr$). Recall from (7.4) that the proper time of a particle falling into a black hole is smooth for all values of r. Instead of setting up the coordinate system

[3] This relation differs from (7.5), because we are now considering lightlike worldlines.

using a static observer far from the gravitational source (as in Schwarzschild co-ordinates), one could describe the Schwarzschild geometry from the viewpoint of an infalling observer. Mathematically, an even simpler choice is to use an infalling photon. While a photon cannot measure a proper time, its traveled distance could serve as a time coordinate: $c\,d\bar{t} = -dr$, which would make the infalling photon's worldline look the same as (7.10) for a flat space. This suggests that we replace the factor $[c\,dt + dr/(1 - r^*/r)]$ in (7.11) with $(c\,d\bar{t} + dr)$. Namely, we introduce a new coordinate time $c\,d\bar{t} = c\,dt + r^*\,dr/(r - r^*)$, so that, in terms of $\bar{t}$, the light geodesic condition $ds^2 = 0$ is satisfied (i.e., (7.11) is solved) by the vanishing in turn of the expressions in the last two parentheses:

$$c\,d\bar{t} = -dr \qquad \text{(incoming)}, \qquad\qquad (7.14)$$

$$c\,d\bar{t} = \frac{r + r^*}{r - r^*}\,dr \qquad \text{(outgoing/incoming)}. \qquad\qquad (7.15)$$

In the new spacetime diagram with $(c\bar{t}, r)$ axes (Fig. 7.3a), incoming light follows straight paths of negative unit slope (for which we rigged the time co-ordinate). Meanwhile, the outgoing lightlike worldlines gradually become steeper

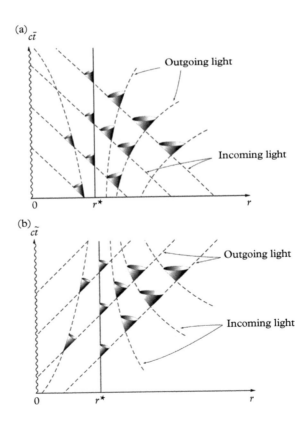

Figure 7.3 *Lightcones tilt over smoo-thly in Eddington-Finkelstein spacetime. (a) A black hole in advanced EF coor-dinates* $(\bar{t}, r)$*: all light geodesics inside the horizon move toward the future $r = 0$ singularity. (b) Reversing the time coor-dinate* $(d\tilde{t} = -d\bar{t})$ *yields a white hole in retarded EF coordinates* $(\tilde{t}, r)$*: all light geodesics inside the horizon move away from the past $r = 0$ singularity.*

as they approach the $r = r^*$ event horizon, which itself is a vertical null line, a lightlike path. Once inside the $r < r^*$ region, the coefficient in (7.15) becomes negative, just like that in (7.14). Thus both lightlike geodesics, (7.14) and (7.15), are incoming; r decreases as $\tilde{t}$ increases. Namely, inside the black hole, all light cones open inward, so all timelike geodesics head for the $r = 0$ singularity. We already saw the tipping of lightcones in Schwarzschild coordinates. The improvement in EF coordinates is that the tipping is smooth. The lightcones do not clam up at the event horizon; some cross over into the black hole. Thus EF coordinates can credibly describe the event horizon.

A stationary ($dr = 0$) point on the $r = r^*$ Schwarzschild surface is the limit of solutions to (7.15) and therefore traces a null (lightlike) path in spacetime, a vertical line in Fig. 7.3(a). Thus the stationary event horizon is a null surface in spacetime, everywhere tangent to inward-pointing lightcones as shown in Fig. 7.4. This is what makes the event horizon special; this is why it is a one-way barrier. A timelike worldline passing through any point on such a null surface can only point inward toward the $r = 0$ singularity.

Black hole vs. white hole One may wonder whether the same procedure can be applied to the last parenthesis in (7.11) to straighten out the outgoing light geodesics. Indeed, we can define a (retarded) EF time $\tilde{t}$ with $c\,d\tilde{t} = dr$ for the outgoing light and $c\,d\tilde{t} = -[(r + r^*)/(r - r^*)]\,dr$ for the incoming/outgoing light. Essentially, we have just reversed the direction of time: $d\tilde{t} = -d\tilde{t}$. Again we see that the lightcones, depicted in Fig. 7.3(b), tilt over smoothly across the Schwarzschild surface. But instead of tipping inward as in Fig. 7.3(a), they lean outward away from the $r = 0$ singularity. That is, while the Schwarzschild geometry depicted in the advanced EF coordinates has a future singularity at $r = 0$, the geometry depicted in retarded EF coordinates contains a past singularity at $r = 0$. Once again, the $r = r^*$ surface is a null surface, a one-way membrane allowing the transmission of particles and light only in one direction—this time outward. Thus we now have a white hole (containing the past singularity). While this time-reversed black hole is a perfectly good solution to Einstein's equation (which of course is covariant under time reversal), we have not found such a thing in our physical universe.

Null surface

Figure 7.4 *A null surface is an event horizon. The lightcones of all points on the null surface are on one side of the surface. All timelike worldlines (samples shown as arrowed lines) are contained inside lightcones and thus can cross the null surface only in one direction. Therefore, a null surface is a one-way barrier.*

Exercise 7.2 Retarded EF coordinates with past $r = 0$ singularity

Above, we obtained the advanced EF coordinates $(\tilde{t}, r)$ with lightlike geodesics, (7.14) and (7.15), defining lightcones tilting over smoothly inward toward a future $r = 0$ singularity. Obtain likewise the corresponding retarded EF coordinates $(\tilde{t}, r)$. Find the outgoing and incoming light geodesics that bound lightcones tilting outward away from a past $r = 0$ singularity as in Fig. 7.3(b).

Metric is singularity-free at $r = r^*$ in Kruskal–Szekeres coordinates
Above, we separately straightened the light geodesics into worldlines of unit slope. Advanced EF coordinates straightened the incoming null geodesics, and retarded EF coordinates the outgoing. One can actually make a coordinate transformation that straightens both at the same time, i.e., that simplifies both parentheses in (7.11). But the metric still has the troublesome prefactor $(1 - r^*/r)$. Kruskal and Szekeres independently found a transformation involving the exponentiation of the coordinates that eliminates the singularity at $r = r^*$, leaving only the genuine one at $r = 0$. As the procedure is somewhat complicated, we shall omit its presentation.[4]

Section 7.1.3 concludes our presentation of the Schwarzschild black hole by discussing the orbits of particles around such a compact source of gravity.

[4] For an elementary introduction with some details worked out, see (Cheng 2010, Section 8.1.3).

7.1.3 Binding energy to a black hole can be extremely large

We are familiar with the fact that thermonuclear fusion, when compared with chemical reactions, is a very efficient process for releasing the rest energy of a particle. Here we show that binding a particle to a compact gravity source like a black hole can be an even more efficient mechanism. The thermonuclear reactions taking place in the sun can be summarized as fusing four protons (hydrogen nuclei, each with a rest energy of 938 MeV) into a helium nucleus (having a rest energy smaller than the sum of the four proton rest energies) with a released energy of 27 MeV, which represents $27/(4 \times 938) = 0.7\%$ of the input energy. Here we discuss the energy that can be released when a particle first falls into stable orbits around a Schwarzschild black hole before it eventually spirals through the event horizon.

Recall that the orbit can be determined from the effective 1D energy balance equation that we studied in the Chapter 6. Equation (6.88) may be written as

$$\frac{1}{2}m\dot{r}^2 + m\Phi_{\text{eff}} = \mathcal{E}, \tag{7.16}$$

Figure 7.5 *Schwarzschild vs. Newtonian effective potential. The solid curve represents a specific choice of angular momentum $(l_0/mc)^2 \simeq 4.6r^{*2}$. For higher l, Φ_{eff} more closely tracks the Newtonian potential before falling off sharply at lower r. In the text it is shown that for l below l_0, $(l_0/mc)^2 = 3r^{*2}$, there are no maxima or minima; the potential is monotonic.*

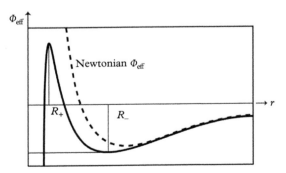

with an effective potential

$$\Phi_{\mathrm{eff}} = -\frac{G_N M}{r} + \frac{l^2}{2m^2 r^2} - \frac{r^* l^2}{2m^2 r^3}. \qquad (7.17)$$

The first term on the right-hand side is clearly the Newtonian gravitational potential, the second term is the rotational kinetic energy (the centrifugal barrier), and the last term is the new GR contribution. It is a small correction for situations such as a planet's motion discussed in Chapter 6, but can be very important when the radial distance r is comparable to the Schwarzschild radius r^* as in the case of an orbit just outside the horizon.

The innermost stable circular orbit   We can find the extrema of this potential by setting $\partial \Phi_{\mathrm{eff}} / \partial r = 0$:

$$\frac{r^* c^2}{2r^2} - \frac{l^2}{m^2 r^3} + \frac{3r^* l^2}{2m^2 r^4} = 0, \qquad (7.18)$$

or

$$r^2 - 2 \left(\frac{l}{mc} \right)^2 \frac{r}{r^*} + 3 \left(\frac{l}{mc} \right)^2 = 0. \qquad (7.19)$$

The solutions R_+ and R_- specify the locations where Φ_{eff} has a maximum and a minimum, respectively (see Fig 7.5):

$$R_{\pm} = \frac{1}{r^*} \left(\frac{l}{mc} \right)^2 \left[1 \mp \sqrt{1 - 3 \left(\frac{r^* mc}{l} \right)^2} \right]. \qquad (7.20)$$

We note the important difference between the Schwarzschild Φ_{eff} and its Newtonian analog, whose centrifugal barrier always dominates in the small-r region ($l^2 / r^3 \to \infty$ as $r \to 0$). This means that a particle in the Newtonian field cannot fall into the $r = 0$ center as long as $l \neq 0$. In the small-r region of the relativistic Schwarzschild geometry, the last (GR correction) term in (7.17) dominates ($-l^2 / r^4 \to -\infty$ as $r \to 0$). When $\mathcal{E} \geq m\Phi_{\mathrm{eff}}(R_+)$, a particle can plunge into the gravity center even if $l \neq 0$. For $\mathcal{E} = m\Phi_{\mathrm{eff}}(R_-)$, just as in the Newtonian case, we have a stable circular orbit with radius R_-. This circular radius cannot be arbitrarily small. In order to have an orbit of any kind and not just plunge into the black hole, a particle must have enough angular momentum to create a sufficient centrifugal barrier. Equation (7.19) must have a solution, so its determinant in the square root of (7.20) must be nonnegative. This fixes a minimum angular momentum l_0:

$$3 \left(\frac{r^* mc}{l_0} \right)^2 = 1, \quad \text{or} \quad \left(\frac{l_0}{mc} \right)^2 = 3 \, (r^*)^2, \qquad (7.21)$$

so that the innermost stable circular orbit (ISCO) has a radius

$$R_0 = \frac{1}{r^*}\left(\frac{l_0}{mc}\right)^2 = 3r^*. \tag{7.22}$$

As we reduce the angular momentum to $l = l_0$, the centrifugal barrier peak in Φ_{eff} (Fig. 7.5) at R_+ falls and the stable orbit trough at R_- rises until they meet, forming a flat point of inflection at R_0, the ISCO. For smaller angular momenta ($l < l_0$), the potential falls monotonically from $\lim_{r\to\infty} \Phi_{\text{eff}} = 0$ to $\lim_{r\to 0} \Phi_{\text{eff}} = -\infty$, so there are no orbits.[5] The plasma in an accretion disc around a black hole settles into stable orbits, but will lose its orbital angular momentum through turbulent viscosity (due to magnetohydrodynamic instability) and eventually, owing to the disappearing centrifugal barrier, spiral into the black hole.

The binding energy of a particle around a black hole   To illustrate the energy of gravitational binding by a Schwarzschild black hole, consider a free particle that falls toward a black hole and ends up bound in the ISCO. Thus, according to (7.22) and (7.21), the particle orbits at a radial distance $r = R_0 = 3r^*$ with angular momentum $l_0 = \sqrt{3}r^* mc$. According to the energy balance equation (7.16) with $\dot{r} = 0$, we have $\mathcal{E} = m\Phi_{\text{eff}} = -mc^2/18$. This solution gives the total energy for the gravitationally bound particle:[6]

$$\frac{E(\infty)}{mc^2} = \frac{\kappa}{c} = \sqrt{\frac{2\mathcal{E}}{mc^2} + 1} = \sqrt{\frac{8}{9}} = 0.94. \tag{7.23}$$

That is, 6% of the rest energy is released—almost ten times larger than the 0.7% from thermonuclear fusion.[7]

7.2 Astrophysical black holes

So far we have concentrated on the Schwarzschild black hole: an idealized, static, spherically symmetric entity. Is it relevant for any astrophysical phenomena? In this section, we shall qualitatively answer this question on two fronts: on the theoretical side we summarize the results of studies of more realistic black hole solutions in GR; on the phenomenological side, we briefly report the present status of our search for black holes in the universe.

7.2.1 More realistic black holes

First, we present some theoretical results applicable to more realistic black holes. The Kerr solution of the Einstein equation generalizes the Schwarzschild solution to rotating sources. Model studies indicate that stellar gravitational collapse can result in rotating black holes.

[5] For a more detailed discussion of Φ_{eff}, see (Wald 1984, pp. 139–143).

[6] Cf. (6.87) and Sidenote 34 in Chapter 6.

[7] We should note that the gravitational binding energy of a particle around a spinning black hole is even greater; it can be as much as 42% of its rest energy!

Rotating black holes

Most stars rotate and thus have only axial symmetry. The simplest of such sources of gravity is characterized not just by its mass M but also by its angular momentum $\mathcal{J}$. The solution of Einstein's equation for the spacetime exterior to such a rotating source was discovered by Roy Kerr (1934–). The **Kerr spacetime** reduces to the Schwarzschild geometry in the limit of $\mathcal{J} = 0$, but has in general a considerably more complicated singularity structure. We shall present only a brief introduction to some of its salient features. The physical singularity is no longer a single point, but a ring perpendicular to the symmetry axis, with a radius proportional to the angular momentum of the source. Similarly, the Kerr black hole has an event horizon, a null surface, that is not spherical but ellipsoidal. While the event horizon of a Schwarzschild black hole coincides with the **surface of infinite redshift**, the Kerr horizon surface is enclosed inside the surface of infinite redshift, which coincides with the **stationary limit surface**, which we will explain below. As the source rotates, GR predicts that the spacetime will be dragged along;[8] see Fig. 7.6. If a particle (or photon) starts with a vanishing angular momentum $l = 0$, we would normally expect it to fall straight toward the center of the gravitational attraction; but with a rotating inertial frame of reference, such a zero-angular-momentum particle would still develop an angular velocity. The stationary limit surface is the boundary of a region where the frame dragging is so strong that no particle (not even light) can be stationary; everything rotates in the same direction as the source—even if it entered with great angular momentum opposing the source rotation.

An interesting feature of the rotating black hole is that one can actually extract energy from it. A physical processes (called the **Penrose process**) taking place in the region (called the **ergosphere**) between the stationary limit surface and the event horizon null surface can send particles to distant observers that carry away the rotational energy of the source. Clearly, a rotating black hole can bring about more complex physical processes than a static Schwarzschild black hole. The mathematics involved is correspondingly more complicated, so we refer interested readers to more advanced texts.[9]

[8] This GR frame-dragging prediction has been tested by the Gravity Probe B experiment. This satellite experiment managed to measure the tiny gyroscopic precession brought about by earth's rotation.

[9] See, e.g., (Hobson et al. 2006, Chapter 13) or (Cheng 2010, Section 8.4).

(a)   (b)

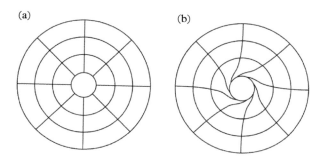

Figure 7.6 *Dragging of the inertial frame: a counterclockwise-rotating source turns (a) some initial spacetime (before source rotation) into (b) a twisted geometry (following the source rotation). Radial geodesics follow those twisted lines; they are swept along with the rotation.*

Gravitational collapse into a black hole

Our discussion of black holes has so far assumed a static source, an eternal black hole that has always existed, with its mass behind its event horizon. Naturally, one would like to know whether GR could allow the creation, through the gravitational collapse of more normal matter, of a region of spacetime with these features? Such an investigation would involve solving the Einstein equation for a nonvanishing energy/momentum source. The simplest model was the original Schwarzschild interior solution for a constant energy density; gradually, more realistic equations of state for the stellar interior were incorporated in such studies. The most influential investigations were carried out by Robert Oppenheimer (1905–1967) and his students around 1939. Their research showed analytically that a cold Fermi gas quickly collapses from a smooth initial distribution to form a black hole with the properties discussed above. The gravitational attraction causes each mass element to follow a geodesic trajectory toward the center. As the interior density increases, ever more exterior space is described by the Schwarzschild metric until all matter passes through the $r = r^*$ surface. The exterior then contains an event horizon, so a black hole is formed.

Nevertheless, the physics community remained skeptical of the reality of black holes. Their reservations were many. For example, it was questioned whether the assumption of spherical symmetry was too much of an idealization. How should one account for realistic complications such as deformation-forming lumps, shock waves leading to mass ejection, effects of electromagnetic and gravitational radiation, etc.? However, numerical calculations years later showed that any multipole distortion to the Schwarzschild metric is quickly shaken off through gravitational radiation; the source relaxes to the exact Schwarzschild black hole.

As for stellar rotation, we have already mentioned the Kerr solution found in 1963. While there does not exist an analytic solution of gravitational collapse for a rotating source analogous to the one discussed above, numerical calculations have again demonstrated that even with large distortions, a collection of matter with nonvanishing angular momentum always collapses into a Kerr black hole.

The revival of theoretical study of black holes since the 1950s was due in large part to the leadership of John Wheeler (1911–2008) in the United States and Yakov Zel'dovich (1914–1987) in the Soviet Union. The final acceptance by physicists of the GR prediction of black holes as the generic end product of gravitational collapse was brought about by the proof in the early 1960s of the singularity theorems[10] by Roger Penrose (1931–). Related to this, we note also a well-known theorem stating that all black holes can be completely characterized by their mass, angular momentum, and electric charge.[11] Their lack of any other features inspired the witty summary "Black holes have no hair."

[10] This set of theorems show in realistic situations the inevitability of the formation of an event horizon, within which always lies a singularity.

[11] The GR solution for an electrically charged source, called the Reissner-Nordström geometry, is thought to be less phenomenologically relevant.

7.2.2   Black holes in our universe

In Chapter 6, we saw that the GR predictions of bending of light and gravitational redshift have all been verified within our solar system. Can we likewise confirm the strong gravity predictions of black holes' existence?

Black holes are faraway, small, black discs in the sky; it would seem rather hopeless to ever observe them. But, by accounting for the gravitational effects of an object on its surroundings (such as gravitational lensing, the orbits of nearby stars, and the accretion of hot gas), one can estimate the mass of the object. If a highly compact source has a mass greater than about $3M_\odot$, it is a strong candidate to be a black hole—since no known mechanism can stop such a massive system from gravitationally collapsing into a black hole.

The Chandrasekhar limit   What is the basis of this $3M_\odot$ limit? In an ordinary star, the gravitational attraction is balanced by the outward pressure from thermonuclear burning in its interior. When the fuel is spent, what else can prevent gravitational collapse? One possibility is the quantum mechanical repulsive force due to Pauli exclusion (called the **degeneracy pressure** or **Pauli blocking**) among particles of half-integer spin (fermions). This is the source of stability for **white dwarfs** (the fermions being electrons) and **neutron stars** (the fermions being neutrons). In 1930, Subrahmanyan Chandrasekhar (1910–1995) used the new quantum mechanics to show that for stellar masses $M > 1.4M_\odot$, the electrons' degeneracy pressure is not strong enough to stop the gravitational contraction. Thus he was the first one to make the radical suggestion that massive enough stars would collapse into black holes (decades before such nomenclature was invented). In 1932, James Chadwick (1891–1974) discovered the neutron, and, soon after, in 1934, Fritz Zwicky (1898–1974) suggested that the remnant of a supernova explosion, associated with the final stage of gravitational collapse, was a **neutron star**. Because of the strong nuclear force, there exists no simple analytic calculation for the corresponding limit for neutron stars, but numerical estimates of the neutron degeneracy pressure all point to a value not much more than $3M_\odot$.

Black holes in X-ray binaries   The majority of stars are members of binary systems orbiting each other. If a black hole is in a binary system with another visible star, then, by observing the Kepler motion of the visible companion, one can obtain some limit on the mass of the invisible companion. If it exceeds $3M_\odot$, it is a black hole candidate. Even better, if the visible star produces significant gas (as in the case of solar flares), the infall of such gas (called **accretion**) into the black hole will produce intense X-rays. A notable example is Cygnus X-1, which is now generally accepted as a black hole binary system, in which the visible companion is a supergiant star[12] of mass $M_{vis} \simeq 30M_\odot$, and the invisible compact object, presumably a black hole, has a mass $M \geq 8M_\odot$. Altogether, close to ten such binary candidate black holes have been identified in our galaxy.

[12] The supergiant star, having a radius of about $20R_\odot$, cannot be the source of the observed X-rays.

Supermassive black holes It has also been discovered (again by detecting the gravitational influence on nearby visible matter) that at the centers of most galaxies are supermassive black holes, with masses ranging from 10^6 to $10^{12} M_\odot$. Even though the initial findings were a great surprise, once this discovery had been made, it was not too difficult to understand why we should expect such supermassive centers. The gravitational interaction between stars is such that they swing-and-fling past each other: one flies off outward while the other falls inward. Thus we can expect stars and dust to be driven inward toward the galactic core, producing a supermassive gravitational aggregate. Some of these galactic nuclei emit huge amounts of X-rays and visible light a thousand times brighter than the stellar light of a galaxy. This is interpreted as light emitted from gas heated as it is funnels into the central black hole.[13] Such galactic centers are called AGN (active galactic nuclei). The well-known astrophysical objects, quasars (quasi-stellar objects), are interpreted as AGN in the early stage of galactic evolution. Observations suggest that an AGN is composed of a massive center surrounded by a molecular accretion disk. It is thought to be powered by a rotating supermassive black hole at the core of the disk. Such huge emissions require extremely efficient mechanisms for releasing the energy associated with the matter surrounding the black hole. Recall our discussion in Section 7.1.3 of the huge gravitational binding energy of particles orbiting close to a black hole horizon. Thus, besides the electromagnetic extraction of rotational energy as alluded to above, an important vehicle is gravitational binding of accreting matter. The gravitational energy is converted into radiation when free particles fall into lower-energy centrally bound states in the formation of the accretion disk around the black hole. From a whole host of such observations and deductions, we conclude that galactic centers contain objects of tens of millions of solar masses. They must be black holes, because no other known object could be so massive and so small.

[13] When stars pass close enough to a massive black hole, tidal forces rip them apart, producing streams of debris that then swirl around and ultimately get swallowed by the black hole.

7.3 Black hole thermodynamics and Hawking radiation

The "no-hair" theorem suggests that black holes, being characterized only by mass, spin, and charge, are extraordinarily simple entities. One might conclude that a black hole has vanishing entropy ($S = 0$). This immediately runs into a contradiction: the process of matter falling into a black hole would then be an entropy-decreasing process! In the early 1970s, Jacob Bekenstein (1947–) pointed out that a black hole must have nonvanishing entropy proportional to its horizon area: $S \propto A^*$. Here we shall restrict ourselves to the simplest case of a Schwarzschild black hole, whose area is simply proportional to M^2:

$$A^* = 4\pi r^{*2} = 16\pi c^{-4} G_N^2 M^2. \tag{7.24}$$

As matter falls into a black hole, the horizon area (and hence the entropy) is clearly ever-increasing, as $(M + dM)^2 > M^2 + dM^2$. The area- and entropy-increasing theorem implies that while two black holes can join to make a bigger black hole, one black hole can never split into two, because $M_1^2 + M_2^2 < (M_1 + M_2)^2$.

7.3.1 Laws of black hole mechanics and thermodynamics

There is in fact a deep analogy between the laws of black hole mechanics and the laws of thermodynamics. After noting the surface gravity of a black hole,[14]

$$\sigma^* = \frac{G_N M}{r^{*2}} = \frac{c^4}{4 G_N M}, \tag{7.25}$$

we list the four laws of black hole mechanics:

0th law: The surface gravity σ^* has the same value everywhere on the event horizon.

1st law: The change in mass of a black hole is proportional to the surface gravity times the change in area:

$$dM = \frac{\sigma^*}{8\pi G_N} \, dA^*. \tag{7.26}$$

2nd law: The surface area of the event horizon of a black hole can only increase, never decrease.

3rd law: It is impossible to lower the surface gravity to zero through any physical process.

> **Exercise 7.3** Change of BH mass is proportional to BH surface gravity and change of area
>
> *Use the definition of BH surface gravity (7.25) and area (7.24) to derive the mass/area relation shown in (7.26).*

These laws of black hole mechanics are closely analogous to the four laws of thermodynamics:

0th law: The temperature T of a system in thermal equilibrium has the same value everywhere in the system.

1st law: The change in energy of a system is proportional to its temperature times the change in entropy: $dE = T \, dS$.

2nd law: The entropy of a system can only increase, never decrease.

3rd law: It is impossible to lower the temperature of a system to zero through any physical process.

[14] The surface gravity of a Schwarzschild black hole is the limit of the weight (per unit mass) of a stationary object near the event horizon, as measured by a distant observer (perhaps holding the object on a long massless string). You can see from (7.25) that it is analogous to the Newtonian acceleration at the event horizon.

Thus we have the following correspondence between black hole physics and the laws of thermodynamics:

Black holes		Thermodynamics
Mass M	$\Longleftrightarrow$	Energy E
Surface gravity at horizon σ^*	$\Longleftrightarrow$	Temperature T
Area of event horizon A^*	$\Longleftrightarrow$	Entropy S

This correspondence, except for that between mass and energy, is apparently mysterious. A black hole is a piece of spacetime geometry, not a container of gas and liquid—why should there be such a correspondence to a thermodynamical system? So far, all one can say is that these two sets of laws appear similar. This suggests the proportionality of entropy to area ($S \propto A^*$) and temperature to surface gravity ($T \propto \sigma^*$), but we do not know their proportionality constants. Such questions were partially answered in 1973 by the discovery of Hawking radiation, from the application of quantum mechanics to black holes.

7.3.2   Hawking radiation: quantum fluctuation around the horizon

Here we shall offer some brief comments on the interplay between black holes and quantum physics. Any detailed discussion of these advanced topics is beyond the scope of this introductory exposition. Our purpose here is merely to alert the readers to the existence of a vast body of knowledge on such topics, which are at the forefront of current research.

The Planck scale   GR is a classical macroscopic theory. For a microscopic description, we would need to combine GR with quantum mechanics into a theory of quantum gravity. The natural scale for such a quantum description of gravity is the Planck scale.

Soon after his 1900 discovery of the eponymous Planck's constant $\hbar$ in fitting the blackbody spectrum, Max Planck (1857–1947) noted that a self-contained system of natural units of mass–length–time can be defined by various combinations of Newton's constant G_N (gravity), Planck's constant $\hbar$ (quantum theory), and the speed of light c (relativity). When we recall from Newtonian theory that $[G_N \cdot (\text{mass})^2 \cdot (\text{length})^{-1}]$ has units of [energy], and from relativistic quantum theory that the natural scale of [energy· length] is $\hbar c$, we can obtain the natural mass scale for quantum gravity, the Planck mass,

$$M_{\text{Pl}} = \left(\frac{\hbar c}{G_N} \right)^{1/2} . \tag{7.27}$$

From this, we can immediately deduce the other Planck scales:

$$\text{Planck energy } E_{\text{Pl}} = M_{\text{Pl}}c^2 = \left(\frac{\hbar c^5}{G_{\text{N}}}\right)^{1/2} = 1.22 \times 10^{22}\,\text{MeV},$$

$$\text{Planck length } l_{\text{Pl}} = \frac{\hbar c}{E_{\text{Pl}}} = \left(\frac{\hbar G_{\text{N}}}{c^3}\right)^{1/2} = 1.62 \times 10^{-35}\,\text{m},$$

$$\text{Planck time } t_{\text{Pl}} = \frac{l_{\text{Pl}}}{c} = \left(\frac{\hbar G_{\text{N}}}{c^5}\right)^{1/2} = 5.39 \times 10^{-44}\,\text{s},$$

$$\text{Planck temperature } T_{\text{Pl}} = \frac{E_{\text{Pl}}}{k_{\text{B}}} = \left(\frac{\hbar c^5}{G_{\text{N}}k_{\text{B}}^2}\right)^{1/2} = 1.42 \times 10^{32}\,\text{K},$$

$$(7.28)$$

where we have also used Boltzmann's constant k_{B} to define a natural temperature scale. Such extreme scales are utterly beyond the reach of any laboratory setup. (Recall that the rest energy of a nucleon is about 1 GeV and that the highest energies probed by the current generation of accelerators are on the order of 10^4 GeV.) The only natural phenomena that can reach such an extreme scale are the physical singularities in GR: the endpoints of gravitational collapse hidden inside black hole horizons and the origin of the cosmological big bang. It is expected that quantum gravity will modify such singularity features of GR.

Quantum field theory vs. quantum gravity   Quantum field theory (QFT) is the union of SR with quantum mechanics. Namely, SR describes classical fields (such as Maxwell's electromagnetic fields); a quantum description of fields (such as quantum electrodynamics) is a QFT. The quanta of a field are generally viewed as particles. For example, the quanta of an electromagnetic field are photons, the quanta of an electron field are electrons, etc. The central claim of QFT is that the essence of reality is a set of fields, subject to the rules of quantum mechanics and SR; all observed phenomena are consequences of the quantum dynamics of these fields. QFT is the natural language to describe interactions that include the possibility of particle creation and annihilation allowed by the relativistic energy and mass relation $E = mc^2$. QFTs of nongravitational interactions need not operate at the extreme Planck scale. Quantum gravity is the quantum theory of the gravitational field. As in other QFTs, the gravitational field has its quantum particle, the graviton. Gravitons interact with the energy–momentum 4-tensor as photons interact with the charge-current 4-vector. GR, with its geometric interpretation and warped spaces, must emerge[15] as the macroscopic (low-energy) limit of quantum gravity in the same way that Maxwell's electromagnetic theory emerges from quantum electrodynamics. But, as mentioned above, the natural scale of quantum gravity is the Planck scale; thus, in this context, all observable phenomena may be macroscopic. Gravity is too weak at our preferred scales to reveal its quantum nature.

[15] There are suggestions that space and time themselves are emergent concepts from quantum gravity.

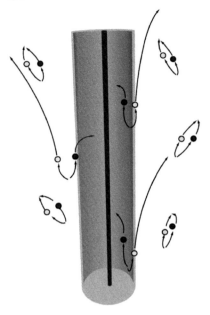

Figure 7.7 *Spacetime diagram of Hawking radiation. The cylinder is the (2+1)-dimensional worldsheet of the Schwarzschild horizon, with the time axis in the vertical direction. Quantum fluctuation causes particles and antiparticles to pop in and out of the vacuum. If one of the particles is inside the horizon and has negative energy, the other one of the pair can reach infinity with a positive energy. The black hole thereby radiates.*

[16] During a fluctuation, one cannot locate any particle to such precision. If one insists on a classical picture with exact particle locations, one can say that after the particles' creation, one of them, during the fluctuation time Δt, travels across the event horizon—or, in quantum mechanical language, one particle tunnels across the horizon.

[17] Recall that just as time and space intervals are components of the same vector (ct, x^i), so are energy and momentum $(E/c, p^i)$.

Hawking radiation

The surprising theoretical discovery by Stephen Hawking (1942–) that a black hole can radiate (contrary to the general expectation that nothing can escape from it) was made in the context of a quantum description of particle fields in a background Schwarzschild geometry. That is, the relevant theoretical framework involves only a *partial* unification of gravity with quantum theory: while the fields of photons, electrons, etc., are treated as quantized systems, gravity is still described by the classical (nonquantum) theory of GR. Thus, the relevant theoretical framework is quantum field theory in a curved spacetime.

The quantum uncertainty principle of energy and time, $\Delta E \Delta t \gtrsim \hbar$, implies that processes can temporarily violate energy conservation, provided they do so for a sufficiently short time interval Δt. Such quantum fluctuations turn empty space into a medium with particle and antiparticle pairs appearing and disappearing. In normal circumstances, such energy-nonconserving processes cannot survive on a macroscopic timescale. Hence the temporarily created and destroyed particles are called virtual particles. However, Hawking showed that if such random quantum fluctuations take place near the event horizon of a black hole, the virtual particles can become real because in such a situation energy conservation can be maintained permanently.

Qualitative explanation Consider the simplest quantum fluctuation: the creation of a particle–antiparticle pair from the vacuum. If the pair are to persist and not promptly annihilate back into the vacuum, energy conservation requires that

$$0 = E(\infty) + \tilde{E}(\infty). \tag{7.29}$$

If both particles could reach $r = \infty$, then $E(\infty)$ and $\tilde{E}(\infty)$ would be their energies measured by observers at infinity. Such energies must be positive, so the equality (7.29) cannot be satisfied. However, if this quantum fluctuation takes place sufficiently close to the event horizon of a black hole, one particle can be outside (and eventually travel to $r = \infty$) and the other particle can be inside[16] the event horizon (and fall into the singularity); see Fig. 7.7. Recall our discussion of the causal structure of the event horizon. The roles of space and time are interchanged at the $r = r^*$ Schwarzschild surface. In the same way, the energy component[17] inside the horizon takes on the properties of momentum; in particular, it is possible for the energy of a particle to be negative. If $\tilde{E}(\infty)$ is negative, then the conservation relation (7.29) can be satisfied on a macroscopic timescale. To a distant observer, the black hole emits a particle with positive energy, while losing a corresponding amount by swallowing its partner with negative energy. This is known as the Hawking effect or Hawking radiation.

Result obtained from QFT in curved spacetime Any radiation field, whether quantum or classical, can be decomposed into plane waves. The (Fourier) coefficients of expansion obey the simple harmonic equation. Thus radiation can be viewed as a collection of harmonic oscillators.[18] A QFT treats these field oscillators according to quantum mechanics, so their energy spectra have discrete, evenly spaced levels, states with a particular number of energy quanta (particles). These quantum states are related by the so-called raising and lowering ladder operators. Thus QFT provides a natural language to describe the creation and annihilation of particles. In a curved spacetime, the natural decomposition is still into plane waves, but they are plane waves with respect to the coordinates that encode the curvature. Near the horizon of a Schwarzschild black hole, for example, the Kruskal–Szekeres coordinates are convenient. The vacuum (i.e., the lowest-energy state) of such a field system, when viewed by a distant observer in flat spacetime, is a state containing a distribution of particles. In this way, Hawking obtained the particle distribution result

[18] For a simple discussion, see (Cheng, 2013, Sections 3.1 and 6.4).

$$|\Psi|^2 = \langle n \rangle = (e^{2\pi cE/\hbar\sigma^*} - 1)^{-1}. \tag{7.30}$$

This has the form of a thermal number distribution, $\langle n \rangle = (e^{E/k_B T} - 1)^{-1}$, with k_B being Boltzmann's constant. In this way, one can identify the black hole's surface gravity σ^* of (7.25) as temperature:

$$k_B T = \frac{\hbar\sigma^*}{2\pi c} = \frac{\hbar c^3}{8\pi G_N M} = \frac{\hbar c}{4\pi r^*}. \tag{7.31}$$

Namely, due to quantum effects in the surrounding space, a black hole of surface gravity σ^* should radiate particles as a perfect blackbody of temperature T, with (7.31) giving us the desired proportionality constant relating σ^* to T. Once derived, this expression for thermal energy appears reasonable on dimensional grounds. It is a relativistic quantum effect, hence the presence of $\hbar c$, which has the dimensions of [energy · length]; the only lengthscale available is the Schwarzschild radius r^*. Equivalently, this Hawking thermal energy can be expressed in terms of the natural quantum gravity unit of Planck energy:

$$\frac{k_B T}{E_{Pl}} = \frac{T}{T_{Pl}} = \frac{l_{Pl}}{4\pi r^*} = \frac{M_{Pl}}{8\pi M}. \tag{7.32}$$

In short, Hawking radiation shows that black holes radiate like blackbodies; smaller and hotter black holes should evaporate completely.

Black hole entropy From the expression for temperature and noting that $E = Mc^2$, we can immediately deduce the black hole's entropy through the definition $dS = T^{-1} dE$ (cf. Exercise 7.4). In this way, we find that the entropy is indeed proportional to the horizon area A^* of (7.24):

$$\frac{S}{k_B} = \frac{A^*}{4l_{Pl}^2}, \tag{7.33}$$

where l_{Pl}^2 is the Planck length squared. This is a shocking result! Entropy is an extensive variable, so one would expect it to be proportional to the volume, not the area, of the system. This result inspired a proposal that in quantum gravity the description of a volume of space is somehow encoded on its boundary; for instance, all the information about a black hole is encoded on its event horizon. Much like Einstein's proposal of the equivalence principle in GR, this idea has been elevated to a fundamental principle of quantum gravity: the holographic principle.[19]

[19] The holographic principle was first proposed by Gerard 't Hooft (1946–) and later formulated in the context of string theory by Leonard Susskind (1940–) and others.

Exercise 7.4 From Hawking temperature to the proportionality of black hole entropy to area

---

By a simple integration of $dS = T^{-1} dE$, derive the proportionality of black hole entropy and area shown in (7.33).

Deciphering the meaning of black hole entropy

One of the great achievements of Ludwig Boltzmann (1844–1906) was to show that the second law of thermodynamics was amenable to precise mathematical treatment. The macroscopic notion of entropy S could be related to a counting of the corresponding microscopic states, the complexions W:

$$S = k_B \ln W. \tag{7.34}$$

What would be the microscopic statistical description of a black hole that corresponds to the just-obtained entropy? The traditional approach would suggest that to do this counting, we need a microscopic theory of quantum gravity. In fact, black hole entropy would provide a check on the viability of any proposed quantum gravity theory. Currently, the most developed theory is superstring theory. Indeed, string theorists have been greatly encouraged by some success in recovering the entropy (7.33) by counting the number of ways a black hole can be formed in superstring theory. Nevertheless, it is still not a total success; the black holes being studied can be properly described only as black-hole-like entities. We are still far from having a realistic quantum description of a black hole.

[20] The reader interested in finding out more about such an approach may wish to start with (Jacobson 1995), (Padmanabhan 2010), and (Verlinde 2010).

[21] As in the case of a rubber band, we are ultimately more interested in the atomic/molecular explanation of its contracting force.

Entropic gravity? Some have suggested[20] that the entropy result is even more fundamental, that gravity is an entropic force. Just like the restoring force of a rubber band, which can be viewed as resulting from entropy maximization, the gravitational force can be derived (by reverse engineering) from extremizing entropy based on some conjectured principle governing the quantized spacetime. But this may not be such an interesting approach if one is ultimately interested in the detailed microscopic description of spacetime that quantum gravity promises to provide.[21]

Review questions

1. What does it mean that the Schwarzschild surface at $r = r^*$ is only a coordinate singularity?

2. What is the event horizon associated with a black hole?

3. At the $r = r^*$ surface of a Schwarzschild black hole, the proper time is finite, while the coordinate time is infinite. To what time measurements do we refer in these two descriptions? In terms of light frequency, what is an alternative (but equivalent) description of this phenomenon of infinite coordinate time dilation?

4. An event horizon is a null 3-surface. What is a null surface? Why does it allow particles and light to traverse only in one direction?

5. What is the basic property of the time coordinate in the advanced EF coordinate system that allows the lightcone to tip over smoothly inward across the $r = r^*$ surface? Answer the same question for the retarded EF system (with outward-tipping lightcones). Such properties of the coordinates allow their respective spacetime diagrams to display the black hole and white hole solutions. What is a white hole?

6. The effective potential for a particle in Schwarzschild spacetime has the form

$$V_{\text{eff}} = -\frac{A}{r} + \frac{Bl^2}{r^2} - \frac{Cl^2}{r^3} \qquad (7.35)$$

with positive coefficients A, B, and C. Use this expression to explain why, unlike in the Newtonian central force problem, a particle can spiral into the center even with nonzero angular momentum $l \neq 0$.

7. Black holes are linked with many of the most energetic phenomena observed in the cosmos. What is the energy source associated with a black hole that can power such phenomena?

8. List three or more astrophysical phenomena that are thought to be associated with black holes.

9. Explain why one expects that stars with a final mass $> 3M_\odot$ after they have burnt out must undergo gravitational collapse into black holes?

10. There is a correspondence between black hole physics and the laws of thermodynamics. While it is not surprising that black hole mass behaves like energy, what properties of the black hole behave like temperature and entropy?

11. Hawking radiation is understood in the context of quantum mechanics applied to black hole physics. But we say it only represents a partial union of quantum theory and GR. Why so?

8

The General Relativistic Framework for Cosmology

- Cosmology treats the universe as one physical system, whose fundamental elements are galaxies. On the present cosmic scale, the only relevant interaction among galaxies is gravity. The general theory of relativity thus provides the natural framework for the study of cosmology.

- Lemaître and Hubble each discovered independently that the universe is expanding, which suggests strongly that everything was once concentrated in a tiny region of extremely high density. The observational estimate of the age of the universe is 12–15 billion years.

- The mass density of (nonrelativistic) matter in the universe, expressed in units of critical density, is $\Omega_M \simeq 0.32$. There is strong evidence that most of this matter does not shine. Out of the matter made up of ordinary atoms (the baryonic matter) $\Omega_B \simeq 0.05$, the luminous part is only $\Omega_{lum} \simeq 0.005$; the rest of the baryonic matter is interstellar and intergalactic media. This still leaves a nonrelativistic, nonluminous, nonbaryonic component $\Omega_{DM} \simeq 0.27$, an exotic dark matter that has no electromagnetic interaction.

- The universe, when observed on distance scales over 300 Mpc, is homogeneous and isotropic. Such a spacetime is said to obey the cosmological principle. It is described by the Robertson–Walker metric in comoving coordinates (the cosmic rest frame).

- The dynamics of the universe are determined by the Friedmann equations, which follow from the Einstein equation with a Robertson–Walker geometry on one side and an ideal fluid source on the other. From the density and pressure of the fluid universe, these equations determine the dynamic scale factor $a(t)$ and the curvature constant k of the metric. The Friedmann equations have simple quasi-Newtonian interpretations.

- Einstein introduced the cosmological constant in his field equation to obtain a static cosmic solution. The cosmological constant may be viewed as a constant vacuum energy density with a negative pressure, which gives rise to a repulsive force that increases with distance. A universe dominated by vacuum energy expands exponentially.

A College Course on Relativity and Cosmology. First Edition. Ta-Pei Cheng.
© Ta-Pei Cheng 2015. Published in 2015 by Oxford University Press.

Cosmology is the study of the whole universe as a physical system:

What is its matter/energy content? How is this content organized?
What is its history? How will it evolve in the future?

Cosmology describes a universe now composed of galaxies spread around evenly at the largest distance scales. On the present cosmic scale, the only relevant interaction among galaxies is gravitation; all galaxies are accelerating under their mutual gravity. Thus the study of cosmology depends crucially on our understanding of the gravitational interaction. For a gravitational system with characteristic mass M and length R, we have learned that the dimensionless ratio $\psi = G_N M/Rc^2$ determines whether Einstein's theory is required or a Newtonian description would suffice. The effects of GR are small for a typical stellar system. On the other hand, we have also considered black holes, astronomical objects so compact that their size is comparable to their Schwarzschild radius: $\psi_{bh} = O(1)$. For the case of cosmology, the mass density ρ is very low. Nevertheless, the distance involved is so large that the total mass ($M = \rho R^3$) is so great that $\psi_{cosmo} = G_N \rho R^2/c^2 = O(1)$. Thus, to describe events on cosmic scales, we must use GR.

The solution of Einstein's equation describes the whole universe, because it describes the whole spacetime. Soon after completing his papers on the foundation of GR, Einstein applied his new theory to cosmology. In 1917, he published his paper "Cosmological considerations in the general theory of relativity," showing that GR can describe an unbounded homogeneous mass distribution.[1] Since then almost all cosmological studies have been carried out in the framework of GR.

8.1 The cosmos observed

Less than a hundred years ago, it was commonly held that the universe was static, infinite in age, and infinite in extent. Actually we could have already noted that this cosmic picture was contradicted by observation: the night sky is dark!

Olbers' paradox

What is the expected brightness of the night sky in a static, infinite universe? The intrinsic brightness of a star, its luminosity $\mathcal{L}$, is its emitted power $\mathcal{L} = dE/dt$. Thus its observed brightness (its flux or radiant power per area) at a distance r is

$$f(r) = \frac{dE}{dt \cdot A} = \frac{\mathcal{L}}{4\pi r^2}. \tag{8.1}$$

The total brightness B of the night sky is the sum of the received flux from all the stars in the universe. So if the stars have uniform luminosity $\mathcal{L}$ and number density n, we integrate over all space (with spherically symmetric volume element $dV = 4\pi r^2\, dr$) to get

$$B = \int nf(r)\, dV = n\mathcal{L} \int_0^\infty dr = \infty. \tag{8.2}$$

[1] Einstein was very much influenced by the teaching of Ernst Mach (1838–1916) in his original motivation for GR. His theory ultimately was unable to incorporate the strong version of Mach's principle (cf. Sidenote 2 in Chapter 1) that the total inertia of any single body is determined by its interaction with all the bodies in the universe. Nevertheless, he viewed his 1917 cosmology paper as the completion of his GR program, because it showed that the matter in the universe determines the entire geometry of the universe.

Namely, while the flux contribution of each star decreases like $1/r^2$, the number of stars (in the distance interval $(r, r + dr)$) increases like r^2; thus the total contribution to brightness from those stars at distances near r would be independent of r. Distant stars out to infinity would therefore make an unbounded contribution, so the total diverges. It is difficult to avoid infinite brightness (or at least a sky hot/bright enough to disrupt the burning of stars) in a static, infinite universe. One might argue that stars have finite angular sizes, and the above calculation assumes no obstruction by foreground stars, or one might argue that perhaps interstellar dust diminishes the intensity of light traveling a long distance, etc. These arguments do not help, because over time the foreground stars and dust particles would be heated and radiate as much as they absorb. Thus Olbers' paradox suggests that our universe is not infinite and static.

In this section, we shall discuss the large recessional motion of galaxies, estimates of the universe's age, and various mass contents. Before doing so, let us first present some of the cosmic distance scales relevant to modern cosmology.

Cosmic distance scales

People long believed that our universe was static, partly because the then-known universe essentially consisted only of our own galaxy, the Milky Way, and the observed stellar motion was relatively slow. The commonly used astronomical distance unit is the parsec (pc), defined as the distance to a star having a parallax of one arcsecond for a baseline of one astronomical unit.[2] It is the characteristic distance between stars. For example, the distance from the sun to its nearest stellar neighbor is 1.2 pc. Our Milky Way is a galaxy containing something like 10^{11} stars in the form of a disk $\simeq 30$ kpc (kiloparsecs) across and $\simeq 2$ kpc in thickness. Galaxies are distributed in a hierarchical structure. The Milky Way and about thirty other galaxies form our Local Group, spanning a distance of about 1 Mpc (megaparsec), whose largest member is the Andromeda Galaxy 0.7 Mpc away. Our Local Group is part of the Virgo Cluster (5 Mpc in size, comprising two thousand galaxies), which in turn is part of our Local Supercluster having a dimension of 50 Mpc. There are other superclusters and voids of comparable size. But on distance scales greater than 300 Mpc, the universe appears to be homogeneous.

[2] A parsec is defined as $1\,\mathrm{pc} \equiv 1\,\mathrm{AU}/1''$. The astronomical unit (AU) is the average radius of the earth's orbit around the sun, 1.50×10^{11} m. An arcsecond ($''$) is one-sixtieth of an arcminute ($'$), which in turn is one-sixtieth of a degree ($°$), so $1'' = \pi/(180 \cdot 60 \cdot 60) \simeq 4.85 \times 10^{-6}$ (radians). Therefore $1\,\mathrm{pc} \simeq 1\,\mathrm{AU}/4.85 \times 10^{-6} \simeq 2.06 \times 10^5\,\mathrm{AU} \simeq 3.09 \times 10^{16}\,\mathrm{m} \simeq 3.26$ light-years (lyr).

8.1.1   The expanding universe and its age

Astronomers have devised a series of techniques to estimate distances ever farther into space. Each new one, although less reliable, furthers our reach into the universe. During the period 1910–1930, this cosmic distance ladder was extended beyond 100 kpc. The great discovery was made that our universe consists of a vast collection of galaxies, each resembling our own. One naturally tried to study the motions of these newly discovered island universes[3] by using the Doppler effect. A galaxy's visible spectrum typically has absorption lines from the relatively cool outer layers of its stars. The observed wavelength of a particular galactic

[3] In early times, a galaxy was often called an "island universe." Such a description of the distant nebulae originated from Immanuel Kant in 1755.

absorption line, λ_{rec}, will differ from the wavelength of the same line produced and measured in the laboratory, λ_{em}. Such a wavelength shift,

$$z \equiv \frac{\lambda_{rec} - \lambda_{em}}{\lambda_{em}}, \tag{8.3}$$

is related to the emitter's movement (relative to the receiver) by the Doppler effect[4] (cf. (3.54)), which for nonrelativistic motion can be stated as

$$z = \frac{\Delta\lambda}{\lambda} \simeq \frac{V}{c}, \tag{8.4}$$

where V is the recession velocity of the emitter (away from the receiver).

The discovery of Hubble expansion

A priori, for a collection of galaxies, one expects a random distribution of velocities and hence of wavelength shifts: some positive (redshift) and some negative (blueshift). This is more or less true for the Local Group. But spectroscopic measurements of some 40 galaxies, taken by Vesto Slipher (1875–1969) over a 10-year period at Arizona's Lowell Observatory, showed that all, except a nearby few in the Local Group, were redshifted. Edwin Hubble (1889–1953) at the Mount Wilson Observatory in California, then correlated these redshift results with the (less easily measured) distances to the galaxies. He found that the redshift was proportional to the distance D to the light-emitting galaxy. In 1929, Hubble announced his result:

$$z = \frac{H_0}{c}D, \tag{8.5}$$

or, substituting in the Doppler interpretation of (8.4),

$$V = H_0 D. \tag{8.6}$$

Namely, we live in an expanding universe.[5] On distance scales greater than 10 Mpc, all galaxies obey Hubble's relation, receding from us with speed linearly proportional to distance. However, the distance actually traveled by light reaching us from remote galaxies will be affected by the dynamics of an expanding universe, as we will see in (10.17). The diagram plotting the luminosity distance to the emitting galaxy (discussed in Box 8.2) versus the redshift, called a Hubble plot, played a central role in Hubble's original discovery as well as in the discovery 70 years later of an accelerating universe (see Figs. 10.5 and 10.6).

The proportionality constant H_0, the Hubble constant, is the recession speed per unit separation (between the receiving and emitting galaxies). Obtaining an accurate value of H_0 has been a great challenge, as it requires one to ascertain great cosmic distances.[6] Only recently have several independent methods begun to yield consistent results. The convergent value[7] is

$$H_0 = (67.80 \pm 0.77 \, \text{km/s})/\text{Mpc}, \tag{8.7}$$

[4] While one traditionally attributes Doppler effects to the relative motion of the source and observer in a given inertial frame, the GR interpretation of the cosmological redshift, as we shall see in (8.32), is that it is due to the expansion of space.

[5] The first to interpret Slipher's redshift result as indicating an expanding universe was the Dutch mathematician and astronomer Willem de Sitter (1872–1934), who was among the first researchers to apply GR to cosmology. His model, the de Sitter universe (discussed in Section 8.4.2), is devoid of matter content, but dominated by a vacuum energy.

[6] See comments in Box 8.1.

[7] Throughout Chapters 8–10, we shall quote the cosmological parameters presented as the "best fit values" by the Planck Collaboration (2014).

where the subscript 0 stands for the present epoch, $H_0 \equiv H(t_0)$. An inspection of Hubble's relation (8.6) shows that $1/H_0$ has the dimension of time; this Hubble time has the value of

$$t_H \equiv H_0^{-1} \simeq 14.46 \pm 0.16 \, \text{Gyr}. \tag{8.8}$$

For comparison, we note that geologists estimate the age of the earth to be around 4.6 Gyr (gigayears). We likewise define a Hubble length $l_H \equiv c t_H \simeq 4400 \, \text{Mpc}$.

Although Hubble is often cited as the discoverer of the expanding universe, this is not an entirely accurate attribution. Hubble deduced the Hubble relation, mainly from Slipher's redshift data, but he never advocated the idea of an expanding universe. It was the Belgian priest and astrophysicist Georges Lemaître (1894–1966) who in 1927 first wrote down Hubble's relation in the context of an expanding universe model based on GR[8] and compared (8.5) with observational data.

[8] See Section 8.2.2 and the historical comments in (Peebles 1984, pp. 23–30).

The same Hubble's law at every galaxy

That all galaxies are receding from us may lead one to suppose erroneously that our location is the center of the universe. The correct interpretation is just the opposite. The key observation is that the Hubble law (8.6) is a linear relation between distance and velocity at each cosmic epoch. Namely, their proportionality factor H_0 is independent of distance and velocity. We will show that such a relation holds consistently for all observers in all galaxies. Namely, observers in every galaxy see all the other galaxies receding according to Hubble's law. In fact, the Hubble relation follows naturally from a straightforward extension of the Copernican principle: neither our earth, nor our sun, nor our galaxy has a privileged position in the universe.

Let us write Hubble's law in vector form:

$$\mathbf{v} = H_0 \mathbf{r}. \tag{8.9}$$

Thus, a galaxy G located at position $\mathbf{r}$ (as in Fig. 8.1) recedes from us (at the origin O) at velocity $\mathbf{v}$ proportional to $\mathbf{r}$. Now consider an observer on another galaxy O' located at $\mathbf{r}'$. According to Hubble's law, O' must recede from us at velocity

$$\mathbf{v}' = H_0 \mathbf{r}', \tag{8.10}$$

with the same Hubble constant H_0. The difference of these two equations yields

$$\mathbf{v} - \mathbf{v}' = H_0 (\mathbf{r} - \mathbf{r}'). \tag{8.11}$$

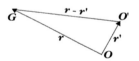

Figure 8.1 *Relative positions of a galaxy G with respect to two observers located at two other galaxies, O and O'.*

But $\mathbf{r} - \mathbf{r}'$ and $\mathbf{v} - \mathbf{v}'$ are the respective location and velocity[9] of G as viewed from O'. Since $\mathbf{v}$ and $\mathbf{v}'$ are in the same direction as $\mathbf{r}$ and $\mathbf{r}'$, the vectors $\mathbf{v} - \mathbf{v}'$ and $\mathbf{r} - \mathbf{r}'$ must also be parallel. That is, the relation (8.11) is just Hubble's law applied to the observer on galaxy O'. Clearly, such a deduction would fail if

[9] This relationship shown in Fig. 8.1 is specific to flat spaces, but (8.32) will show that Hubble's law is valid in Robertson–Walker spacetime, which describes our universe.

the velocity–distance relation, at a given cosmic time, were nonlinear (i.e., if H_0 depended on position and/or on velocity).

An often-noted 2D analog[10] of an expanding universe is a balloon inflating. Every point on its surface recedes from every other point as the balloon enlarges.

The big bang beginning of the universe and the estimate of its age

If all the galaxies are now rushing away from each other, presumably they must have been closer in the past. Unless there was some new physics involved (cf. Box 8.1), one could extrapolate back in time to a moment, the big bang, when all objects were concentrated at one point of infinite density. This is taken to be the origin of the universe, as first suggested in 1927 by Lemaître, who called it the "primeval atom."

How much time has passed since this fiery beginning? What is the age of our universe? As the universe expands, its matter and energy content should slow this expansion through mutual gravitational attraction, leading to a monotonically decreasing expansion rate $V(t) = D(t)/t$ of the distance $D(t)$ between two galaxies—a decelerating universe. Only in an empty universe do we expect the expansion rate V to be constant throughout its history: $V(t) = D(t)/t = V_0$. In that case, the age t_0 of the empty universe is given by the Hubble time,

$$[t_0]_{\text{empty}} = \frac{D_0}{V_0} = \frac{1}{H_0} = t_{\text{H}}. \tag{8.12}$$

For a decelerating universe, the expansion rate would have been larger in the past: $V(t) > V_0$ for $t < t_0$. Thus the age of the universe would be shorter than that of the empty universe: $t_0 < t_{\text{H}}$. Nevertheless, we shall often use the Hubble time as a rough benchmark value for the age of the universe.

Olbers' paradox is resolved in our expanding universe, because the received light flux from distant galaxies is diminished by redshift, and because the cosmic age is not infinite, so the integral in (8.2) no longer diverges.

Phenomenologically, we can estimate the age of the universe from observational data. For example, from astrophysical calculations, we know the relative abundance of nuclear elements produced in a star. Since they have different decay rates, their present relative abundance will differ from its initial value. This difference is a function of time. From the decay rates and the initial and observed relative abundances, we can estimate the time elapsed since the star's formation. Typically, such calculations give the ages of the oldest stars to be around 13.5 Gyr. This gives only an estimate of time passed since stars were first formed, and thus only a lower bound for the age of the universe. However, we currently understand that stars started forming about a hundred million years after the big bang; thus such a lower limit still serves as a useful estimate of t_0.

Further insights into the universe's age comes from studying systems of 10^5 or so old stars known as globular clusters. These stars are located in the halo

[10] Another familiar analog, this one 3D, is a rising pudding whose embedded raisins separate like galaxies.

rather than the disk of our galaxy. The halo lacks the interstellar gas for star formation, so these stars must have been created in the early epochs after the big bang. Stars spend most of their lifetimes undergoing nuclear burning. From the observed brightness (flux) and the distance to the stars, one can deduce their intrinsic luminosity (energy output per unit time). From such properties, astrophysical calculations based on established models of stellar evolution yield the stars' ages:

$$12\,\text{Gyr} \lesssim [t_0]_{\text{gc}} \lesssim 15\,\text{Gyr}. \tag{8.13}$$

Box 8.1   The steady-state universe

We have already mentioned the difficulty of accurately deducing the Hubble constant, as it requires one to measure great cosmic distances. In fact, Hubble originally overestimated H_0 by a factor of more than three. The standard candles (sources of known intrinsic luminosity) he used were Cepheid variable stars, which have a direct relationship between their luminosity and pulsation period. As it turns out, there are different classes of Cepheids, with different luminosity–period relations. Hubble's initial erroneous calculation of the cosmic expansion rate brought about a significant underestimate of the cosmic age (with a Hubble time smaller than even the known age of the earth).

This cosmic age problem was one of the issues that prompted Hermann Bondi (1919–2005), Thomas Gold (1920–2004), and Fred Hoyle (1915–2001) to propose in the late 1940s an alternative cosmology, called the **steady-state universe** (SSU). They suggested that, consistent with the Robertson–Walker description of an expanding universe, cosmological quantities (aside from the scale factor)—the Hubble constant, spatial curvature, matter density, etc.—are all time-independent. Thus one of the motivations for the SSU was its strong aesthetic appeal: it obeys the perfect cosmological principle[11]—the universe is homogeneous not only in space but also in time. A steady state means that the universe did not have a hot beginning; it existed forever; hence there cannot be a cosmic age problem.

To have a constant mass density in an expanding universe requires the continuous, energy-nonconserving, creation of matter. To the advocates of the SSU, this spontaneous mass creation is no more peculiar than the creation of all matter and energy at the instant of the big bang. In fact, the name "big bang" was Hoyle's offhand description of the competing cosmology. The absence of a hot beginning led to the downfall of the SSU when cosmic thermal relics, particularly the cosmic microwave background (CMB) radiation, the afterglow of the big bang, were discovered in the 1960s.

[11] The cosmological principle will be discussed in Section 8.2.

8.1.2   Mass/energy content of the universe

Critical mass density   It is useful to express mass densities in terms of a benchmark value called the critical density ρ_c. For a universe with relative expansion rate given by the Hubble constant H, this critical density is

$$\rho_c \equiv \frac{3H^2}{8\pi G_N}. \tag{8.14}$$

A density can thus be expressed as a **density ratio parameter**

$$\Omega \equiv \frac{\rho}{\rho_c}, \tag{8.15}$$

which, as will be discussed in (8.49), determines the spatial geometry of the universe. Since the Hubble constant is a function of cosmic time, the critical density also evolves with time. We denote the values for the present epoch with the subscript 0. For example, $\rho(t_0) \equiv \rho_0$, $\rho_c(t_0) \equiv \rho_{c,0}$, $\Omega(t_0) \equiv \Omega_0$, etc. For the present Hubble constant H_0 given in (8.7), the critical density is

$$\rho_{c,0} = (0.87 \pm 0.04) \times 10^{-29} \text{ g/cm}^3 \tag{8.16}$$

or, equivalently, the critical energy density[12] is

$$\rho_{c,0}c^2 \simeq 0.78 \times 10^{-9} \text{ J/m}^3 \simeq 4.9 \text{ keV/cm}^3. \tag{8.17}$$

Exercise 8.1 Dimension of critical density

Check the right-hand side of (8.14) to ensure that it has the correct dimensions of [mass]/[volume].

[12] One may note that such a density (about one nucleon per 200 liters) is more dilute than the best vacuum one can achieve in a modern laboratory. In the natural unit system of quantum field theory, this critical energy density can be written as $\rho_c c^2 \simeq (2.5 \times 10^{-3} \text{ eV})^4/(\hbar c)^3$, where $\hbar$ is Planck's constant (over 2π) with $\hbar c \simeq 1.97 \times 10^{-5}$ eV·cm.

Baryonic matter

Regular matter is made up of atoms, which in turn are composed of nucleons (protons and neutrons) and electrons. Since a nucleon is about 2000 times more massive than an electron, the mass of ordinary matter resides mostly in nucleons, which in particle physics are classified as baryons; hence ordinary matter is referred to in cosmology as **baryonic matter**. Shining stars are composed of baryonic matter; their cosmic mass density ρ_{lum} can be deduced from the luminosity-to-mass ratio (established through some well-studied portions of the cosmos) and the luminosity density measured from all-sky surveys. The resultant density parameter of luminous matter is

$$\Omega_{lum} = \frac{\rho_{lum}}{\rho_{c,0}} \simeq 0.005. \tag{8.18}$$

There is also baryonic matter in the form of interstellar and intergalactic gas that does not shine, but is subject to electromagnetic interactions and can thus be detected by its absorption spectra.[13] As it turns out, we have methods to deduce directly the total baryonic abundance $\Omega_B = \Omega_{lum} + \Omega_{gas}$, regardless of whether it is now luminous or nonluminous. The light nuclear elements (helium, deuterium, etc.) were produced predominantly in the early universe at the cosmic time $O(10^2 \text{ s})$; cf. Section 9.2. Their abundance (in particular that of deuterium) is sensitive to the baryonic abundance. From such considerations, we find that

$$\Omega_B = 0.049 \pm 0.002, \tag{8.19}$$

which is confirmed by the CMB anisotropy measurements, as well as by gravitational lensing. We see that $\Omega_B \gg \Omega_{lum}$, which means that most of the ordinary matter does not shine. Theoretical studies, backed up by detailed simulation calculations, confirm this conclusion that most baryonic matter is in the form of neutral interstellar and intergalactic gas.

Dark matter

There is compelling evidence that the vast majority of matter does not have electromagnetic interactions. Hence we cannot see it through its emission or absorption of radiation. Its presence can, however, be deduced from its gravitational attraction. This **dark matter** is more than five times as abundant as baryonic matter, giving a total matter density of

$$\Omega_M = \Omega_B + \Omega_{DM} \simeq 0.05 + 0.27 \simeq 0.32. \tag{8.20}$$

Zwicky's discovery That there might be a significant amount of dark matter in the universe was first proposed in the 1930s by Fritz Zwicky (1898–1974). He noted that given the observed velocities of the galaxies, the combined mass of the visible stars and gases in the Coma Cluster is simply not enough to gravitationally hold them together.[14] He concluded that there must be a large amount of invisible mass, the dark matter, that holds this cluster of galaxies together. However, Zwicky's work did not attract close attention from the astronomical community for decades. The modern era began in the 1970s when the accumulated dynamical evidence for dark matter was again noted. For instance, Jeremiah Ostriker (1937–) and James Peebles (1935–) found[15] that a disk-like galaxy cannot be stable unless it is embedded in a roughly spherical dark mass distribution of comparable magnitude. Outside this inner portion (a sphere covering the visible disk), further evidence, gathered by them and by many other astronomers, all pointed to an even larger dark mass distribution. The inner and outer portions together appeared to have a dark matter mass almost ten times the mass that can be attributed to the visible part of the galaxy.

[14] The proper mathematical formulation uses the **virial theorem**, which statistically determines the mass needed to gravitationally contain the observed (average) motion. This is comparable to the much simpler situation of a spherical mass M, which cannot confine a particle having a velocity greater than the escape velocity $v_{esc} = \sqrt{2G_N M/r}$.

[15] (Ostriker and Peebles 1973)

Galactic rotation curve In the late 1970s and early 1980s, Vera Rubin (1928–) and W. Kent Ford (1931–) and their collaborators, using more sensitive

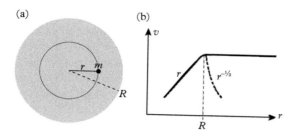

(a)   (b)

Figure 8.2 *Evidence of dark matter from galactic rotation curves. (a) A test mass m is displaced by a distance r from the center of a spherical mass distribution (shaded) of radius R. The circle passing through m represents the spherical Gaussian surface discussed prior to (8.21). (b) The solid line is the observed rotation velocity curve v(r). It does not fall as expected like $r^{-1/2}$ beyond R, the edge of the visible portion of a galaxy. Its constancy suggests a halo of dark matter extending much further out.*

techniques, were able to extend the rotation curve measurements[16] far beyond the visible edge of galaxies. Primed by the supporting evidence assembled in the previous decade, the general astronomy community readily accepted these latest data as straightforward and definitive evidence for dark matter. Consider in Fig. 8.2(a) the gravitational force that a spherical mass distribution exerts on a mass m in a circular orbit of radius r around the center of a galaxy. Since the contribution from outside the Gaussian sphere (radius r) cancels out, only the interior mass $M(r)$ enters into the Newtonian formula for gravitational attraction. The object is held by this gravity in circular motion with centripetal acceleration v^2/r; hence

[16] The observed rotational velocity as a function of the radial distance from the galactic center. See, e.g., (Rubin et al. 1980).

$$v(r) = \sqrt{\frac{G_N M(r)}{r}}. \tag{8.21}$$

Thus, the tangential velocity inside a galaxy is expected to rise linearly with the distance from the center ($v \sim r$) if the mass density is approximately constant. For a mass orbiting outside the galactic mass distribution (radius R), the velocity is expected to decrease as $v \sim 1/\sqrt{r}$; see Fig. 8.2(b). The velocities of particles located at different distances (the rotation curves) can be measured through the 21 cm line of the hydrogen atoms. The surprising discovery was that beyond the visible portion of the galaxies ($r > R$), instead of falloff, $v(r)$ is observed to stay at its constant peak value $v(R)$ (as far out as it can be measured). The gravitational pull of the luminous matter cannot account for this; hence it constitutes evidence for the existence of dark matter.

Many subsequent studies of galactic rotation curves confirmed this discovery. The general picture has emerged of a disk of stars and gas embedded in a large halo of dark matter. In our simple representation (Fig. 8.3), we take the halo to be spherical. In reality, the dark matter halo may not be spherical, and its distribution may not be smooth. There are theoretical and observational grounds to expect that the dark matter spreads out as an inhomogeneous web, and that the visible stars congregate in its potential wells. For instance, a mammoth filament of dark matter stretching between two galaxy clusters has been detected through gravitational lensing; see, e.g., (Dietrich et al. 2012).

Cold vs. hot dark matter Neutrinos, having only weak interaction, are an example of dark matter. In fact, they are called **hot dark matter**, because (being of

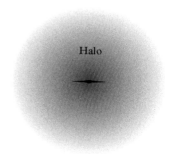

Halo

Figure 8.3 *The dark matter halo around the luminous portion of the galaxy.*

extremely low mass) they typically move with relativistic velocities. Studies of galactic structure formation have concluded that most dark matter is in the form of **cold dark matter** (CDM),[17] i.e., made up of heavy nonrelativistic particles. No known particles fit this description. It is commonly held that there are undiscovered elementary particles beyond those in the Standard Model that may be hundreds or thousands of times more massive than the nucleons. Such **WIMPs** (weakly interacting massive particles) may form the CDM content of the universe. Many physicists guess the most likely WIMP candidates are the lightest of the **supersymmetric** (SUSY) particles. Theories of the early universe contain mathematical symmetries that call for each known particle, such as the electron, photon, and quarks, to have a (yet undiscovered) massive partner. Just after the big bang, almost all of the heavy SUSY particles supposedly decayed or annihilated, but the lightest SUSY particles, unable to decay any further, were stable and survived. Nevertheless, all efforts to date in search of such dark matter particles have not produced a confirmed positive result.

In summary, cosmic matter has two components: the familiar baryonic matter and the dark matter. Dark matter is some unknown (collisionless) material that interacts very weakly compared with the usual electromagnetic interactions.[18]

[17] More recent research suggests a window of possibility for the dark matter to be **warm** and composed of **sterile neutrinos** with a mass in the region of $7\,\mathrm{keV}/c^2$, so that they were relativistic (hot) when produced in the early universe but became cold as the universe expanded and cooled down. A sterile neutrino is a spin-1/2 particle that does not even have a weak interaction charge but interacts (through mixing) with other (active) neutrinos.

[18] As Max Tegmark (2014) suggests, a more apt name would be "invisible matter," since it looks more transparent than dark and can pass through our bodies without our noticing.

Bullet Cluster as direct empirical evidence of dark matter In all the above discussion, the presence of dark matter was inferred from its gravitational effects. One might wonder whether it is possible to explain the observations, instead of by postulating the existence of a new form of matter, by modifying the law of gravity. Here we present a piece of evidence for dark matter that simply cannot be evaded in this way. Two images of the galaxy cluster 1E0657-558, nicknamed the **Bullet Cluster**, are displayed in Fig. 8.4. The picture (a) shows galaxies that make up a few percent of the cluster's mass; the picture (b) is an X-ray image from the Chandra telescope showing where the bulk of the hot gas is located. Superimposed on these two pictures are the mass contours derived from gravitational lensing. These contour lines show two mass peaks that, while more-or-less tracking the locations of observed galaxies, are situated far from the atomic gas. This observation cannot be explained by any modified law of gravity, but is consistent with the interpretation that the Bullet Cluster resulted from a collision

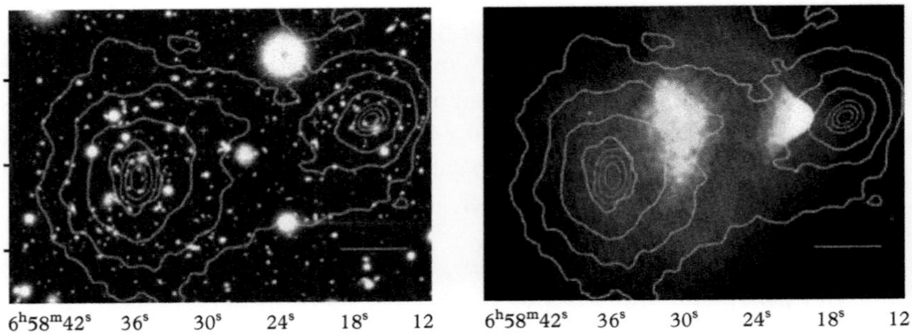

Figure 8.4 *Images of the Bullet Cluster 1E0657-558, produced by the collision of two galactic clusters, resulting in the separation of hot gas from dark matter (with the embedded stars and galaxies). (a) Optical image from the Hubble Space Telescope. (b) X-ray image from the Chandra telescope. The mass density contours from gravitational lensing reconstruction, superposed on both panels, show two mass peaks separated from the regions of hot gas. The figure is reproduced with permission from (Clowe et al. 2006). ©2006 American Astronomical Society.*

of two clusters of galaxies. The dark matter and (X-ray producing) baryonic gas were separated, because the weakly interacting dark matter passed through like a bullet, leaving the baryonic gases behind.

Matter densities in the universe: a summary Dark matter consists of non-relativistic particles having only gravitational and weak interactions. It does not emit or absorb electromagnetic radiation. Its presence can be inferred from the velocity distributions of gravitationally bound systems as well as from gravitational lensing. The most direct empirical evidence is the mass distribution of the Bullet Cluster. On an even larger scale, the abundance of dark matter can be quantified from the study of large cosmic structure and the CMB. All this leads to a total mass density just under a third of the critical density:

$$\Omega_M = \Omega_B + \Omega_{DM} \simeq 0.32. \tag{8.22}$$

The total baryonic (atomic) density can be deduced from the observed amounts of light nuclear elements and primordial nucleosynthesis theory (or from the observed CMB temperature anisotropy):

$$\Omega_B \simeq 0.05, \tag{8.23}$$

the bulk of which is in the intergalactic medium and has been detected through its electromagnetic absorption lines. Luminous matter (stars and galaxies) makes up only a small part of this baryonic density:

$$\Omega_B = \Omega_{gas} + \Omega_{lum} \quad \text{with} \quad \Omega_{gas} \gg \Omega_{lum} \simeq 0.005. \tag{8.24}$$

Thus the luminous matter associated with the stars we see in galaxies represents under 2% of the total mass content. The bulk of this total ($\Omega_M \simeq 0.32$) is nonrelativistic, weakly interacting dark matter ($\Omega_{DM} \simeq 0.27$), suspected to be composed mostly of exotic particles such as WIMPs. The exact nature of these exotic nonbaryonic CDM particles remains one of the major unsolved problems in physics.

8.2 The homogeneous and isotropic universe

The cosmological principle Modern cosmology usually adopts the working hypothesis called the **cosmological principle**: at each epoch (i.e., each fixed value of the cosmological time t), the universe is homogeneous and isotropic. It presents the same aspects (except for local irregularities) at each point; the universe has no center and no edge. The absence of any privileged location is, in essence, the fundamental generalization of the Copernican principle. The hypothesis has been confirmed by modern astronomical observations, especially of the cosmic back-ground radiation, which is homogeneous to one part in 10^5 in every direction. The observed irregularities (i.e., the structure) in the universe—stars, galaxies, clusters of galaxies, voids, etc.—are assumed to have arisen from gravitational clumping around some initial density unevenness. Various mechanisms for seed-ing such density perturbations have been explored. Most efforts have focused on the idea that quantum fluctuations in the earliest moments were inflated[19] to astro-physical size, thereby establishing the initial lumpiness required for the formation of the observed cosmic structure.

[19] The universe in its earliest moments passed through a phase of extraordinarily rapid expansion, the cosmic inflationary epoch, which will be discussed in Section 10.1.

8.2.1 Robertson–Walker metric in comoving coordinates

The cosmological principle depicts a universe filled with a cosmic fluid. The fun-damental particles of this fluid are galaxies. Each fluid element has a volume that contains many galaxies, yet is small compared with the whole universe. Thus, the motion of a cosmic fluid element is the average (smeared-out) motion of its con-stituent galaxies. This motion is determined by the gravitational interaction of the entire system, the self-gravity of the universe. This means that each element is in free fall; all elements follow geodesic worldlines. In reality, the random motion of the galaxies relative to the fluid (peculiar motion ignored by our homogeneous model) is small, on the order of 10^{-3}.

Comoving coordinate system   The picture of the universe discussed above allows us to pick a privileged coordinate frame, the comoving coordinate system, whose time and position coordinate are chosen to be

$$t \equiv \text{the proper time of each fluid element,}$$
$$x^i \equiv \text{the spatial coordinates carried by each fluid element.}$$

The comoving coordinate time can be synchronized over the whole homogeneous spacetime. A comoving observer flows in free fall with a cosmic fluid element—the Hubble flow. Because each fluid element carries its own fixed position label, the comoving coordinate frame may be regarded as the cosmic rest frame. But we must remember that in GR the coordinates do not directly measure distance, which is a path-dependent[20] combination of the coordinates and the metric as in (5.14). As we shall detail below, this comoving coordinate system describes the expansion of the universe (with all galaxies rushing away from each other), not by changing position coordinates, but by an increasing metric. Since the metric describes the geometry of space, we are modeling the expanding universe not as something exploding in space, but as the expansion of space itself.[21]

[20] This is an important point to keep in mind. There is no well-defined distance or length (or even interval or proper time) between two separate (not infinitesimally close) points in a curved space (or space-time) unless you specify a path (e.g., the geodesic, the radial path, or a stationary worldline).

[21] The recession of galaxies away from each other is described in comoving coor-dinates as the expansion of space. But in other coordinate systems, galaxies could be moving in space. GR allows for all these equivalent descriptions.

The Robertson–Walker metric   Just as Schwarzschild showed that spherical symmetry restricts the metric to $g_{\mu\nu} = \text{diag}(g_{00}, g_{rr}, r^2, r^2 \sin^2 \theta)$ with only two scalar functions, g_{00} and g_{rr}, here we discuss the geometry resulting from the cosmological principle for a homogeneous and isotropic universe. The resulting metric, when expressed in comoving coordinates, is the Robertson–Walker metric.[22] Because the coordinate time is chosen to be the proper time of the fluid elements, we must have $g_{00} = -1$. The fact that spacelike slices for fixed t can be defined means that the spatial axes are orthogonal to the time axes: $g_{0i} = g_{i0} = 0$. This implies that the 4D metric that satisfies the cosmological principle is block-diagonal:

[22] Named after the American physicist H.P. Robertson (1903–1961) and the English mathematician A.G. Walker (1909–2001).

$$g_{\mu\nu} = \begin{pmatrix} -1 & 0 \\ 0 & g_{ij} \end{pmatrix}, \qquad (8.25)$$

where g_{ij} is the 3×3 spatial part of the metric. Namely, the invariant interval is

$$ds^2 = -c^2\, dt^2 + g_{ij}\, dx^i\, dx^j \equiv -c^2\, dt^2 + dl^2. \qquad (8.26)$$

Because of the requirement of the cosmological principle (i.e., no preferred direction or position in the 3D space), the time dependence in g_{ij} must be given by an overall scale factor $a(t)$ with no dependence on any of the spatial coordinates:

$$dl^2 = a^2(t)\, d\bar{l}^2, \qquad (8.27)$$

where the scale factor is normalized at the present epoch by $a(t_0) = 1$, so the fixed (comoving) length element $d\bar{l}$ is the present proper spatial length.

One can picture the universe as a three-dimensional map like Fig. 8.5 with cosmic fluid elements labeled by the fixed (t-independent) comoving coordinates $\tilde{x}_i$. Time evolution enters entirely through the time dependence of the map scale, as $a(t)$ determines the size of the grids per (8.27) and is independent of the map coordinates $\tilde{x}_i$. As the universe expands (i.e., $a(t)$ increases), the relative distance relations (i.e., the shapes of things) are not changed.

Because the 3D space is homogeneous and isotropic, one naturally expects this space to have at a given time the same curvature everywhere. In Box 6.1, we wrote down the metric[23] for the 3D spaces with constant curvature in two coordinate systems with the dimensionless radial coordinates $\chi = r/R_0$ and $\xi = \rho/R_0$ and the differential solid angle $d\Omega^2 = d\theta^2 + \sin^2\theta\, d\phi^2$. We insert the intervals of the 3D spaces of constant curvature into (8.27):

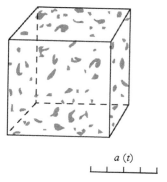

Figure 8.5 *A 3D map of the cosmic fluid with elements labeled by t-independent comoving coordinates $\tilde{x}_i$. The time dependence of any distance comes entirely from the t-dependent scale factor of the map.*

[23] While our deduction in Box 6.1 of the metric for a 3D space of constant curvature is only heuristic, it can be derived from the cosmological principle assumption of homogeneity and isotropy (constant scalar curvature); see (Cheng 2010, Section 14.4.1).

- Equation (6.16) for the comoving polar coordinates (χ, θ, ϕ):

$$dl^2 = R_0^2 a^2(t)\left[d\chi^2 + k^{-1}(\sin^2\sqrt{k}\chi)\, d\Omega^2\right]. \qquad (8.28)$$

- Equation (6.17) for the comoving cylindrical system (ξ, θ, ϕ):

$$dl^2 = R_0^2 a^2(t)\left(\frac{d\xi^2}{1 - k\xi^2} + \xi^2\, d\Omega^2\right). \qquad (8.29)$$

As discussed in Box 6.1, the curvature signature k can take the values $\pm 1, 0$, with $k = +1$ for a positively curved 3-sphere called a **closed universe**; $k = -1$ for a negatively curved 3-pseudosphere, an **open universe**; and $k = 0$ for a 3D flat[24] (Euclidean) space, a **flat universe**. For $k = +1$, the constant distance scale R_0 is the present radius of the 3-spherical universe, whose size varies with $a(t)$. As expected from GR theory, the mass/energy content of the universe determines its geometric form and (as we will see) its dynamical scale.

8.2.2 Hubble's law follows from the cosmological principle

[25] Namely, the length of the shortest path from emitter to receiver on the spatial surface of fixed time t.

In an expanding universe with a space that may be curved, we must be very careful in any treatment of distance. Nevertheless, the time dependence of distance measurement is simple. The (proper) distance[25] from a comoving emitter at $\tilde{x}_{em}$ to a comoving receiver at $\tilde{x}_{rec}$ at cosmic time t, according to (8.27), is

$$D(\tilde{x}_{em}, \tilde{x}_{rec}, t) = a(t) D(\tilde{x}_{em}, \tilde{x}_{rec}, t_0), \tag{8.30}$$

where $D(\tilde{x}_{em}, \tilde{x}_{rec}, t_0)$ is the fixed (comoving) distance at the present time t_0. This implies a proper relative velocity of

$$V(t) = \frac{dD(t)}{dt} = \frac{\dot{a}(t)}{a(t)} D(t). \tag{8.31}$$

Evidently, the relative velocity of two comoving points is proportional to the separation between them. This is Hubble's law, (8.6), with Hubble's constant identified as

$$H(t) = \frac{\dot{a}(t)}{a(t)}, \tag{8.32}$$

so $H_0 = \dot{a}(t_0)$. Recall that the overall scale factor in the spatial part of the Robertson–Walker metric follows from our imposition of the homogeneity and isotropy conditions. The result in (8.31) confirms our expectation that in any geometric description of a dynamic universe that satisfies the cosmological principle, Hubble's law emerges automatically. This result was obtained by Lemaître and, independently, by Hermann Weyl (1885–1955),[26] before Hubble's announcement in 1929.

Redshift and the scale factor   The scale factor $a(t)$ is the key to describing the time evolution of the universe. In fact, because $a(t)$ is usually a monotonic function, it can serve as a kind of cosmic clock. How can the scale factor be measured? The observable quantity that has the simplest relation to $a(t)$ is the redshift of a light signal.

The spectral shift, according to (8.3), is

$$z \equiv \frac{\Delta\lambda}{\lambda_{em}} = \frac{\lambda_{rec}}{\lambda_{em}} - 1. \tag{8.33}$$

We expect that the wavelength (in fact any length) scales as $a(t)$:

$$\frac{\lambda_{\text{rec}}}{\lambda_{\text{em}}} = \frac{a(t_{\text{rec}})}{a(t_{\text{em}})}. \tag{8.34}$$

For signals we receive in the present epoch, $a(t_0) = 1$, so we have the relation

$$1 + z = \frac{1}{a(t_{\text{em}})}. \tag{8.35}$$

For example, at redshift $z = 1$, the universe was half as large as at present. In fact a common practice in cosmology is to refer to the redshift of an era instead of its cosmic time. For example, the photon decoupling time, when the universe became transparent to light (cf. Section 9.3), is said to occur at redshift $z_\gamma = 1100$, etc.

Box 8.2 Hubble plot of luminosity distance vs. redshift

The Hubble diagram (or Hubble plot) is a graph of the distance to the emitting source against its redshift. We will use Hubble diagrams (Figs. 10.5 and 10.6) in Section 10.2 to explain the discovery that the expansion of the universe is accelerating. The distance plotted is not the proper distance D_{p} but the luminosity distance D_{L}, which is more directly related to observational measurements. In a static universe, the measured flux (energy per unit time per cross-sectional area) is related to the proper distance from the source as in (8.1):

$$f_{\text{static}} = \frac{\mathcal{L}}{4\pi D_{\text{p}}^2}, \tag{8.36}$$

where $\mathcal{L}$ is the intrinsic luminosity (emitted power) of the source. Since the luminosity distance D_{L} is defined by

$$f \equiv \frac{\mathcal{L}}{4\pi D_{\text{L}}^2}, \tag{8.37}$$

we see that in a static universe $D_{\text{L}} = D_{\text{p}}$. In an expanding universe, this observed flux, $f \sim dE/(dt \cdot A)$, is reduced by a factor of $(1 + z)^2$: one power of $(1 + z)$ comes from energy reduction due to lengthening wavelength of the emitted light and another power from the increasing time interval between emission and reception. Let us explain:

1. The ratio of the emitted energy to that received over a given coordinate time period is the inverse of the corresponding wavelengths:[27]

$$\frac{E_{\text{em}}}{E_0} = \frac{\lambda_0}{\lambda_{\text{em}}} = \frac{1}{a(t_{\text{em}})} = 1 + z, \tag{8.38}$$

continued

[27] Recall from elementary quantum mechanics that the energy of a photon is inversely proportional to its wavelength, which gets stretched like any proper length by the expanding universe. The same result could be derived using classical electromagnetic waves as well, see Sidenote 10 in Chapter 4.

Box 8.2 *continued*

where we have used $a(t_0) = 1$ and converted the scale factor to the redshift at emission with (8.35).

2. Wavelength is related to proper time rate: $\lambda \sim c\,dt$. Thus time intervals must be increased by $dt_0 = dt_{em}(1 + z)$, thereby reducing the proper energy transfer rate by another power of $(1 + z)$

Thus the observed flux in an expanding universe, in contrast to the static universe result (8.36), is given by

$$f = \frac{\mathcal{L}}{4\pi D_p^2 (1 + z)^2}.$$

(8.39)

A comparison of (8.37) and (8.39) leads to

$$D_L = (1 + z)D_p.$$

(8.40)

8.3 Time evolution in FLRW cosmology

Above, we have studied the kinematics of the standard model of cosmology. Requiring a homogeneous and isotropic space forces the spacetime to have the Robertson–Walker metric in comoving coordinates. This geometry is specified by a curvature signature k and a time-dependent scale factor $a(t)$. Here we study the dynamics of the model universe. Its unknown quantities k and $a(t)$ are determined by the Einstein field equation with the cosmic fluid as its matter/energy source. The theory of an expanding universe filled with matter was written down in 1922 by the Russian physicist and mathematician Alexander Friedmann (1888–1925). As noted before, Lemaître was the first to connect the expansion predicted by GR with the galactic redshift observation. Thus this GR model universe[28] is often referred to as the Friedmann–Lemaître–Robertson–Walker (FLRW) cosmology.

[28] This FLRW model does not include the cosmological constant to be discussed in Section 8.4.

8.3.1 Friedmann equations and their simple interpretation

The Einstein equation (6.37) relates spacetime geometry on one side to the mass/energy distribution on the other:

$$\underbrace{G_{\mu\nu}}_{a(t),\,k} = \underbrace{\kappa}_{G_N} \underbrace{T_{\mu\nu}}_{\rho(t),\,p(t)} .$$

(8.41)

The cosmic spacetime must have the Robertson–Walker metric in comoving coordinates. This means that the Einstein curvature tensor $G_{\mu\nu}$ on the geometry side of (8.41) is expressed[29] in terms of k and $a(t)$. The source term should also be compatible with a homogeneous and isotropic space. The simplest plausible choice for the energy–momentum tensor $T_{\mu\nu}$ is that of an ideal fluid, specified by two scalar functions: mass density $\rho(t)$ and pressure $p(t)$ (as discussed in Box 3.5). Thus specified, Einstein's equation yields the dynamical equations for cosmology, called the **Friedmann equations**. Because of the symmetry assumed from the cosmological principle, there are only two (of the original six)[30] independent component equations. One component of the Einstein equation becomes the **first Friedmann equation**,

$$\frac{\dot{a}^2(t)}{a^2(t)} + \frac{kc^2}{R_0^2 a^2(t)} = \frac{8\pi G_{\mathrm{N}}}{3}\rho, \qquad (8.42)$$

where R_0 is a constant distance factor, the present radius of curvature for a closed 3-spherical universe.[31] Another component yields the **second Friedmann equation**,

$$\frac{\ddot{a}(t)}{a(t)} = -\frac{4\pi G_{\mathrm{N}}}{c^2}\left(p + \frac{1}{3}\rho c^2\right). \qquad (8.43)$$

As one expects the pressure and density to be positive, the second derivative $\ddot{a}(t)$ is expected to be negative: the expansion should decelerate owing to mutual gravitational attraction among the cosmic fluid elements.

[29] Given the expression for the Robertson–Walker metric $g_{\mu\nu}$, this involves a straightforward, but tedious, calculation of the Christoffel symbols $\Gamma^{\mu}_{\nu\lambda}$ and then the Riemann tensor $R^{\mu}_{\nu\lambda\rho}$, which is contracted to obtain the Ricci tensor $R_{\mu\nu}$ and scalar R. They are finally combined into the Einstein tensor $G_{\mu\nu}$. See, e.g., (Wald 1984, p. 97) for a more complete derivation of the two Friedmann equations.

[30] See Sidenote 12 in Chapter 6.

[31] More generally, $K_0 = k/R_0^2$ is the present Gaussian curvature (6.8).

Exercise 8.2 Friedmann equations and energy conservation

Show that a linear combination of the Friedmann equations (8.42) and (8.43) leads to

$$\frac{d}{dt}(\rho c^2 a^3) = -p\frac{da^3}{dt}, \qquad (8.44)$$

which, having the form of the first law of thermodynamics $dE = -p\, dV$, is the statement of energy conservation. Since it has such a simple physical interpretation, we shall often use (8.44) instead of (8.43). By "Friedmann equation," one usually means the first Friedmann equation (8.42).

Equation-of-state parameter Because there are only two independent equations but three unknown functions $a(t), \rho(t)$, and $p(t)$, we need one more relation. This is provided by the **equation of state**, relating the pressure to the density of the system. Usually such a relation is rather complicated. However, since cosmology deals mostly with a dilute gas, the equation of state can usually be written simply as

$$p = w\rho c^2, \tag{8.45}$$

which defines the **equation-of-state parameter** w that characterizes the material content of the system. For example, for nonrelativistic matter, the pressure is negligibly small compared with the rest energy of the material particles, so $w_M = 0$; for radiation,[32] we have $w_R = \frac{1}{3}$, etc.

[32] The radiation energy density $u_R \equiv T^{00}$ can't be expressed as ρc^2 as in (3.84), because radiation has no mass density. We should properly write the equation of state (8.45) for radiation as $p_R = w_R u_R = u_R/3$. There are a number of ways to deduce $p_R = u_R/3$. For a thermodynamical derivation, see, e.g., (Cheng 2013, Section 3.5.2).

Critical density of the universe   The first Friedmann equation (8.42) can be rewritten as

$$-k = \left(\frac{\dot{a}R_0}{c}\right)^2 \left(1 - \frac{\rho}{\rho_c}\right), \tag{8.46}$$

where the critical density ρ_c is defined as in (8.14), with Hubble's constant $H = \dot{a}/a$ as in (8.32). Denoting the density ratio by $\Omega = \rho/\rho_c$ as in (8.15), we can rewrite (8.46) as

$$-\frac{kc^2}{\dot{a}^2 R_0^2} = 1 - \Omega. \tag{8.47}$$

In particular, at $t = t_0$ with $H_0 = \dot{a}(t_0)$ and $\Omega_0 = \rho_0/\rho_{c,0}$, it becomes

$$\frac{kc^2}{R_0^2} = H_0^2(\Omega_0 - 1). \tag{8.48}$$

This displays the GR connection between matter/energy distribution (Ω_0) and geometry (k):

$$\begin{array}{llll}
\Omega_0 > 1 & \longrightarrow & k = +1 & \text{closed universe;} \\
\Omega_0 = 1 & \longrightarrow & k = 0 & \text{flat universe;} \\
\Omega_0 < 1 & \longrightarrow & k = -1 & \text{open universe.}
\end{array} \tag{8.49}$$

Thus, the critical density is a cutoff value that separates a positively curved, high-density universe from a negatively curved, low-density universe. From the measured value $\Omega_0 = \Omega_{M,0} \simeq 0.3$ of (8.22), it would seem that we live in a negatively curved, open universe. In Section 8.4, we shall discuss this topic further, ultimately concluding that we need to modify Einstein's equation (by including the cosmological constant). As we shall discuss in Chapters 9 and 10, theoretical considerations, together with new observational evidence, now suggest that in fact we live in a $k = 0$ flat universe, whose energy/mass density just equals the critical value.

We now show that the Friedmann equations (8.42) and (8.43), although derived from the Einstein equation of GR, actually have rather simple Newtonian interpretations.

Applicability of Newtonian interpretation   We argued previously that GR must be the proper framework to study cosmology. Indeed, the Friedmann equations determine the scale factor $a(t)$ and the curvature signature k, which are the fundamental parameters of the curved spacetime describing gravity in a homogeneous and isotropic universe. Nevertheless, we shall presently show that these equations have rather simple Newtonian interpretations when supplemented by global geometric concepts. It is not contradictory that cosmological equations must be fundamentally relativistic and yet may have Newtonian interpretations. The cosmological principle states that every part of the universe, large or small, behaves in the same way. When we concentrate on a small region, a Newtonian description should be valid, because the gravity involved is not strong, and a small space can always be approximated by a flat geometry. Thus, we should be able to understand the cosmological equations with a Newtonian approach, carried out in an overall GR framework.

Quasi-Newtonian interpretations of the Friedmann equations

Equation (8.44) is clearly the statement of energy conservation as expressed in the form of the first law of thermodynamics $dE = -p\,dV$. That is, the change in energy E (with the energy per unit volume ρc^2) is equal to the work done on the system, whose volume V is proportional to a^3.

Next we shall demonstrate that the Friedmann equation (8.42) can be plainly interpreted as the usual energy balance equation for a central force problem: total energy is the sum of kinetic and potential energy. Recall that in our homogeneous and isotropic cosmological models, we ignore any peculiar motion of the galaxies relative to the cosmic fluid (i.e., any changes in the galaxies' comoving coordinates). The only dynamic effect we need to consider is the change in separation due to the change of the scale factor $a(t)$. Namely, the only relevant dynamical question is the time dependence of the separation between any two fluid elements, the Hubble flow.

To be specific, let us consider a cosmic fluid element with mass m (i.e., an element composed of a collection of galaxies) in the homogeneous and isotropic fluid (Fig. 8.6) at a distance $r(t) = a(t)r_0$ (cf. (8.30)) from an arbitrarily selected comoving coordinate origin O. We wish to study the effect of gravitational attraction on this element m by the whole fluid, which may be treated as spherically symmetric[33] around O. The gravitational attractions due to all the mass outside the sphere (radius r) mutually cancel. To understand this, you can imagine the outside region to be composed of a series of concentric spherical shells; the interior gravitational field inside each shell vanishes. This is a familiar Newtonian result. Here we must use GR, because the gravitational attraction from the mass shells at large distances is not Newtonian. But it turns out that this familiar nonrelativistic solution is also valid in GR; it follows from Birkhoff's theorem (stated at the end of Section 6.4.1). Consequently, the mass element m feels only the total mass M inside the sphere.

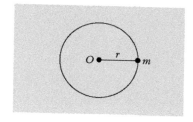

Figure 8.6 *The effect on the separation r between two galaxies due to the gravitational attraction by all the mass in the cosmic fluid: the net force on m is as if all the mass inside the sphere were concentrated at the center O. All the mass outside the sphere has no net effect.*

[33] Homogeneity and isotropy implies spherical symmetry with respect to every point.

This is a particularly simple central force problem, as we have only the radial motion, $\dot{r} = \dot{a}r_0$. The energy balance equation (6.88) has no orbital angular momentum term:

$$\frac{1}{2}m\dot{r}^2 - \frac{G_N mM}{r} = E_{tot},\tag{8.50}$$

which may be rewritten as

$$\frac{1}{2}m\dot{a}^2 r_0^2 - \frac{G_N mM}{ar_0} = E_{tot}.\tag{8.51}$$

The expansion of the universe implies an increasing $a(t)$ and hence an increasing (i.e., less negative) potential energy. This necessarily implies a decreasing $\dot{a}(t)$, namely, a slowdown of the expansion. The total mass inside the (flat-space) sphere is $M = \rho 4\pi a^3 r_0^3/3$, so we get

$$\dot{a}^2 - \frac{8\pi G_N}{3}\rho a^2 = \frac{2E_{tot}}{mr_0^2}.\tag{8.52}$$

This is to be compared with the Friedmann equation (8.42). Remember that this calculation is carried out for an arbitrary center O. Different choices of a center correspond to different values of r_0 and thus different E_{tot}. The assumption of a homogeneous and isotropic space leads to the GR conclusion that the right-hand side of (8.52) is a constant with respect to any choice of O. This constant is related to the curvature parameter (k) on the left-hand side of (8.42). We accordingly identify E_{tot} by

$$\frac{2E_{tot}}{mr_0^2} = \frac{-kc^2}{R_0^2}.\tag{8.53}$$

Exercise 8.3 Newtonian interpretation of the second Friedmann equation

---

Show that the second Friedmann equation (8.43) can be viewed as the $F = ma$ equation in a Newtonian system (i.e., a system with negligible pressure $p = 0$).

Mass density also determines the fate of the universe

With this interpretation of the Friedmann equation (8.42) as the energy balance equation, and with the identification of the curvature signature k as being proportional to the total energy in (8.53), it is clear that the value of k, and hence that of the density ρ through (8.46), determines not only the geometry of the 3D space but also the fate of cosmic evolution.[34] Recall that in the central force problem, whether the motion of the test mass m is bound or not is determined

[34] Even though this connection between density and the outcome of time evolution is broken when Einstein's equation is modified by a cosmological constant term as discussed in Section 8.4, the following presentation still provides some insight into the meaning of the critical density.

by the sign of the total energy E_{tot}. An unbound system allows $r \to \infty$, where the potential energy vanishes and the total energy is given by the kinetic energy, which must be positive: $E_{tot} > 0$ so $k < 0$, cf. (8.53). Equation (8.50) also shows that the sign of E_{tot} reflects the relative size of the positive kinetic energy compared with the negative potential energy. We can phrase this relative size question in two equivalent ways:

1. Compare the kinetic energy with a given potential energy: the escape velocity Given a source mass (i.e., the potential energy), one can determine whether the kinetic energy term (i.e., test particle velocity) is big enough to have an unbound system. To facilitate this comparison, we write the potential energy term in the form

$$G_N \frac{mM}{r} \equiv \frac{1}{2} m v_{esc}^2,$$ (8.54)

which defines the escape velocity

$$v_{esc} = \sqrt{\frac{2 G_N M}{r}}.$$ (8.55)

The energy equation (8.50) then takes the form

$$G_N \frac{mM}{r} \left(\frac{v^2}{v_{esc}^2} - 1 \right) = E_{tot}.$$ (8.56)

When $v < v_{esc}$, and thus $E_{tot} < 0$, the test mass m is bound and can never escape.

2. Compare the potential energy with a given kinetic energy: the critical mass Given the test particle's velocity (i.e., the kinetic energy), one can determine whether the potential energy term (i.e., the amount of mass M) is big enough to overcome the kinetic energy to bind the test mass m. We write the kinetic energy term as

$$\frac{1}{2} m \dot{r}^2 \equiv G_N \frac{mM_c}{r},$$ (8.57)

which defines the critical mass

$$M_c = \frac{r \dot{r}^2}{2 G_N} = \frac{a \dot{a}^2 r_0^3}{2 G_N}.$$ (8.58)

The energy equation (8.50) then takes the form

$$\frac{1}{2} m \dot{r}^2 \left(1 - \frac{M}{M_c} \right) = E_{tot}.$$ (8.59)

When $M > M_c$, and thus $E_{tot} < 0$, the test mass m is bound and can never escape.

The analogous question of whether, given an expansion rate $H(t)$, a test galaxy m is bound by the gravitational attraction of the cosmic fluid is determined by whether there is enough mass in the arbitrary sphere (on whose surface m lies) to prevent m from escaping completely. Namely, the value of $k \sim -E_{\text{tot}}$ determines whether the universe will expand forever or its expansion will eventually slow down and recollapse. Since the sphere is arbitrary, what matters is the density of the cosmic fluid. We will divide the critical mass (8.58) by the volume of the sphere and use the Hubble constant relation $H = \dot{a}/a$ of (8.32) to obtain

$$\rho_c = \frac{M_c}{V} = \frac{a\dot{a}^2 r_0^3 / 2 G_{\text{N}}}{4\pi a^3 r_0^3 / 3} = \frac{3H^2(t)}{8\pi G_{\text{N}}}, \tag{8.60}$$

which checks with the definition of ρ_c given in (8.14). Substituting E_{tot} from (8.53) into the energy balance equation (8.59) and replacing M/M_c by ρ/ρ_c or Ω yields the Friedmann equation as written in (8.46) and (8.47), respectively.

8.3.2   Time evolution of model universes

We now use the Friedmann equations (8.42), (8.44), and the equation of state (8.45) to find the time dependence of the scale factor $a(t)$ for a definite value of the curvature k. Although in realistic situations we need to consider several different energy/mass components $\rho = \Sigma_w \rho_w$ with their respective pressure terms $p_w = w \rho_w c^2$, we shall at this stage consider mostly single-component systems. To simplify the notation, we shall omit the subscript w in the density and pressure functions.

Scaling of the density function

Before solving for $a(t)$, we shall first study the scaling behavior of density and pressure as dictated by the energy conservation condition (8.44). Carrying out the differentiation in this equation, we have

$$\dot{\rho} c^2 = -3(\rho c^2 + p)\frac{\dot{a}}{a}, \tag{8.61}$$

which, after using the equation of state (8.45), turns into

$$\frac{\dot{\rho}}{\rho} = -3(1 + w)\frac{\dot{a}}{a}. \tag{8.62}$$

This can be solved by integrating $d\rho/\rho = -3(1 + w)\, da/a$ and then exponentiating the result to yield

$$\rho = \rho_0 a^{-3(1+w)}. \tag{8.63}$$

For a matter-dominated universe (MDU), $w_{\text{M}} = 0$, and for a radiation-dominated universe (RDU), $w_{\text{R}} = \frac{1}{3}$, with the respective densities scaling as

$$\rho_{\text{M}} = \rho_{\text{M},0} a^{-3} \quad \text{and} \quad \rho_{\text{R}} = \rho_{\text{R},0} a^{-4}. \tag{8.64}$$

The first equation displays the expected scaling behavior of an inverse volume. The second relation can be understood because radiation energy is inversely proportional to wavelength, and hence scales as a^{-1}, which is then divided by the volume factor a^3 to get the density. For the special case of negative pressure $p = -\rho c^2$ with $w = -1$, (8.63) leads to a constant energy density $\rho = \rho_0$ even as the universe expands. Einstein's cosmological constant and the newly discovered dark energy (discuss in Sections 8.4.1 and 10.2, respectively) seem to have just this property.

Model universe with $k = 0$

We proceed to solve (8.42) for the time evolution of the scale factor $a(t)$ for some simple situations. We first consider a class of model universes with a flat geometry $k = 0$. As we shall see, a spatially flat geometry is particularly relevant to the universe we live in. Substituting (8.63) into the Friedmann equation (8.42) yields

$$\left(\frac{\dot{a}}{a}\right)^2 = \frac{8\pi G_N}{3}\rho_0 a^{-3(1+w)} - \frac{kc^2}{R_0^2 a^2}, \tag{8.65}$$

which for the $k = 0$ case is

$$\left(\frac{\dot{a}}{a}\right)^2 = \frac{8\pi G_N}{3}\rho_0 a^{-3(1+w)}. \tag{8.66}$$

Assuming a power-law growth for the scale factor,

$$a(t) = \left(\frac{t}{t_0}\right)^x, \tag{8.67}$$

we have

$$\frac{\dot{a}}{a} = \frac{x}{t}. \tag{8.68}$$

With this power law assumption, we can immediately relate the age of the universe t_0 to the Hubble time $t_H = H_0^{-1}$:

$$H_0 \equiv \left(\frac{\dot{a}}{a}\right)_{t_0} = \frac{x}{t_0}, \quad \text{or} \quad t_0 = x t_H. \tag{8.69}$$

Furthermore, by substituting (8.68) and (8.67) into (8.66), we see that in order to match the powers of t on both sides, the following relation must hold:

$$x = \frac{2}{3(1 + w)}. \tag{8.70}$$

For the matter-dominated and radiation-dominated flat universes, (8.70), (8.67), and (8.69) lead to

$$\text{MDU} \;(w = 0): \quad x = \tfrac{2}{3}, \quad a = \left(\frac{t}{t_0}\right)^{2/3}, \quad t_0 = \tfrac{2}{3} t_{\text{H}},$$

$$\text{RDU} \;(w = \tfrac{1}{3}): \quad x = \tfrac{1}{2}, \quad a = \left(\frac{t}{t_0}\right)^{1/2}, \quad t_0 = \tfrac{1}{2} t_{\text{H}}. \tag{8.71}$$

We note from (8.70) that for the cosmic age to be less than the Hubble time $t_0 < t_{\text{H}}$ (i.e., $x < 1$), we must have $w > -\tfrac{1}{3}$; contrariwise, $t_0 > t_{\text{H}}$ if $w < -\tfrac{1}{3}$. Equation (8.70) also informs us that x is singular for the $w = -1$ cosmic fluid, which possesses, as we have already noted, negative pressure and constant energy density. Thus, in this case, the scale factor cannot be a power function of t; that is, the assumption (8.67) cannot apply.

We have considered the specific case of a flat geometry with $k = 0$. But we note that the result also correctly describes the early epoch even in a universe with curvature $k \neq 0$. This is so because in the $t \to 0$ limit, the curvature (total energy) term in the Friedmann equation (8.42) is negligible compared with the $\dot{a}^2$ (kinetic energy) term, which blows up as some negative power of the cosmic time.

Time evolution in single-component universes

We now consider the time evolution of toy universes with only one component of energy/matter but with no restriction on k. They can be treated as approximations to a more realistic multicomponent universe in which the energy content is dominated by one component. Plugging (8.64) for the scaling of RDU energy density into the Friedmann equation (8.42) yields

$$\dot{a}^2(t) = \frac{A^2}{a^2(t)} - \frac{kc^2}{R_0^2}, \tag{8.72}$$

where the constant $A = [(8\pi G_{\text{N}}/3)\rho_0]^{1/2}$. This has the solution for a RDU

$$a^2(t) = 2At - \frac{kc^2}{R_0^2} t^2. \tag{8.73}$$

Using the MDU scaling from (8.64) changes only the power of $a(t)$ under the constant A^2 in (8.72). The MDU solution is more complicated, but the qualitative behavior of $a(t)$ is essentially the same[35] as that depicted in Fig. 8.7 for an RDU. For subcritical density $\rho < \rho_{\text{c}}$, the expansion of the open universe ($k = -1$) will continue forever; for $\rho > \rho_{\text{c}}$, the closed universe ($k = +1$) will slow its expansion to a stop and then start to recollapse—all the way to $a = 0$ (the big crunch); for the critically dense flat universe ($k = 0$), the expansion will slow down, but not enough to stop. Thus for both the matter- and radiation-domination cases discussed above, the density ratio Ω (of density to the critical density) determines not only the geometry of the universe (whether it is positively or negatively curved) but also its fate (whether it expands forever or ultimately contracts).

[35] See (Wald 1984, pp. 98–102).

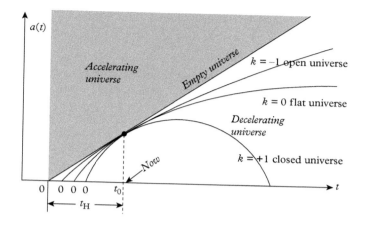

Figure 8.7 *Time dependence of the scale factor a(t) for open, flat, and closed universes. The qualitative features of these curves are the same for radiation- and matter-dominated universes. All models must have the same scale $a_0 \equiv 1$ and slope $\dot{a}_0$ at t_0 in order to match the observed Hubble constant $H_0 = \dot{a}_0/a_0$ at the present epoch. The origin of the cosmic time $t = 0$ (the big bang) is different for each curve (plotted thus to make the present times coincide for universes of different ages).*

8.4   The cosmological constant Λ

One of Einstein's great contributions to cosmology was his discovery that GR can accommodate a form of cosmic gravitational repulsion by including a new term in the field equation.

Before the late 1920s and the discovery by Lemaître and by Hubble of an expanding universe, just about everyone, Einstein included, believed that we lived in a static universe. (Recall that the then-observed universe consisted essentially of stars within the Milky Way galaxy.) In order to obtain a static universe solution from GR, Einstein altered his field equation to include a repulsion component. This could, in principle, balance the usual gravitational attraction to yield a static cosmic solution.

Einstein discovered that the geometry side (the metric's second derivatives) of his field equation (8.41) could accommodate an additional term. Like $G_{\mu\nu}$, it must be another symmetric rank-2 tensor with a vanishing covariant divergence constructed out of the metric.[36] He noted that the metric tensor $g_{\mu\nu}$ itself possessed all these properties, in particular $D^\mu g_{\mu\nu} = 0$, (11.27). Thus the geometric side of the field equation (8.41) can be modified by the addition of a new term proportional to $g_{\mu\nu}$:

$$G_{\mu\nu} + \Lambda g_{\mu\nu} = \kappa T_{\mu\nu}. \tag{8.74}$$

Such a modification will, however, alter its nonrelativistic limit to differ from Newton's equation $\nabla^2 \Phi = 4\pi G_N \rho$. As we shall explain, the new term induces a force that increases with distance. In order for this alteration to be compatible with known phenomenology, it must have a coefficient Λ so small as to be unimportant except on truly large cosmic scales. Hence, this additional constant Λ has come to be called the **cosmological constant**.

[36] The covariant derivative D_μ is the derivative that, when acting on a tensor, does not spoil its tensor transformation properties (i.e., it yields another tensor). D_μ can be expressed in terms of ordinary derivatives ∂_μ and Christoffel symbols (derivatives of the metric) as in (11.24)–(11.25). The requirement that the geometric side of Einstein's equation must be covariantly constant ($D_\mu [\hat{O}g]^{\mu\nu} = 0$) follows from energy/momentum conservation ($D_\mu T^{\mu\nu} = 0$) in a gravitational system. In Sections 11.1.1 and 11.2.3, we show that $D_\mu g^{\mu\nu} = 0$ and $D_\mu G^{\mu\nu} = 0$.

8.4.1 Λ as vacuum energy and pressure

The cosmological constant may be interpreted as a constant vacuum energy density and negative pressure. We shall presently show that energy conservation demands that such a negative pressure accompany a constant energy density. Moreover, the negative pressure gives rise to a cosmologically significant gravitational repulsion.

While we introduced this Λ term as an additional geometric term, we could just as well move it to the right-hand side of (8.74) and view it as an additional source term. In particular, when the regular energy–momentum tensor is absent ($T_{\mu\nu} = 0$ in the vacuum state), the field equation becomes

$$G_{\mu\nu} = -\Lambda g_{\mu\nu} \equiv \kappa T^{\Lambda}_{\mu\nu}. \tag{8.75}$$

Thus we interpret the new term as the vacuum's energy–momentum tensor,

$$T^{\Lambda}_{\mu\nu} = -\kappa^{-1}\Lambda g_{\mu\nu} = \frac{-c^4 \Lambda}{8\pi G_{\mathrm{N}}}\begin{pmatrix} -1 & \\ & g_{ij} \end{pmatrix}, \tag{8.76}$$

where we have displayed the Robertson–Walker metric of (8.25). Just as $T_{\mu\nu}$ for an ideal fluid depends on two functions of energy density ρ and pressure p, this vacuum energy–momentum tensor $T^{\Lambda}_{\mu\nu}$ can be similarly parameterized, cf. (3.84), by a vacuum energy density ρ_{Λ} and vacuum pressure p_{Λ}:

$$T^{\Lambda}_{\mu\nu} = \begin{pmatrix} \rho_{\Lambda} c^2 & \\ & p_{\Lambda} g_{ij} \end{pmatrix}. \tag{8.77}$$

From a comparison of (8.76) and (8.77), we see that ρ_{Λ} and p_{Λ} are related one another and to Λ. For a positive cosmological constant $\Lambda > 0$, the vacuum energy[37] per unit volume,

$$\rho_{\Lambda} = \frac{\Lambda c^2}{8\pi G_{\mathrm{N}}} > 0, \tag{8.78}$$

is a positive constant (in space and in time); the vacuum pressure,

$$p_{\Lambda} = -\rho_{\Lambda} c^2 < 0, \tag{8.79}$$

is negative, corresponding to an equation-of-state parameter $w_{\Lambda} = -1$ as defined in (8.45).

[37] In nonrelativistic physics, only the relative value of energy is meaningful; the motion of a particle with potential energy $V(x)$ is exactly the same as one with $V(x) + C$, where C is a constant. In GR, since the whole energy–momentum tensor is the source of gravity, the actual value of the energy makes a difference.

Consistency with energy conservation   What is a negative pressure? Consider the simple case of a piston chamber filled with ordinary matter and energy, which exerts a positive pressure by pushing out against the piston. If it is filled with this Λ energy, as in Fig. 8.8, it will exert a negative pressure by pulling in the piston. Physically, this is sensible, because its energy per unit volume $\rho_{\Lambda} c^2$ is constant, so the change in the system's energy is strictly proportional to its volume change: $dE = \rho_{\Lambda} c^2\, dV$. The system tends to lower its energy by contracting its volume (pulling in the piston). When we increase the volume of the chamber

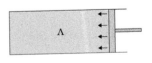

Figure 8.8 *The Λ energy in a chamber has negative pressure and thus pulls in the piston.*

$(dV > 0$ and hence $dE > 0$) by pulling out the piston, we have to do positive work to overcome the pulling by the Λ energy. Energy is conserved because the negative pressure $p < 0$ is just what is required by the first law of thermodynamics, $dE = -p\,dV$, when both dE and dV have the same sign. In fact, the first law clearly implies (8.79) if energy density is a constant, $dE = \rho c^2\,dV = -p\,dV$, so $p = -\rho c^2$; the pressure must equal the negative of the energy density.

Exercise 8.4

Use the Friedmann equation (8.44), which is the cosmological version of the first law of thermodynamics, to show that a constant ρ_Λ implies $w_\Lambda = -1$.

Λ as a gravitational repulsion that increases with distance   Consider the second Friedmann equation (8.43): $\ddot{a} \sim -(\rho c^2 + 3p)$. For an ideal matter fluid ($p_M = 0$ and $\rho_M c^2 > 0$), the right-hand side is negative, and we have decelerated expansion; for the Λ vacuum, $\ddot{a} \sim -(\rho_\Lambda c^2 + 3p_\Lambda) = +2\rho_\Lambda c^2$, resulting in an accelerated expansion. Perhaps it would be more instructive to look at the Newtonian limit of the field equation (6.40), which shows that not only mass/energy density but also pressure is a source of gravity. Plugging the expressions (8.78) and (8.79) for ρ_Λ and p_Λ into (6.40) yields $\nabla^2\Phi = -c^2\Lambda$, which has the solution $\Phi_\Lambda(r) = -\frac{1}{6}\Lambda c^2 r^2$. Between any two mass points, this potential corresponds to a repulsive force (per unit mass) that increases with separation $\mathbf{r}$,

$$\mathbf{g}_\Lambda = -\nabla\Phi_\Lambda = +\frac{\Lambda c^2}{3}\mathbf{r}, \tag{8.80}$$

in contrast to the familiar gravitational attraction $\mathbf{g}_M \sim -\mathbf{r}/r^3$. With this pervasive repulsion that increases with distance, even a small Λ can have a significant effect on truly large cosmic dimensions.

Box 8.3 The static universe and Einstein's "biggest blunder"

We now consider the Friedmann equations (8.42) and (8.43) with a nonvanishing cosmological constant:

$$\frac{\dot{a}^2 + kc^2/R_0^2}{a^2} = \frac{8\pi G_N}{3}(\rho_M + \rho_\Lambda), \tag{8.81}$$

$$\frac{\ddot{a}}{a} = -\frac{4\pi G_N}{c^2}\left[(p_M + p_\Lambda) + \frac{1}{3}(\rho_M + \rho_\Lambda)c^2\right]. \tag{8.82}$$

The right-hand side of (8.82) need not necessarily be negative, because of the presence of the negative pressure term $p_\Lambda = -\rho_\Lambda c^2$. For nonrelativistic matter ($p_M = 0$), we have

continued

Box 8.3 *continued*

$$\frac{\ddot{a}}{a} = -\frac{4\pi\,G_{\mathrm{N}}}{3}\,(\rho_{\mathrm{M}} - 2\rho_{\Lambda})\,. \qquad (8.83)$$

The static condition $\ddot{a} = 0$ (imposed by Einstein to reconcile GR cosmology with his belief that the universe was static) leads to the constraint

$$\rho_{\mathrm{M}} = 2\rho_{\Lambda} = \frac{\Lambda c^2}{4\pi\,G_{\mathrm{N}}}\,. \qquad (8.84)$$

The mass density ρ_{M} of Einstein's static universe is fixed by the cosmological constant. The other static condition $\dot{a} = 0$ implies, through (8.81), the relation (with $a = a_0 = 1$)

$$\frac{kc^2}{R_0^2} = 8\pi\,G_{\mathrm{N}}\rho_{\Lambda} = \Lambda c^2\,. \qquad (8.85)$$

Since the right-hand side is positive, we must have $k = +1$. Namely, the static universe has positive curvature (it is closed) and finite size. The radius of the universe is also determined, according to (8.85), by the cosmological constant: $R_0 = 1/\sqrt{\Lambda}$. Thus, the basic features of such a static universe, its density and size, are determined by the arbitrary input parameter Λ. Not only is this a rather artificial arrangement, but also the solution is, in fact, unstable. That is, a small variation will cause the universe to deviate from this static point. A slight increase in separation will cause the regular gravitational attraction to decrease and the vacuum repulsion to increase. This causes the separation to increase even further—the system blows up. A slight decrease in separation will increase the gravitational attraction and decrease the vacuum repulsion, causing the separation to decrease further until the whole system collapses.

Einstein's "biggest blunder"?

Having missed the chance to predict an expanding universe before its discovery,[38] Einstein came up with a solution that did not really solve the perceived difficulty (his static solution is unstable). It has often been said that later in life Einstein considered the introduction of the cosmological constant to be the biggest blunder of his life! This originated from a characterization by George Gamow in his autobiography (Gamow, 1970):

> *Thus, Einstein's original gravity equation was correct, and changing it was a mistake. Much later, when I was discussing cosmological problems with Einstein, he remarked that the introduction of the cosmological term was the biggest blunder he ever made in his life.*

[38] Einstein was initially hostile to Lemaître's expanding universe cosmology. Lemaître recalled Einstein saying to him: "Your calculations are correct, but your physics is atrocious" (Deprit 1984, p. 370). But once Einstein accepted the expanding universe discovery of Hubble, he quickly and publicly endorsed Lemaître's cosmology, helping its wider reception.

Then Gamow went on to say,

> *But this blunder, rejected by Einstein, is still sometimes used by cosmologists even today, and the cosmological constant Λ rears its ugly head again and again and again.*

What we can conclude for sure is that Gamow himself considered the cosmological constant "ugly" (because this extra term makes the field equation less simple). Generations of cosmologists kept on including it, because the quantum vacuum energy[39] gives rise to such a term, and there is no physical principle one can invoke to exclude it. In fact, the discovery of the cosmological constant as the source of a new cosmic repulsive force must be regarded as one of Einstein's great achievements. As it has turned out, Λ is the key ingredient of inflationary cosmology and of the cosmic dark energy.

[39] See Box 10.4 for a discussion of how the quantum vacuum state has an energy with just the properties of the cosmological constant.

8.4.2   Λ-dominated universe expands exponentially

Let us consider the behavior of the scale factor $a(t)$ in a model with $\Lambda > 0$ when the matter density can be ignored. In such a vacuum-energy-dominated universe, the behavior of the expansion rate $\dot{a}(t)$ is such that[40] we can neglect the curvature signature k term; (8.81) then becomes

[40] Not only is $a(t)$ increasing exponentially, so too is its derivative $\dot{a}(t)$.

$$\frac{\dot{a}^2}{a^2} = \frac{8\pi G_N}{3}\rho_\Lambda = \frac{\Lambda c^2}{3}. \tag{8.86}$$

Thus $\dot{a}$, the derivative of the scale factor a, is proportional to a itself.[41] Namely, we have the familiar rate equation. It can be solved to yield an exponentially expanding universe (called the **de Sitter universe**):

$$a(t_2) = a(t_1)e^{(t_2-t_1)/\Delta\tau} \tag{8.87}$$

with the time constant

$$\Delta\tau = \sqrt{\frac{3}{\Lambda c^2}} = \sqrt{\frac{3}{8\pi G_N \rho_\Lambda}}, \tag{8.88}$$

where we have expressed the cosmological constant in terms of the vacuum energy density $\rho_\Lambda c^2$ as in (8.78). The reason for this exponential expansion is easy to see: the repulsion is self-reinforcing. The energy density ρ_Λ is constant, so the more the space expands, the greater are the vacuum energy and the volume of negative pressure, causing the space to expand even faster. In fact, we can think of this Λ repulsive force as residing in the space itself, so as the universe expands, the push from this Λ energy increases as well.[42]

[41] Once we can neglect curvature (as well as any normal matter and radiation), the Hubble constant $H = \dot{a}/a$ is truly constant, as are the vacuum energy density and pressure. The universe reaches a steady state (in the sense of Hoyle's expanding steady-state universe).

[42] Because Λ represents a constant energy density, it will be the dominant factor $\rho_\Lambda \gg \rho_M$ at later cosmic time, as $\rho_M \sim a^{-3}$. This dominance means that it is possible for the universe to be geometrically closed ($\Omega > 1$ and $k = +1$), but not stop expanding (although, as the discussion leading up to (8.86) shows, the exponentially expanding 3-sphere becomes practically flat, and $\Omega \to 1$). Namely, with the presence of a cosmological constant, the mass/energy density Ω (hence the geometry) no longer determines the fate of the universe in a simple way. A universe with a nonvanishing Λ, regardless of its geometry, will expand forever unless the matter density is so large that the universe starts to contract before ρ_Λ becomes the dominant term.

Review questions

1. What does it mean that Hubble's law is a linear relation? What is the significance of this linearity? Support your statement with a proof.

2. What are the definition and value of the Hubble time t_H? Under what condition is it equal to the age of the universe t_0? In a universe full of matter and radiation energy, which of these two quantities would be greater ($t_H > t_0$ or $t_H < t_0$)? What is the lower bound for t_0 deduced from the observational data on globular clusters?

3. What are galactic rotation curves? What feature would we expect if the luminous matter were a good representation of the total mass distribution? What feature of the rotation curve told us that there was a significant amount of dark matter associated with galaxies and clusters of galaxies?

4. What is baryonic matter? The bulk of the baryonic matter resides in the intergalactic medium (IGM) and does not shine. Why do we not count it as part of dark matter?

5. What are the present values of the total mass density Ω_M, the luminous matter density Ω_{lum}, and the baryonic matter density Ω_B? From this estimate the dark matter density Ω_{DM}.

6. What is the cosmological principle? What are comoving coordinates?

7. Write out the Robertson–Walker metric for two possible coordinate systems. What is the input (i.e., the assumption) used in the derivation of this metric? What are the physical meanings of the scale factor $a(t)$ and the parameter k in this metric?

8. How is the epoch-dependent Hubble constant $H(t)$ related to the scale factor $a(t)$? Derive the relation between $a(t)$ and the redshift z.

9. Describe the relation of the Friedmann equation (8.42) to the Einstein equation and give its Newtonian interpretation. Why can we use nonrelativistic Newtonian theory to interpret this GR equation in cosmology? Also, in what sense is it only quasi-Newtonian?

10. In what sense can the critical density be understood as akin to the more familiar concept of escape velocity?

11. Why do we expect the energy density of radiation to scale as a^{-4}. Why should the energy of the universe be radiation-dominated in its earliest moments?

12. What is the equation-of-state parameter w for radiation? For nonrelativistic matter? If we know that the universe has a flat geometry, what is the time dependence of the scale factor $a(t)$ in a radiation-dominated universe (RDU)? In a matter-dominated universe (MDU)? How is the age of the universe t_0 related to the Hubble time t_H in an RDU and in an MDU? Justify the approximation that the age of our universe is two-thirds of the Hubble time.

13. Draw a schematic diagram showing the behavior of the scale factor $a(t)$ for various values of k in cosmological models (with zero cosmological constant). (It is suggested that all $a(t)$ curves be drawn to meet at the same point $a(t_0)$ with the same slope $\dot{a}(t_0)$.) Also mark the regions corresponding to a decelerating universe, an accelerating universe, and an empty universe.

14. Write out the Einstein equation with the Λ term. Why can such a term be interpreted as the vacuum energy–momentum source of gravity?

15. Demonstrate that the first law of thermodynamics implies a negative pressure in a system of constant positive energy density.

16. Solve the Newtonian field equation with $\nabla^2 \Phi$ equal to a negative constant to show that it yields a repulsive force that increases with distance. Give a simple argument that a cosmological-constant-dominated system expands exponentially.

Big Bang Thermal Relics

- The universe began hot and dense (the big bang), and thereafter expanded and cooled. The early universe underwent a series of phase changes between thermal equilibria—each one a cosmic soup composed of a different mix of particles. This cosmic history left behind telltale thermal distributions of relic particles: light nuclei, neutrinos, and photons.

- The observed abundances of the light nuclear elements (helium, deuterium, etc.) support the view that they are products of big bang nucleosynthesis around a hundred seconds after the big bang.

- When the universe was around 380 000 years old, photons ceased to interact with matter and began to pass freely though the universe. This primordial light remains today as the cosmic microwave background (CMB), with a blackbody spectrum of temperature $T = 2.725$ K.

- Similarly, at an earlier cosmic epoch, neutrinos decoupled to form a cosmic neutrino background. Thus the most abundant particles in our universe are relic photons and neutrinos, with a density of about 400 of each per cubic centimeter.

- The CMB is not perfectly uniform. Its dipole anisotropy is primarily determined by our motion in its rest frame (the comoving Robertson–Walker frame); higher multipoles contain much information about the geometry and the matter/energy content of the universe, as well as the initial density perturbation, from which grew the cosmic structure we see today.

During the epochs immediately after the big bang, the universe was much more compact, and the energy associated with the random motions of matter and radiation was much larger. Thus we say that the universe was much hotter. Space was filled with a plasma, in which various particles could be in thermal equilibrium through high-energy interactions. As the universe expanded, it also cooled. As thermal energy lowered, particles (and antiparticles) either disappeared through annihilation, combined into various composites of particles, or decoupled (i.e., ceased to interact) to become free particles. Consequently, there remain today different kinds of thermal relics left behind by the hot big bang.

One approach to studying the universe's history is to start with some initial state that may be guessed based on our knowledge of (or speculation about) particle physics. Then we can evolve this proposed universe forward in the hope of ending up with something like the observed universe today. That we can speak of the early universe with any sort of confidence rests with the idea that the universe passed through a series of equilibria. At such stages, its properties were determined, independently of the details of the interactions, by a few parameters such as the temperature, density, and pressure.

The thermodynamical investigation of cosmic history was pioneered by Richard Tolman (1881–1948). This approach of extracting observable consequences from big bang cosmology was vigorously pursued in the late 1940s and early 1950s by Alexander Friedmann's PhD student George Gamow (1904–1968) and his collaborators, Ralph Alpher (1921–2009) and Robert C. Herman (1914–1997).

Here, we shall first give an overview of the thermal history of the universe, in particular the scale dependence of radiation temperature.

9.1 The thermal history of the universe

[1] The Boltzmann equation relates the number of a particular species of particle to the difference between its production and disappearance rates.

Once again, it should be pointed out that the calculations carried out in this chapter are rather crude and for illustration only—to give us a flavor of how in principle cosmological predictions can be made. Realistic calculations would typically use the Boltzmann equation,[1] involving many reaction rates with numerous conditions.

[2] Except for photons, there is no strong theoretical ground to set the chemical potential μ to zero. Nonetheless, since there is nothing requiring a sizable μ, we shall for simplicity assume $|\mu| \ll k_B T$.

9.1.1 Scale dependence of radiation temperature

For the radiation component of the universe, we can neglect particle masses and chemical potentials[2] much smaller than $k_B T$ (where k_B is Boltzmann's constant). The number density (per unit volume) distributions with respect to the energy E (of the bosons (minus sign in $\pm$) and fermions (plus sign)) in a radiation gas are

[3] Particles with mass and spin s have $2s + 1$ spin states (e.g., spin-$\frac{1}{2}$ electrons have two spin states), but massless particles (e.g.,spin-1 photons or spin-2 gravitons) have only two spin states. Note that antiparticles are counted separately, so the electron and positron have four spin states between them. Moreover, Standard Model neutrinos have only one spin state, because only left-handed states interact; they come in three flavors, making six states of left-handed neutrinos and right-handed antineutrinos.

$$dn = \frac{g}{2\pi^2 (\hbar c)^3} \frac{E^2 \, dE}{e^{E/k_B T} \pm 1}, \tag{9.1}$$

where $\hbar$ is Planck's constant (with dimensions of energy×length for $\hbar c$) and g is the number of spin states[3] of the particles making up the radiation. We integrate to get the number densities of the bosons and fermions:

$$n_b = \frac{4}{3} n_f = \zeta(3) \frac{g}{\pi^2} \left(\frac{k_B T}{\hbar c} \right)^3, \tag{9.2}$$

[4] The Riemann zeta function is defined as

$$\zeta(n) = \sum_{l=1}^{\infty} \frac{1}{l^n} = \frac{1}{(n-1)!} \int_0^{\infty} dx \, \frac{x^{n-1}}{e^x - 1}.$$

In particular,

$$\zeta(2) = \pi^2/6 \simeq 1.645,$$
$$\zeta(3) \simeq 1.202,$$
$$\zeta(4) = \pi^4/90 \simeq 1.082.$$

where $\zeta(3) = 1.202$ is the Riemann zeta function.[4] We can derive the thermodynamic relation (the **Stefan–Boltzmann law**) between radiation energy density and temperature by integration, $u = \int E \, dn \sim T^4$:

$$\rho_R c^2 \equiv u_R = \frac{g^*}{2} a_{SB} T^4, \tag{9.3}$$

where the radiation energy density has been written as $\rho_R c^2$ (cf. Sidenote 31 in Chapter 8) and a_{SB} is the Stefan–Boltzmann constant,

$$a_{SB} = \frac{3!\zeta(4)}{\pi^2}\frac{k_B^4}{(\hbar c)^3} = \frac{\pi^2 k_B^4}{15(\hbar c)^3} = 4.722 \text{ keV m}^{-3} \text{ K}^{-4}. \tag{9.4}$$

We have summed the energy contributions of all the constituent radiation particles, so that g^* is the effective number of spin states of all the particles making up the radiation:

$$g^* = \sum_i (g_b)_i + \frac{7}{8}\sum_i (g_f)_i, \tag{9.5}$$

where $(g_b)_i$ and $(g_f)_i$ are the spin multiplicities of the ith species of boson and fermion radiation particles, respectively. The factor 7/8 reflects the different integral values for the fermion distribution, with a plus sign in (9.1), vs. the boson distribution, with a minus sign.

From the number and energy densities, we can also compute the average energies $\bar{E} = \rho_R c^2/n$ with $\bar{E}_b = \frac{6}{7}\bar{E}_f$ for bosons and fermions in the radiation. In particular, for photons,

$$\bar{E}_\gamma = \frac{3\zeta(4)k_B T}{\zeta(3)} = \frac{\pi^4 k_B T}{30\zeta(3)} = 2.701 k_B T. \tag{9.6}$$

Scale dependence of temperature Combining the Stefan–Boltzmann law (9.3) $\rho_R \sim T^4$ with our previously derived relation (8.64) for a radiation-dominated system $\rho_R \sim a^{-4}$, we deduce the scaling property for the radiation temperature:

$$T \propto a^{-1}. \tag{9.7}$$

This expresses, in precise scaling terms, our expectation that temperature is high when the universe is compact, so it cools as it expands. Under this temperature-scaling law, the total number distributions $dN = V\,dn$ (for some volume $V \sim a^3$) from (9.1) are unchanged. Because the radiation energy varies inversely with the wavelength, $E \sim \lambda^{-1} \sim a^{-1}$, the combinations $VE^2\,dE$ and $E/k_B T$ are invariant under scale changes. Thus, as the universe expands and the temperature falls, the form of the blackbody spectrum is maintained.

Remark 9.1 *In the context of the Newtonian interpretation of the cosmological (Friedmann) equations, we can understand energy conservation in an expanding universe as follows: while the total number of radiation particles $N = nV$ does not change during expansion, the total radiation energy ($\sim Nk_B T$) scales as a^{-1}. This loss of radiation energy as the scale a increases is balanced by the increase in the gravitational energy of the universe. The gravitational potential energy is also inversely proportional to the scale, but is negative. Thus, it increases (becomes less negative) with an increase in a.*

Example 9.1 Relation between radiation temperature
and cosmic time

The early universe was dominated by radiation, so it grew like $a \propto t^{1/2}$, cf. (8.71). We can drop the curvature (k) term in the Friedmann equation (8.42), and replace $\dot{a}/a$ by $(2t)^{-1}$, so that the radiation energy density is related to cosmic time by

$$\rho_{\mathrm{R}} c^2 = \frac{3}{32\pi} \frac{c^2}{G_{\mathrm{N}}} t^{-2}. \tag{9.8}$$

The left-hand side can be expressed in terms of the thermal energy by the Stefan–Boltzmann law (9.3):

$$\rho_{\mathrm{R}} c^2 = \frac{g^* \pi^2}{30} \frac{(k_{\mathrm{B}} T)^4}{(\hbar c)^3}. \tag{9.9}$$

Thus, in a radiation-dominated universe, time is related to temperature by

$$t = \left(\frac{45 \hbar^3 c^5}{16 \pi^3 g^* G_{\mathrm{N}} k_{\mathrm{B}}^4 T^4} \right)^{1/2} \tag{9.10}$$

$$= \frac{0.3012}{\sqrt{g^*}} \left(\frac{T_{\mathrm{Pl}}}{T} \right)^2 t_{\mathrm{Pl}} = \frac{3.26 \times 10^{20}\ \mathrm{K}^2 \cdot \mathrm{s}}{\sqrt{g^*} T^2},$$

where we have used the Planck time and temperature from (7.28). For an effective multiplicity[5] of $g^* = 10\frac{3}{4}$, this gives an easy-to-remember numerical relation between the cosmic time in seconds and temperature in kelvins:

$$t\,(s) \simeq \frac{10^{20}}{[T(K)]^2}. \tag{9.11}$$

From this estimate we can see that the big bang nucleosynthesis at temperature $T_{\mathrm{bbn}} \simeq 10^9$ K (cf. (9.21)) took place about a hundred seconds after the big bang: $t_{\mathrm{bbn}} = O(10^2)$ s.

[5] While electrons and positions contribute to the radiation, (9.5) gives $g^* = 2(\text{photons}) + \frac{7}{8} \times 4(\mathrm{e}^+\ \&\ \mathrm{e}^-) + \frac{7}{8} \times 6(\nu\ \&\ \bar{\nu}) = 10\frac{3}{4}$; (9.11) is good to $\simeq 1\%$. When the electrons and positrons mostly vanish at reheating when $k_{\mathrm{B}} T \simeq m_e c^2 \simeq 0.5\,\mathrm{MeV}$, g^* drops to $7\frac{1}{4}$. Moreover, reheating (Exercise 9.1) increases subsequent temperatures by a factor of $(11/4)^{1/3}$ from what they would be, so $t(T)$ must go up by a factor of $(11/4)^{1/3} \simeq 1.96$. Thus (9.11) is low at later epochs: $t(s) \simeq 2.37 \times 10^{20}/[T(K)]^2$.

9.1.2   Different thermal equilibrium stages

After the big bang (the inflationary epoch to be discussed in Chapter 10), the cooling of the universe allowed the existence of different mixtures of particles in the equilibrium plasma. Composites such as nucleons, nuclei, and atoms were able to form and survive. When the falling thermal energy $k_{\mathrm{B}} T$ could no longer produce various types of particle–antiparticle pairs, the antiparticles annihilated with particles and disappeared.[6] Particles eventually ceased to interact with (decoupled from) the cosmic soup.

Determining what reactions took place to maintain each thermal equilibrium involves dynamical calculations, taking into account reaction rates in an expanding and cooling medium. The basic requirement for a given particle interaction to be significant at a given epoch is that the time interval between particle scatterings be much shorter than the age of the cosmos. This can be expressed as the Gamow condition that the reaction rate Γ must be faster than the expansion rate of the universe as measured by Hubble's constant:

$$\Gamma \geq H. \tag{9.12}$$

The reaction rate Γ of a particle is the product of the number density n of the particles with which it interacts, their relative velocity v, and the reaction cross section σ:[7]

$$\Gamma = nv\sigma. \tag{9.13}$$

Particle velocity enters because nv is the flux of the interacting particles (the number of particles passing through unit area in unit time). The velocity distribution is determined by the thermal energy of the system. The cross section can be measured in a laboratory or predicted by theory. The condition for a transition to a new equilibrium phase is that $\Gamma = H$. Since the cosmic age $\sim H^{-1}$, one can think of this condition as requiring on average one interaction since the beginning of the universe. This condition can be used to solve for the thermal energy and the redshift value at which a new equilibrium stage started. These different thermal stages (ordered by their time since the beginning of the universe and their average thermal energies) can be summarized as follows:

A chronology of the universe

- $\lesssim 10^{-43}$ s ($k_B T \gtrsim 10^{19}$ GeV) Planck epoch. Quantum gravity is expected to be relevant at such high energy scales.

- $\simeq 10^{-35}$ s ($k_B T \simeq 10^{16}$ GeV) Inflation. The big bang is described as an exponential expansion in which the scale factor of space grew by something like 30 orders of magnitude owing to vacuum energy during a phase transition likely associated with the grand unification of particle physics. This enormous expansion left a homogeneous, flat universe with the seeds for subsequent formation of cosmic structure.

- $\simeq 10^{-35}$–10^{-14} s ($k_B T \simeq 10^{16}$–10^2 GeV) Early stages of a radiation-dominated universe with all the fundamental Standard Model species present. Any phase transitions or nontrivial dynamics would be due to physics beyond the Standard Model. It is often speculated that this early

[6] Why there was an excess of particles over antiparticles is currently an area of active research in particle physics and cosmology.

[7] You might imagine the cross section σ as the cross-sectional area of one of the interacting particles if the other particle were a point. Properly, the reaction rate should be thermally averaged: $\Gamma = n\overline{\sigma v}$. Since the product σv is nearly constant (except at a resonance or cutoff threshold), this average for our purpose can be trivially done. **Exercise:** check the dimensions of reaction rate as given here.

universe was also populated with yet undiscovered particles such as the supersymmetric particles.

- $\simeq 10^{-14}$–10^{-10} s ($k_B T \simeq 100\,\text{GeV}$) **Electroweak symmetry breaking** of the Standard Model. Up to this time, the gauge bosons of the weak interaction were massless, just like the photons of electromagnetism. After this phase transition, W and Z bosons, as well as all quarks and leptons, gained their masses.

- $\simeq 10^{-5}$ s ($k_B T \simeq 200\,\text{MeV}$) **Quark/gluon confinement.** Strongly interacting particles (previously deconfined in a plasma) were bound into hadrons. Unstable hadrons (e.g., pions) shortly fell out of equilibrium, leaving only nucleons.

- $\simeq 1$ s ($k_B T \simeq 1\,\text{MeV}$) **Neutron–proton freeze-out and neutrino decoupling.** Nucleons ceased interconverting into each other through the weak interaction involving neutrinos, thus fixing their ratio (although neutrons would very slowly decay into protons until bound later into stable nuclei). Another consequence of ending weak interactions is that neutrinos decoupled, becoming free-streaming to form the cosmic neutrino background.

- $\simeq$ **4–8 s** ($k_B T \simeq 0.5\,\text{MeV}$) **Positron disappearance and photon reheating.** The early universe contained comparable numbers of electrons and positrons, with a slight excess ($\simeq 10^{-9}$) of electrons. When the thermal radiation energy fell below the rest energies, they could no longer be produced; they annihilated with one another, leaving only the few leftover electrons present in matter today. This annihilation also boosted the temperature of the cosmic photons (compared with the previously decoupled cosmic neutrinos).

- $\simeq$ **200 s** ($k_B T \simeq 0.1\,\text{MeV}$) **Nucleosynthesis.** Protons and neutrons combined into charged ions of light nuclei: helium, deuterium, etc.

- $\simeq$ **70 kyr** ($k_B T \simeq 1\,\text{eV}$) **Radiation–matter equality.** The radiation energy density fell below that of nonrelativistic matter (which falls more slowly), thereby changing the dynamics (power-law expansion) of the universe.

- $\simeq$ **380 kyr** ($k_B T \simeq 0.3\,\text{eV}$) **Photon decoupling.** The thermal energy of radiation dropped too low to ionize the just-formed neutral atoms. Photons could free-stream to form the cosmic microwave background observed today.

- $\simeq$ **100 Myr** Formation of stars and galaxies.

- $\simeq$ **8 Gyr** Transition to an accelerating universe; $\simeq$ **10 Gyr** Equality of matter and dark energy. The matter energy fell below the ever-constant dark energy density. As the dark energy came to dominate, the expansion of the universe stopped slowing down and began to accelerate.

- $\simeq$ **14 Gyr** The present epoch.

Thermal relics of light nuclei, neutrinos, and CMB

In the following sections, we shall discuss two particular epochs in the history of the universe that left observable relics in our present-day cosmos. In Section 9.2, we study the epoch of big bang nucleosynthesis at $t_{bbn} \simeq 200$ s. In Section 9.3, we study the epoch at $t_\gamma \simeq 380\,000$ years when the photons decoupled to form the CMB radiation.

In Box 9.2 in Section 9.3.1, we shall also briefly comment on epochs prior to the two mentioned above. About a second after the big bang, neutrinos decoupled to form the first half of the main thermal relic particles of the universe. This was followed by reheating (Exercise 9.1): the disappearance of positrons (and most of the electrons) from the cosmic soup. Because this reheating took place after neutrino decoupling, the relic neutrinos (as yet undetected) should have a cooler thermal distribution than the CMB photons, which decoupled at a later epoch.

9.2 Primordial nucleosynthesis

When we look around our universe, we see mostly hydrogen but very little of the heavy elements. The observed amounts of the heavy elements can all be satisfactorily accounted for by the known nuclear reactions taking place inside stars and supernovae. On the other hand, everywhere we look, besides hydrogen, we also see a significant amount of helium. (The helium abundance can be deduced from measurements of the intensities of spectral lines of helium-4 (^{4}He) in stars, planetary nebulae, and galactic as well as extragalactic HII regions.) The data indicate a ^{4}He mass fraction close to a quarter:

$$y \equiv \left(\frac{^4\text{He}}{\text{H} + ^4\text{He}} \right)_{\text{mass}}, \qquad \text{with} \quad y_{\text{obs}} \simeq 0.24. \tag{9.14}$$

Similarly, we observe much smaller uniform abundances of other light elements: deuterium (D), helium-3 (^{3}He), and lithium-7 (^{7}Li). These light nuclear elements are theorized to have been synthesized in the early universe by the path described below.

Proton–neutron equilibrium and freeze

By the time of 10^{-5} s after the big bang (when the thermal energy was about 200 MeV), quarks had coalesced into nucleons, and unstable particles had vanished from the rapidly cooling universe. The cosmic soup was then composed of protons (p^+), neutrons (n), electrons (e^-), positrons (e^+), three flavors of neutrinos (ν) and antineutrinos ($\bar{\nu}$), and photons. Up till the first second after the big bang, neutrons and protons could interconvert by weak interactions such as

$$n + e^+ \rightleftarrows p^+ + \bar{\nu} \quad \text{and} \quad p^+ + e \rightleftarrows n + \nu. \tag{9.15}$$

Neutrons and protons were in thermal equilibrium, so their number ratio $\lambda = n_n/n_p$ was governed by the Boltzmann distribution $\exp(-E/k_B T)$:

$$\lambda = \exp\left[-\left(\frac{E_n - E_p}{k_B T}\right)\right] \simeq \exp\left[-\left(\frac{m_n - m_p}{k_B T}\right)c^2\right] \simeq \exp\left(\frac{-1.3\,\text{MeV}}{k_B T}\right) \tag{9.16}$$

for nonrelativistic neutrons and protons with a difference in rest energy of 1.3 MeV.

When thermal energy fell below $k_B T_{fr} \simeq 1$ MeV ($T_{fr} \simeq 10^{10}$ K at time $t_{fr} \simeq 1$ s per (9.11)), the interconversion rate Γ fell below the rate of relative expansion H. We call this the freeze-out time, because when the neutron and protons left thermal equilibrium, their ratio was frozen at $\lambda_{fr} \simeq 1/6$. In fact, their ratio would subsequently drop owing to neutron beta decay, $n \longrightarrow p + e + \bar{\nu}$, but not too rapidly, since the mean lifetime of a free neutron, $\tau_n \simeq 880$ s, is considerably longer than the then age of the universe.

Epoch of primordial nucleosynthesis

Protons and neutrons tend to join (through strong interactions) into bound nuclear states.[8]. However, during this epoch, as soon as they were formed, they were blasted apart by energetic photons (photodissociation):

$$p^+ + n \rightleftarrows D^+ + \gamma. \tag{9.17}$$

As the universe cooled, there were fewer photons energetic enough to photodissociate the deuteron (the reaction proceeding from right to left), so deuterons accumulated. The following nucleon-capture reactions could then build up heavier elements:

$$D^+ + n \rightleftarrows {}^3H^+ + \gamma, \quad D^+ + p^+ \rightleftarrows {}^3He^{2+} + \gamma, \tag{9.18}$$

and

$$ {}^3H^+ + p^+ \rightleftarrows {}^4He^{2+} + \gamma, \quad {}^3He^{2+} + n \rightleftarrows {}^4He^{2+} + \gamma. \tag{9.19}$$

These reversible nuclear reactions would not go further to form heavier nuclei; because helium-4 is particularly tightly bound, the formation of nuclei with five nucleons is not energetically favored. Lacking a stable $A = 5$ nucleus, the synthesis of lithium with mass numbers six or seven from stable helium-4 requires the much less abundant deuterons or tritium:

$$ {}^4He^{2+} + D^+ \rightleftarrows {}^6Li^{3+} + \gamma \quad \text{and} \quad {}^4He^{2+} + {}^3H^+ \rightleftarrows {}^7Li^{3+} + \gamma. \tag{9.20}$$

Big bang nucleosynthesis could not progress further to produce even heavier elements ($A > 7$), because there is no stable $A = 8$ element. Beryllium-8 almost

[8] A nucleus is composed of Z (the atomic number) protons and N neutrons, giving it the mass number $A = Z + N$. Since chemical properties are determined by the number of protons, we can identify Z from the name of the element; e.g., hydrogen has $Z = 1$ and helium $Z = 2$. Nuclei having the same Z but different numbers of neutrons are isotopes. From the mass number, usually denoted by a superscript on the left side of the nucleus symbol, we can figure out the number of neutrons. The most abundant helium isotope is helium-4 (${}^4He$) with two neutrons, followed by helium-3 (${}^3He$) with one neutron. Hydrogen isotopes have specific names: the deuteron has one proton and one neutron (${}^2H \equiv D$), and the tritium nucleus (${}^3H$) has two neutrons. Superscript + to the right of the nucleus symbols denotes positively charged ions.

immediately disintegrates into a pair of helium-4's. (Only by packing very large concentrations of helium-4 at high temperatures can mature stars proceed to form stable carbon-12.)

According to a detailed rate calculation, below the thermal energy

$$k_B T_{bbn} \simeq O(0.1) \text{ MeV}, \tag{9.21}$$

photons were no longer energetic enough to photodissociate the bound nuclei. This corresponds to a temperature on the order of $T_{bbn} \simeq 10^9$ K and a cosmic age (cf. (9.11)) of $t_{bbn} \simeq O(10^2)$ s. The net effect of the above reactions from (9.17) to (9.19) was to bind almost all the neutrons into helium-4 nuclei, because there were more protons than neutrons:

$$2n + 2p^+ \longrightarrow {}^4He^{2+} + \gamma. \tag{9.22}$$

We can then conclude that the resultant number density n_{He} for helium-4 must equal half of the neutron density n_n. The number density of hydrogen n_H (the protons leftover after all the others were bound with neutrons into helium ions) should equal the proton number density minus that of the neutrons. The helium mass m_{He} is about four times the nucleon mass m_N. This yields a helium mass fraction of

$$y \equiv \left(\frac{{}^4He}{H + {}^4He} \right)_{mass} = \frac{n_{He} m_{He}}{n_H m_H + n_{He} m_{He}}$$

$$= \frac{(n_n/2) \cdot 4m_N}{(n_p - n_n)m_N + (n_n/2) \cdot 4m_N} = \frac{2\lambda}{1 + \lambda}, \tag{9.23}$$

where λ is the neutron-to-proton ratio, n_n/n_p. Between freeze-out $t_{fr} \simeq 1$ s and nucleosynthesis $t_{bbn} \simeq O(100)$ s, the ratio dropped owing to free-neutron decay[9] from $\lambda_{fr} \simeq 1/6$ to its ultimate value $\lambda \simeq 1/7$. This yields a primordial helium-4 mass fraction very close to the observed ratio of 0.24:

$$y = \frac{2\lambda}{1 + \lambda} \simeq \frac{2/7}{8/7} = \frac{1}{4}. \tag{9.24}$$

[9] Neutrons are stable once bound into nuclei.

In summary, once deuterium was formed by the fusion of protons and neutrons, this chain of fusion reactions proceeded rapidly, so that by about 180 s after the big bang, nearly all the neutrons were bound into helium. Since these reactions were not perfectly efficient, trace amounts of deuterium and helium-3 were left over. (Any leftover tritium would also decay into helium-3.) Formation of nuclei beyond helium progressed slowly; only small amounts of lithium-6 and -7 were synthesized in the big bang. Again, we must keep in mind that the crude calculations presented here are for illustrative purposes only. They are meant to give us a simple picture of the physics involved in such cosmological deductions. Realistic calculations often include many simultaneous reactions. The detailed computations leading to theoretical predictions such as (9.24) must also consider the following:

1. The Standard Model assumes three flavors of light neutrinos (ν_e, ν_μ, and ν_τ), which we include in the effective degrees of freedom g^* in (9.5) for the radiation energy density. Additional neutrinos or other light particles would increase the expansion rate, making freeze-out earlier and hotter, thereby increasing the neutron/proton ration and the helium-4 mass fraction. The observed abundance thus constrain new theories that include more neutrinos or any light exotic particles beyond the Standard Model.

2. The baryon mass density ρ_B affects the cooling rate of the universe. Deuterium is particularly sensitive to ρ_B. Thus we can use the observed abundance of deuterium (one in every 300 000 hydrogens!) to constrain the baryon density.[10] The best fit, as shown in plots like Fig. 9.1, is at $\rho_B \simeq 0.5 \times 10^{-30}$ g/cm^3, or, as a fraction of the critical density:

$$\Omega_B = 0.0475 \pm 0.0006. \tag{9.25}$$

As we already pointed out in Section 8.1.3, when compared with the total mass density $\Omega_M = \Omega_B + \Omega_{DM} \simeq 0.31$, this shows that ordinary atomic matter (baryons) is only a small part of the matter in the universe. Furthermore, we can obtain an estimate of baryon number density n_B by dividing the baryon energy density $\rho_B c^2 = \Omega_B \cdot \rho_c c^2 \simeq 220$ MeV/m^3 by the energy

[10] As can be seen from (9.17)–(9.19), the production of all light elements passed through deuterium. The remnant abundance observed today of such an intermediate state is very sensitive to the reaction and cooling rates. This Ω_B value deduced from nucleosynthesis is confirmed by other cosmological observations, notably the CMB anisotropy. In fact, the quoted value in (9.25) is from the CMB summary by the Planck Collaboration (2014).

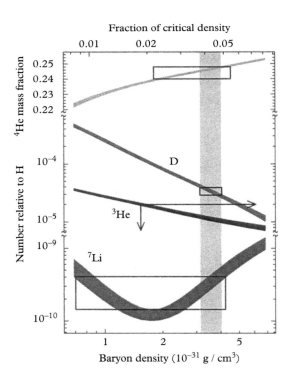

Figure 9.1 *The abundances of light nuclear elements vs. the baryon mass density ρ_B of the universe. The curves are big bang nucleosynthesis predictions, and the boxes are observational results: the vertical heights represent uncertainties in observation, and the horizontal widths the ranges of ρ_B for which theory can accommodate observation. The shaded vertical column represents the range of ρ_B for which theory and observation agree for all four elements. Its uncertainty (the width of the column) is basically determined by the deuterium abundance, which is well measured and strongly dependent on ρ_B. The graph is reproduced with permission from (Burles, Nollett, and Turner 2001). ©2001 American Astronomical Society.*

of each nucleon, whose average can be taken to be the rest energy of a nucleon (939 MeV), because these particles are nonrelativistic:

$$n_B \simeq 0.23/\text{m}^3. \qquad (9.26)$$

9.3 Photon decoupling and cosmic microwave background

The early universe after nucleosynthesis contained a plasma of photons and charged matter coupled by their mutual electromagnetic interactions. As the universe expanded and cooled, baryonic matter (ions and electrons) congealed into neutral atoms, so the cosmic soup lost its ability to entrap the photons. These free thermal photons survived as the CMB radiation we see today. The nearly uniformly distributed relic photons obey a blackbody spectrum. Their discovery gave strong support to the hot big bang theory of the beginning of our universe, as it is difficult to think of any other alternative to account for the existence of such a physical phenomenon on the cosmic scale. Furthermore, its slight temperature fluctuation, the CMB anisotropy, is a picture of the early universe. Careful study of this anisotropy has furnished and will continue to provide us with detailed information about the history and composition of the universe. This is a major tool for quantitative cosmology.

9.3.1 Universe became transparent to photons

The epoch when charged nuclear ions and electrons combined into neutral atoms is called the **photon-decoupling time**[11] t_γ. This took place when the thermal energy of the photons dropped below the threshold required to ionize the newly formed atoms. Namely, the dominant reversible reaction during the age of ions,

$$e^- + p^+ \rightleftarrows H + \gamma, \qquad (9.27)$$

ceased to proceed from right to left when the photon energy fell below the ionization energy. All the charged electrons and ions were swept up and bound themselves into stable neutral atoms.

One would naturally expect the thermal energy at the decoupling time, $k_B T_\gamma$, to have been comparable to the typical atomic binding energy $O(\text{eV})$. In fact, a detailed calculation yields the thermal energy and redshift of this cosmic epoch respectively as

$$k_B T_\gamma \simeq 0.26 \text{ eV} \quad \text{and} \quad z_\gamma \simeq 1100. \qquad (9.28)$$

Dividing out Boltzmann's constant k_B, this energy corresponds to a photon temperature of

$$T_\gamma \simeq 3000 \text{ K}. \qquad (9.29)$$

[11] The photon-decoupling time is also referred to in the literature as the recombination time. We do not use this terminology often, as up to this time ions and electrons had never been combined. The name has been used because of the analogous situation in the interstellar plasma, where such atomic formation is indeed a recombination.

Also, we note that the average photon energy (cf. (9.6)) at decoupling was

$$\bar{E}_\gamma = 2.7 \times 0.26\,\text{eV} \simeq 0.70\,\text{eV}. \tag{9.30}$$

From (8.35), this redshift factor in (9.28) also tells us that the universe was about a thousand times smaller in linear dimension, and by (8.64) a trillion times denser, on average, than it is today.

After the cosmic time t_γ, the decoupled photons could travel freely through the universe, but they kept the blackbody spectrum, whose shape was unchanged as the universe expanded. These relic photons cooled according to the scaling law $T \propto a^{-1}$. Thus, the big bang cosmology predicts that everywhere in the present universe there should be a sea of primordial photons following a blackbody spectrum.

What should the photon temperature be now? From the estimates of $T_\gamma \simeq 3000\,\text{K}$ and $z_\gamma \simeq 1100$, we can use (9.7) and (8.35) to deduce

$$T_{\gamma,0} = a_\gamma T_\gamma = \frac{T_\gamma}{1 + z_\gamma} \simeq 2.7\,\text{K}. \tag{9.31}$$

A blackbody spectrum of this temperature $T_{\gamma,0}$ has its maximal intensity (power per unit area per unit wavelength)[12] at the wavelength $\lambda_{\max}$ such that $\lambda_{\max} T_{\gamma,0} \simeq 0.290\,\text{cm}\cdot\text{K}$ (known as the Wien displacement constant). Thus $T_{\gamma,0} \simeq 2.7\,\text{K}$ implies a thermal spectrum with the maximal energy density at $\lambda_{\max}$ on the order of a millimeter—namely, a relic background radiation in the microwave range.[13]

[12] The reader should be cautioned not to confuse the maximal energy density per unit wavelength, discussed here, with the maximal energy density per unit frequency, as shown for example in Fig. 9.2. They are related but not the same.

[13] While this electromagnetic radiation is outside the visible range, we can still "see" it, because such a microwave noise constitutes a percentage of the (cathode-ray) television snow between channels.

[14] As we shall discuss below, the universe ceased to be radiation-dominated way before the recombination time. Also, the dark energy became significant only recently (on the cosmic timescale)—see the "cosmological coincidence problem" to be discussed in Section 10.2.2 in particular (10.32).

Example 9.2 An estimate of the recombination time

The redshift can be translated into the cosmic age with the following estimate. To the extent that we can ignore the repulsive effect of dark energy, the universe has been matter-dominated since the photon-decoupling time,[14] so the time dependence of the scale factor is $a \propto t^{2/3}$, cf. (8.71):

$$\frac{a_0}{a_\gamma} = \left(\frac{t_0}{t_\gamma}\right)^{2/3} = \frac{1 + z_\gamma}{1 + z_0}. \tag{9.32}$$

As it turns out, the age of the universe is fairly close to the Hubble time $t_0 \simeq 14\,\text{Gyr}$. We get

$$t_\gamma = (1100)^{-3/2} \times 14\,\text{Gyr} \simeq 3.8 \times 10^5\,\text{years} \tag{9.33}$$

In summary, photons in the early universe were tightly coupled to ionized matter (especially electrons) through Thomson scattering. Such interactions stopped around a redshift of $z_\gamma \simeq 1100$, when the universe had cooled sufficiently to form neutral atoms (mainly hydrogen). Ever since this last-scattering time, the photons have traveled freely through the universe, and have redshifted to microwave frequencies as the universe expanded. This primordial light should appear today as the CMB thermal radiation with a temperature of about 3 K.

Box 9.1 The discovery of CMB radiation

The observational discovery of the CMB radiation was one of the great scientific events of the modern era. It made the big bang cosmology much more credible, as it is difficult to see how else such thermal radiation could have been produced. The discovery and its interpretation also constitute an interesting story. Gamow (1946, 1948), Alpher, and Herman (1948) first predicted that a direct consequence of the big bang model is the presence of a relic background of radiation with a temperature of a few degrees. However, their contribution was not widely appreciated, and no effort was mounted to detect such a microwave background. Only in 1964 did Robert Dicke (1916–1997) at Princeton University rediscover this result and lead a research group (including James Peebles, Peter Roll, and David Wilkinson) to detect this CMB. While they were constructing their apparatus, Dicke was contacted by Arno Penzias (1933–) and Robert W. Wilson (1936–) at the nearby Bell Laboratories. Penzias and Wilson had used a horn-shaped microwave antenna over the preceding year to do astronomical observations. This Dicke radiometer had originally been used in a trial satellite communication experiment and was known to have some excess noise. Not content to ignore it, they made a careful measurement of this background radiation, finding it to be independent of direction, time of day, and season of the year. While puzzling over the cause of such radiation, they were informed by one of their colleagues of the Princeton group's interest in the detection of a cosmic background radiation. (Peebles had given a colloquium on this subject at another university.) This resulted in the simultaneous publication of two papers: one in which Penzias and Wilson (1965) announced their discovery, the other by the Princeton group (Dicke et al. 1965) explaining the cosmological significance of the discovery.

Because of microwave absorption by water molecules in the atmosphere, it is desirable to carry out CMB measurements at locations having low humidity and/or at high altitude. Thus some of the subsequent observations were done with balloon-borne instruments launched in Antarctica (low temperature, low humidity, and high altitude)—or, even better, above the atmosphere from a satellite. Satellite observations were first accomplished in the early 1990s by the Cosmic Background Explorer (COBE) satellite observatory,

continued

Box 9.1 *continued*

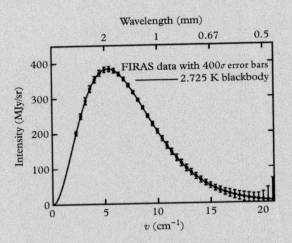

Figure 9.2 *The cosmic background radiation spectrum measured by the FIRAS instrument on the COBE satellite, showing a perfect fit to the blackbody distribution. The horizontal axis is the frequency ~ 1/(wavelength in cm). The vertical axis is the power per unit area per unit frequency per unit solid angle in megajanskies per steradian. In order to make the error bars visible, these estimated uncertainties have been multiplied by a factor of 400. The graph is based on data from (Fixsen et al. 1996). ©1996 American Astronomical Society.*

led by John Mather (1946–) and George Smoot (1945–), which obtained results (Fig. 9.2) showing that the CMB radiation followed a perfect blackbody spectrum with a temperature

$$T_{\gamma,0} = 2.725 \pm 0.002 \, \text{K}. \tag{9.34}$$

The COBE observations not only confirmed that the thermal nature of the cosmic radiation was very uniform (the same temperature in every direction), but also discovered a minute anisotropy at the microkelvin level. This has been interpreted as resulting from the matter density perturbations, that, through subsequent gravitational clumping, gave rise to the cosmic structures we see today: galaxies, clusters of galaxies, voids, etc.[15]

[15] This will be discussed further in Section 9.3.2.

The photon and baryon numbers

Knowing the CMB photon temperature $T_{\gamma,0} = 2.725$ K, we can calculate the relic photon number density via (9.2):

$$n_{\gamma,0} = \frac{2.404}{\pi^2} \left(\frac{k_B T_{\gamma,0}}{\hbar c} \right)^3 \simeq 410/\text{cm}^3. \tag{9.35}$$

That is, there are now in the universe, on average, 410 photons for every cubic centimeter. Clearly this density is much higher than the baryon number density obtained in (9.26). The baryon-to-photon number ratio, with more precise inputs, comes out to be

$$\frac{n_B}{n_\gamma} = (6.05 \pm 0.07) \times 10^{-10}. \tag{9.36}$$

For every proton or neutron, there are almost two billion photons. This explains why the average energy at decoupling (0.70 eV from (9.30)) fell so far below the hydrogen ionization energy of 13.6 eV before ionization stopped. There was such a high density of photons that even though the average photon energy was only 0.70 eV, there remained up to that time a sufficient number of high-energy photons (at the tail end of the distribution) to hold off the transition to a new equilibrium phase.

This ratio (9.36) should hold all the way back to the photon-decoupling time, because not only was the number of free photons unchanged, but also the baryon number, since all the interactions in this low-energy range (in fact, all the Standard Model interactions we have ever observed) respect the law of baryon number conservation. In Exercise 9.2, this density ratio will be used to estimate the cosmic time at which the universe switched from being radiation-dominated to matter-dominated.

Box 9.2 The cosmic neutrino background

We have already discussed the cosmic microwave background formed from decoupled photons. There is another cosmic background of comparable abundance, formed by decoupled neutrinos. Neutrinos have only weak interaction. Their collision cross section with other particles (e.g., electrons) is small. This cross section has a strong energy dependence. In the early universe, high-energy neutrinos interacted strongly enough with the cosmic particle soup to be in thermal equilibrium. They played a role in keeping neutrons and protons in equilibrium (9.15) until freeze-out at $k_B T_{\text{fr}} \simeq 1$ MeV. Shortly thereafter, when the temperature fell to about $k_B T_\nu \simeq 0.3$ MeV, they decoupled and become free-streaming. Thus the cosmic neutrino background was formed considerably earlier than the CMB (at 0.26 eV), prior even to primordial nucleosynthesis (at 0.1 MeV). While neutrinos are about as numerous as photons, the present numbers of all other particles, including dark matter particles, are very much smaller—about one per every few billion photons and neutrinos, cf. (9.36).

continued

Box 9.2 *continued*

Cosmic neutrino temperature

According to (9.7), the radiation temperature scales as a^{-1}, regardless of whether or not the particles are coupled. Thus one would expect the temperature of the cosmic neutrino background to be the same as that of the CMB. This is not the case, because in between the neutrino and photon decoupling times, the photon temperature got a boost from electron–positron annihilation (into photons). As a result, the photon temperature is somewhat higher than the neutrino temperature (see Exercise 9.1):

$$T_\gamma = \left(\frac{11}{4}\right)^{1/3} T_\nu. \tag{9.37}$$

Given that the CMB has at the present epoch a temperature $T_\gamma = 2.7\,\text{K}$, the cosmic neutrino background should have a temperature of $1.9\,\text{K}$.

Neutrino number density

From the neutrinos' temperature, one can fix their number density via (9.2). Because neutrinos are fermions and photons are bosons, we have

$$\frac{n_\nu}{n_\gamma} = \frac{3}{4}\left(\frac{T_\nu}{T_\gamma}\right)^3 \frac{g_\nu}{g_\gamma}. \tag{9.38}$$

As mentioned in Sidenote 3, there are two photon states and six neutrino states (left-handed neutrinos and right-handed antineutrinos for each of the three lepton flavors). Plugging in these multiplicities and the temperature ratio from (9.37) yields

$$\frac{n_\nu}{n_\gamma} = \frac{9}{11}. \tag{9.39}$$

Neutrino contribution to the radiation energy density

From the Stefan–Boltzmann law for the radiation energy density (9.3), $u_R \propto g^* T^4$, with g^* being the effective spin degrees of freedom, as shown in (9.5), $g_\gamma^* = 2$ and $g_\nu^* = \frac{7}{8} \times 6$. Therefore,

$$\frac{\rho_\nu}{\rho_\gamma} = \frac{u_\nu}{u_\gamma} = \frac{g_\nu^*}{g_\gamma^*}\left(\frac{T_\nu}{T_\gamma}\right)^4 = \frac{21}{8}\left(\frac{4}{11}\right)^{4/3} = 0.68, \tag{9.40}$$

where to reach the numerical result we have again used the temperature ratio of (9.37).

Exercise 9.1 Photon temperature boost by e^+e^- annihilation

Since neutrinos and photons were once coupled and in thermal equilibrium, their temperatures were the same: $T'_\nu = T'_\gamma$. The reaction $e^+ + e^- \leftrightarrows \gamma + \gamma$ ceased to proceed from right to left when the photon energy fell below 0.5 MeV.[16] The disappearance of positrons increased the photons' number and hence their temperature. This temperature boost can be calculated through the entropy conservation condition. Entropy S is related to energy U as $dS = (1/T)\, dU = (V/T)\, du$, where V and u are respectively the volume and energy density. Given $u \propto g^* T^4$ from (9.3), the key entropy dependences can be identified as $S \propto g^* V T^3$. By comparing the volume and photon temperature change as required by the entropy conservation condition $S'_{e^+} + S'_{e^-} = S_\gamma - S'_\gamma$ in the annihilation reaction and the corresponding volume and neutrino temperature change for the uncoupled neutrinos $S'_\nu = S_\nu$, show that the final photon and neutrino temperatures are related by (9.37): $T_\gamma = (11/4)^{1/3}\, T_\nu$.

[16] Recall that the rest energy of an electron or positron is about 0.5 MeV.

Exercise 9.2 The radiation–matter equality time

The early universe was radiation-dominated; it then gave way to a matter-dominated system. The radiation–matter equality time t_{RM} is defined to be the cosmic time at which the energy densities of radiation and matter were equal:

$$1 = \frac{\rho_R(t_{RM})}{\rho_M(t_{RM})} = \frac{\Omega_R(t_{RM})}{\Omega_M(t_{RM})}. \tag{9.41}$$

Calculate t_{RM} by the following steps:

(a) *From the scaling behavior of the radiation and matter densities, relate the scale factor a_{RM} at the radiation–matter equality time to the matter-to-radiation density ratio now, $\Omega_M(t_0)/\Omega_R(t_0)$.*

(b) *This density ratio can be calculated to have the value of 3300 with the following inputs: radiation is composed of photons and neutrinos, so the total radiation energy density $\Omega_R(t_0)$ is related to the photon density by (9.40); the matter content $\Omega_M(t_0)$ can be deduced from the baryon fraction of matter $\Omega_B(t_0)/\Omega_M(t_0) \simeq 0.05/0.32$ and the photon-to-baryon number ratio n_γ/n_B. The energy of baryonic matter $E_B(t_0)$ can be calculated by adding up the nucleon rest energies, while the photon energy $\bar{E}_\gamma(t_0)$ can be deduced from its value of 0.7 eV at redshift $z_\gamma \simeq 1100$.*

(c) *Follow the worked Example 9.2, using the result for a_{RM} from parts (a) and (b) and a cosmic age $t_0 = 14$ Gyr to show that the radiation–matter dominance transition happened approximately 73 000 years after the big bang.*

9.3.2   CMB anisotropy as a baby picture of the universe

The CMB shows a high degree of isotropy. After subtracting off the Milky Way foreground radiation, one obtains in every direction the same blackbody temperature. However, the isotropy is not perfect. The great achievement of COBE (improved by the better angular resolution of NASA's Wilkinson Microwave Anisotropy Probe (WMAP), and now of the European Space Agency's Planck satellite)[17] was the first detection of slight variations of temperature: first at the 10^{-3} level associated with the motion of our Local Group of galaxies, then at the 10^{-5} level, which, as we shall explain, holds the key to our understanding the origin of structure in the universe—how the primordial plasma evolved into stars, galaxies, and clusters of galaxies. Furthermore, the CMB fluctuation provides us with another means to measure the matter/energy content of the universe, as well as many cosmological parameters. The free-streaming photons of the CMB were created around $t_\gamma \simeq 3.8 \times 10^5$ years, and the universe has an approximate age of $t_0 \simeq 1.4 \times 10^{10}$ years. If one were to regard the universe as a one-hundred-year-old person, the CMB anisotropy image would be like her one-day-old baby picture.

The dipole anisotropy   Although each point on the sky has a blackbody spectrum, in one half of the sky the spectrum corresponds to a slightly higher temperature, while the other half is slightly cooler with respect to the average background temperature: $\delta T/T \simeq 1.237 \times 10^{-3} \cos\theta$, where θ is measured from the hottest spot on the sky. The dipole distortion is a simple Doppler shift, caused by the net motion of our own galaxy due to the gravitational attraction resulting from the uneven distribution of masses in our cosmic neighborhood. Namely, it shows directly that the Local Group is traveling toward the Virgo Cluster of galaxies at about 600 km/s. This peculiar motion[18] is measured with respect to the frame in which the CMB is isotropic.

The existence of such a CMB rest frame does not contradict special relativity. SR only says that no internal physical measurements can detect absolute motion. Namely, physics laws must be covariant; they may not single out an absolute rest frame. Covariance does not mean that we cannot compare motion relative to a cosmic structure such as the microwave background. Space may not be a thing, but the CMB certainly is. Nor does the CMB rest frame violate the isotropy assumption underlying Robertson–Walker spacetime. Each point in an isotropic spacetime must have an observer (defined to be at rest in the comoving frame) to whom the universe appears the same in every direction. But there is no reason to expect the earth (or the sun, the Milky Way, or the Local Group) to be at rest. There is certainly no expectation that every observer sees the same thing every way he looks. At the same time, there is no definitive explanation why the CMB rest frame defines the inertial frames for us. However, Mach's principle would certainly suggest that this is not a coincidence.[19]

[17] WMAP was preceded by other groups such as the Boomerang and Maxima high-altitude balloon observations in the late 1990s. See the further discussion in Section 10.1, where we discuss the CMB evidence for a flat universe.

[18] The quoted number represents the observational result after subtracting out the orbital motion of COBE around the earth (~ 8 km/s) and the seasonal motion of the earth around the sun (~ 30 km/s). The measured value is the vector sum of the orbital motion of the solar system around the galactic center (~ 220 km/s), the motion of the Milky Way around the center of mass of the Local Group of galaxies (~ 80 km/s), and the motion of the Local Group (630 ± 20 km/s) in the general direction of the constellation Hydra. The last, the peculiar motion of our small galaxy cluster toward the large mass concentration in the neighboring part of the universe, reflects the gravitational attraction by the very massive Virgo Cluster at the center of our Local Supercluster, which is in turn accelerating toward the Hydra–Centaurus Supercluster.

[19] This paragraph ties together several fundamental principles motivating GR and cosmology. Review Section 1.1 for a discussion of covariance and Mach's principle (in particular Sidenote 2 in Chapter 1 and Sidenote 1 in Chapter 8), Section 4.1.2 for Einstein's conviction that "space is not a thing," and Section 8.2.1 for the assumption of an isotropic universe.

Exercise 9.3 Temperature dipole anisotropy as Doppler effect

By converting temperature variation to that of light frequency, show that the Doppler effect implies that an observer moving with a nonrelativistic velocity **v** *through an isotropic CMB would see a temperature dipole anisotropy* $\delta T/T = (v/c)\cos\theta$, *where the angle* θ *is measured from the direction of the motion.*

$$-200 \qquad T(\mu K) \qquad +200$$

Figure 9.3 *The temperature fluctuation of the CMB is a snapshot of the baby universe at the photon decoupling time. The foreground emission of the Milky Way has been subtracted out. The data are from nine-year observations by the WMAP Collaboration, reproduced with permission from (Bennett et al. 2013). ©2013 American Astronomical Society.*

Physical origin of the temperature inhomogeneity Aside from this 10^{-3}-level dipole anisotropy, the background radiation is seen to be quite isotropic. The CMB is a snapshot of the early universe, so its observed isotropy is direct evidence that our working hypothesis of a homogeneous and isotropic universe is essentially valid as far back as the photon decoupling time. Nevertheless, this isotropy should not be perfect. The observed universe has all sorts of structure; some of the superclusters of galaxies and the largest voids are as large as 100 Mpc across. Such a basic feature of our universe must be reflected in the CMB in the form of small temperature anisotropies. There must have been some matter density nonuniformity at t_γ, which would have brought about temperature inhomogeneity; photons traveling from denser regions would be gravitationally redshifted and therefore arrive cooler, while photons from less-dense regions would do less work and arrive warmer. Such small temperature variations ($\simeq 30\,\mu K$) coming from different directions were finally detected, providing evidence for a primordial density nonuniformity that, under gravitational attraction, grew into the structures of stars, galaxies, and clusters of galaxies that we observe today.

Dark matter forms the cosmic scaffolding of matter distribution The temperature variation $\delta T/T = O(10^{-5})$ is smaller than expected based on the observed structure of baryonic matter. But this discrepancy can be resolved by the existence of dark matter. Dark matter has no electromagnetic interaction; its density inhomogeneity does not directly produce temperature inhomogeneities in the cosmic fluid that can be seen in the CMB anisotropy. In fact, we expect that the gravitational clumping of the dark matter in forming structure took place first. The corresponding clumping of the baryonic matter (having electromagnetic interaction) was countered by radiation pressure until after photons were decoupled. Once the baryonic structure was formed, it tended to fall into the gravitational potential wells of the already-formed dark-matter scaffolding. This extra early growth of density perturbation for the nonbaryonic dark matter means that less baryonic inhomogeneity at t_γ is needed to produce the structure seen today. Thus a lesser $\delta T/T$ in the baryonic matter at t_γ (and hence in the CMB) is needed if the bulk of the matter has no electromagnetic interaction.

Cosmic inflation, primordial gravitational waves, and CMB polarization Where did the primordial density perturbations come from? As we shall discuss in Chapter 10, the favored theory of the big bang is that at the cosmic time $O(10^{-35}\,\mathrm{s})$, the fluctuation of some (scalar) quantum field led to a state having a large cosmological constant, which drove an exponential expansion of the universe. This stretched the quantum fluctuations of the density to macroscopic sizes, seeding subsequent structure formation, and brought about the CMB temperature anisotropy discussed above. Furthermore, this anisotropic radiation would in turn lead, through Compton scattering, to CMB polarization (to be discussed in Box 10.3). Of particular interest is the theoretical prediction that such an inflationary epoch would generate tensor perturbations of the spacetime metric (i.e., gravitational waves), which would give rise to a unique pattern of polarization (called *B*-mode polarization). Its definitive detection would provide us with direct evidence of gravitational waves and of the inflationary theory of the big bang.

Box 9.3 The statistical study of CMB anisotropy

Cosmological theories can be checked by a statistical study of the CMB temperature $T(\theta, \phi)$ across the celestial sphere. With an average

$$\langle T \rangle = \frac{1}{4\pi} \int T(\theta, \phi) \sin\theta\, d\theta\, d\phi = 2.725\,\mathrm{K}, \tag{9.42}$$

the temperature fluctuation

$$\frac{\delta T}{T}(\theta, \phi) \equiv \frac{T(\theta, \phi) - \langle T \rangle}{\langle T \rangle} \tag{9.43}$$

has a root-mean-square value of $\sqrt{\langle (\delta T/T)^2 \rangle} = 1.1 \times 10^{-5}$. How do we describe such a temperature variation and connect it to the underlying cosmological theory? Recall that for a function of one variable, a useful approach is Fourier expansion of the function in a series of sine waves with frequencies that are integral multiples of the fundamental frequency (of the wave with the largest wavelength). Similarly, we expand the temperature fluctuation in terms of spherical harmonics (think of them as vibration modes on the surface of an elastic sphere):

$$\frac{\delta T}{T}(\theta, \phi) = \sum_{l=0}^{\infty} \sum_{m=-l}^{l} a_{lm} Y_l^m(\theta, \phi). \tag{9.44}$$

The multipole number l represents the number of nodes (locations of zero amplitude) between equator and poles, while m is the longitudinal node number. For a given l, there are $2l+1$ values for m: $-l, -l+1, \ldots, l-1, l$. The series can of course be inverted so that the expansion coefficients a_{lm} are expressed in terms of the temperature fluctuation.

Cosmological theories predict statistical information about CMB temperature fluctuation. The most useful statistic is the 2-point correlation. Consider two points on a unit sphere at $\hat{\mathbf{n}}_1$ and $\hat{\mathbf{n}}_2$, separated by θ. We define the correlation function

$$C(\theta) \equiv \left\langle \frac{\delta T}{T}(\hat{\mathbf{n}}_1) \frac{\delta T}{T}(\hat{\mathbf{n}}_2) \right\rangle_{\hat{\mathbf{n}}_1 \cdot \hat{\mathbf{n}}_2 = \cos\theta}, \tag{9.45}$$

where the angle brackets denote the averaging over an ensemble of realizations of the fluctuation.[20] The inflationary cosmology predicts that the fluctuation is Gaussian[21] (i.e., maximally random) and is thus independent of the a_{lm}. Namely, the multipoles a_{lm} are uncorrelated. For different values of l and m,

$$\langle a_{lm} \rangle = 0, \qquad \langle a_{lm}^* a_{l'm'} \rangle = C_l \delta_{ll'} \delta_{mm'}, \tag{9.46}$$

which defines the power spectrum C_l as a measure of the relative strength of the spherical harmonics in the decomposition of the temperature fluctuations. Namely, it measures the typical size of the temperature irregularity on a given angular scale. The lack of m dependence reflects the (azimuthal) rotational symmetry of the underlying cosmological model. When we plug (9.44) into (9.45) with the conditions (9.46), the expansion is simplified[22] to

$$C(\theta) = \frac{1}{4\pi} \sum_{l=0}^{\infty} (2l+1) C_l P_l(\cos\theta), \tag{9.47}$$

where $P_l(\cos\theta)$ is the Legendre polynomial. Namely, the information carried by $C(\theta)$ in the angular space can be represented by C_l in the space of multipole number l. For the large multipole, the Legendre polynomial has the

[20] In principle, this means averaging over many universes. Since we have only one universe, this ensemble averaging is carried out by averaging over multiple moments with different m, which in theory should be equal because of spherical symmetry. But for small l, there are fewer m available, which makes the average more uncertain. Figure 9.4 exhibits this *cosmic variance* for low l.

[21] If the temperature fluctuation were not Gaussian, higher-order correlations would contain additional information.

[22] We also need to use the addition theorem of spherical harmonics:

$$\sum_m Y_l^{*m}(\hat{\mathbf{n}}_1) Y_l^m(\hat{\mathbf{n}}_2) = \frac{2l+1}{4\pi} P_l(\cos\theta_{12}).$$

continued

[23] The exact limit form is

$$P_l(\cos\theta) = \sqrt{\frac{2}{\pi l \sin\theta}} \cos\left[\left(l + \frac{1}{2}\right)\theta - \frac{\pi}{4}\right].$$

Figure 9.4 *The angular power spectrum of the CMB temperature anisotropy. The theoretical curve follows from the* ΛCDM *(dark energy* Λ*–cold dark matter) model. The first major peak at multipole number* $l \simeq 200$ *is evidence for a flat universe. The theoretical uncertainty for low multipoles is due to cosmic variance (cf. Sidenote 20). The data are from nine-year observations by the WMAP Collaboration, reproduced with permission from (Bennett et al. 2013). ©2013 American Astronomical Society.*

Box 9.3 *continued*

asymptotic form[23] $P_l(\cos\theta) \simeq \cos l\theta$. At peak locations in this regime, there is a correspondence between angular separation and multipole number as

$$l \simeq \frac{\pi}{\theta}. \tag{9.48}$$

Large l's correspond to small angular scales, with $l \simeq 10^2$ corresponding to degree-scale separation; cf. Box 10.2.

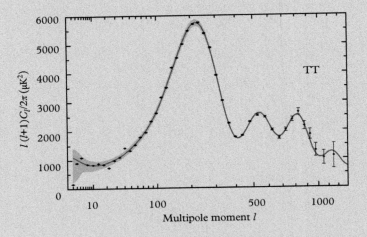

The power spectrum C_l can be measured by observations and compared with theoretical predictions. An example is displayed in Fig. 9.4. We note the prominent features:

(i) a fairly flat region of lower l values, the Sachs–Wolfe plateau, due to the variation of gravitational potential on the last-scattering surface;

(ii) a series of oscillations, the acoustic peaks, coming from the complex motions of the cosmic fluid when the CMB was first created;

(iii) the damping tail in the large-l region, reflecting the fact that the last-scattering surface has finite thickness.

Theoretical predictions are made by selecting cosmological parameters that yield the best match with the observed power spectrum. The cosmological parameters used in our presentation are mainly from the CMB power spectra obtained by the WMAP and Planck satellites.

In other words, when one looks at the anisotropy distribution shown in Fig. 9.3, one sees spots of many different sizes. This pattern can be translated into the power spectrum of Fig. 9.4, which encodes the key cosmological information that can be compared with theoretical predictions.

Review questions

1. Give an argument for the scaling behavior of the radiation temperature: $T \sim a^{-1}$. Show that under such a scaling law, the shape of the blackbody radiation spectrum is unchanged as the universe expands; i.e., a redshifted blackbody spectrum is simply a colder blackbody spectrum.

2. What is the condition (called the Gamow condition) for any particular set of interacting particles to be in thermal equilibrium during a given epoch of the expanding universe?

3. Cosmic helium synthesis combines two protons and two neutrons into a helium nucleus. The Boltzmann distribution at a thermal energy of the order of MeV yields a neutron-to-proton number density ratio $n_n/n_p \simeq 1/7$. From this, how would you estimate the cosmic helium mass fraction?

4. How can one use the theory of big bang nucleosynthesis and the observed abundance of light elements such as deuterium and helium to deduce the baryon number density Ω_B, and that the number of neutrino flavors should be three? (The reader is not asked why there are three neutrino flavors, but how the astrophysical observation is only compatible with three flavors of light neutrinos.)

5. What physical process took place around the photon-decoupling time t_γ? What are the average thermal energy and temperature at t_γ? Given the redshift $z_\gamma \simeq 10^3$, calculate the expected photon temperature now.

6. What was the cosmic time at which the universe made the transition from a radiation-dominated to a matter-dominated system. How does it compare with the nucleosynthesis and photon-decoupling times?

7. The abundance of cosmic background neutrinos should be comparable to the CMB. How come we have not detected them?

8. Why would the peculiar motion of our galaxy show up as a CMB dipole anisotropy?

9. Besides the dipole anisotropy, how does the CMB temperature anisotropy reflect the origin of cosmic structure?

10. Why would the presence of a significant amount of dark matter reduce the baryonic matter inhomogeneity (and hence the CMB temperature inhomogeneity) required to account for the observed structure in the universe?

10

Inflation and the Accelerating Universe

- The inflationary theory of cosmic origin posits that in its earliest moments the universe underwent a huge expansion due to phase change into a field state with a large vacuum energy. This scenario solves two serious issues regarding the initial conditions for the standard FLRW cosmology: the flatness and horizon problems.

- This primordial inflation answers Hoyle's "big bang" objection by explaining not only the origin of all the matter/energy in the universe, but also the unevenness in its initial distribution that led to the subsequent formation of the observed cosmic structures.

- Interestingly, inflation, which ties up so many loose ends in cosmology, emerges naturally from grand unified theories of particle physics.

- This primordial inflation would have left behind a flat universe, which can be reconciled with the observed subcritical matter density and with a cosmic age greater than 9 Gyr (our estimate for a flat matter-dominated universe) if there remains a small but nonvanishing cosmological constant—a dark energy. This would imply that the expansion of the universe is now accelerating.

- The measurement of supernovae at high redshifts provided direct evidence of an accelerating universe. Such data, together with other observational results, especially the anisotropy of the cosmic microwave background and large-scale structure surveys, gave rise to a concordant cosmological picture of a spatially flat universe ($\Omega_{tot} = 1$) whose primary component is dark energy, $\Omega_\Lambda \simeq 0.69$. Nearly all the balance is nonrelativistic matter energy, $\Omega_M \simeq 0.31$. As discussed earlier, most of this matter is some yet-unknown dark matter, $\Omega_{DM} \simeq 0.26$; ordinary (baryonic) matter comprises only $\Omega_B \simeq 0.05$. The cosmic age comes out to be $t_0 \simeq 14$ Gyr, close to the Hubble time.

- The lack of a physical basis for the observed dark energy is the cosmological constant problem. It suggests a need for new fundamental physics.

A College Course on Relativity and Cosmology. First Edition. Ta-Pei Cheng.
© Ta-Pei Cheng 2015. Published in 2015 by Oxford University Press.

We related in Section 8.4 how Einstein introduced the cosmological constant in the hope of obtaining a static-universe solution of the GR field equation; this gravitational repulsion can counteract the familiar attraction. The subsequent discovery that the universe is expanding removed this original motivation. Nevertheless, we shall demonstrate in this chapter that this repulsive gravity is needed not only to account for the initial explosion of the big bang (inflationary epoch), but also to explain how the expansion of the universe could in the present epoch be accelerating (dark energy). In fact, we now regard the cosmological constant as one of Einstein's great contributions to modern cosmology. It allows GR to incorporate a form of cosmic antigravity[1].

[1] By antigravity, we simply mean a repulsion in GR opposing the familiar gravitational attraction.

10.1 The cosmic inflation epoch

The FLRW cosmology is a very successful model of the evolution and composition of the universe: how the universe expanded and cooled after the big bang; how the light nuclear elements were formed; how in an expanding universe matter inhomogeneities congealed to form stars, galaxies, and clusters of galaxies. In short, it describes very well the aftermath of the big bang. However, the model says very little about the nature of the big bang itself—how this explosion of space came about. It assumes that all the matter existed from the very beginning, with just the right seeds for the later development of cosmic structure. These extraordinary initial conditions just clamor for an explanation.

Amazingly, cosmic inflation theory, which emerges naturally from grand unified theories of particle physics, not only answers this but also fixes a couple of other nagging problems in cosmology. It explains so much so well that it allows us to speak with considerable confidence about a universe so young and so hot that its energies outstripped by many orders of magnitude anything we can achieve in our largest accelerator experiments.

10.1.1 Initial condition problems of FLRW cosmology

Here we concentrate on the flatness and horizon problems, the need for two fine-tuned (apparently unnatural) initial conditions in the successful FLRW cosmology:

The flatness problem

Because of the gravitational attraction among matter and energy, we would expect the expansion of the universe to slow down; cf. (8.43). Such a deceleration $\ddot{a}(t) < 0$ would imply that the time derivative of the scale factor, $\dot{a}(t)$, was an **ever-decreasing** function of cosmic time. Recall that the Friedmann equation can be written in terms of the mass density parameter Ω as in (8.47):

$$1 - \Omega(t) = \frac{-kc^2}{[\dot{a}(t)]^2 R_0^2}. \qquad (10.1)$$

This displays the connection between geometry (k) and matter/energy (Ω). If $k = 0$ (a flat geometry), we must have the exact density ratio $\Omega = 1$. When $k \neq 0$ (a curved, expanding universe and thus a nonvanishing right-hand side); the magnitude of $1 - \Omega(t)$ on the left-hand side must be ever-increasing in a matter- or radiation-dominated universe, because the denominator $\dot{a}^2$ on the right-hand side is ever-decreasing. Thus, the value $\Omega = 1$ is an unstable equilibrium point; if Ω ever deviates from 1, this deviation will increase with time. Or, we may say, gravitational attraction always enhances any initial curvature. In light of this property, it is puzzling that the present mass density is observed (see Section 8.1.3) to be not too different from the critical density: $1 - \Omega_0 = O(1)$. This means that Ω must have been extremely close to unity (extremely flat) in the cosmic past. Such a fine-tuned initial condition requires an explanation.

This flatness problem can be quantified by considering the deviation of the density parameter from its flat-universe value of $\Omega = 1$. Ever since the radiation–matter equality time $t_{RM} \simeq 73$ kyr with a scale factor $a_{RM} = O(10^{-4})$, cf. Exercise 9.2, the evolution of the universe has been dominated by nonrelativistic matter:[2] $a(t) \sim t^{2/3}$, so $\dot{a} \sim t^{-1/3} \sim a^{-1/2}$. We can then estimate the relative deviation from flat geometry using (10.1):

$$\frac{1 - \Omega(t_{RM})}{1 - \Omega(t_0)} = \left[\frac{\dot{a}(t_0)}{\dot{a}(t_{RM})} \right]^2 = \frac{a_{RM}}{a_0} = O(10^{-4}). \qquad (10.2)$$

[2] Specifically, we shall use the results shown in (8.71), which assume perfect flatness $\Omega = 1$. Nonetheless, in our calculation, $1 - \Omega$ is so small that they will suffice to produce rough estimates.

Namely, the universe at t_{RM} must have been very flat: $|1 - \Omega(t_{RM})| \lesssim O(10^{-4})$. The successful prediction of light-element abundances by primordial nucleosynthesis provides direct evidence for the validity of the FLRW cosmology back to the big bang nucleosynthesis time $t_{bbn} = O(10^2)$ s. One can then show (see Exercise 10.1) that the density ratio at the primordial nucleosynthesis epoch must have been equal to unity to an accuracy of $1 - \Omega(t_{bbn}) = O(10^{-14})$. That the FLRW cosmology requires such an extraordinary initial condition constitutes the flatness problem.

Exercise 10.1 Estimate $|1 - \Omega(t_{bbn})|$ from $|1 - \Omega(t_{RM})|$

Given that the universe was radiation-dominated during the period from t_{bbn} to t_{RM}, estimate the magnitude of $|1 - \Omega(t_{bbn})|$ from $|1 - \Omega(t_{RM})|$, as in (10.2). The estimate will depend on knowing the scale factor $a(t_{nbb}) = O(10^{-9})$, which can be deduced from the temperature ratio $T_{bbn}/T_\gamma = O(10^6)$ and the temperature evolving as the inverse scale factor.

The horizon problem

Our universe is observed to be very homogeneous and isotropic at large scales—in agreement with the cosmological principle. In fact, the observed universe appears to be *too* homogeneous and isotropic. Consider two different parts of the universe that appear to be outside of each other's horizons. They are so far apart that no light signal sent from one at the beginning of the universe should have reached the other. They are observed to have similar properties, suggesting that they were in thermal contact at some time in the past. But their present separation seems to preclude them from interacting, much less achieving thermal equilibrium. How can one resolve this contradiction?

This horizon problem can be stated most precisely in terms of the observed isotropy of the CMB radiation. We measure the same blackbody temperature in all directions (up to one part in 10^5, after subtracting out the dipole anisotropy due to our peculiar motion). This observed isotropy indicates that at the cosmic photon-decoupling age $t_\gamma = O(10^5)$ years, regions then separated by much further than the naive horizon distance $O(10^5)$ light-years were strongly correlated. How could such distant regions communicate with each other to arrive at the same thermal properties? This is the horizon problem of the FLRW cosmology.

In short, cosmic physics suggests that the universe should be highly curved and heterogeneous, yet we observe it to be homogeneous and very nearly flat. This can be modeled by the FLRW cosmology with a particular set of initial conditions. Still, these initial values are so finely tuned that one might wish for a more natural explanation. As we shall detail below, the inflationary theory of the big bang can provide such a justification.

10.1.2   The inflationary scenario

Inflationary cosmology attempts to give an account of the big bang itself, back to an extremely short instant (something like 10^{-37} s) after the $t = 0$ cosmic singularity.[3] During this primordial inflation, the universe expanded in a burst, increasing its scale factor by more than 30 orders of magnitude;[4] see Fig. 10.1. This explosive expansion could have happened if there existed then a large cosmological constant Λ, which would have supplied a huge repulsion leading to a period of exponential expansion; see Section 8.4.2. This solves the flatness problem, as any initial curvature would have been stretched flat by the burst of expansion. It also solves the horizon problem, because such an expansion can cause points in space to separate with superluminal speed. Regions that today (or at photon decoupling) appear to be outside each other's horizons could thus have been connected and in thermal equilibrium prior to the inflationary epoch.

Around 1980, Alan Guth (1947–) was researching grand unified theories (GUTs) of particle physics, which postulate that at energies above 10^{16} GeV, strong and electroweak forces, which are distinct in the Standard Model, are

[3] This may be compared with the even earlier period, comparable to the Planck time $t_{Pl} = O(10^{-43})$ s, which requires quantum gravity for a proper description; see Section 7.3.2.

[4] These numbers are for illustrative purposes only. Inflation as motivated by grand unified theories suggests such magnitudes, but there is at present no observational evidence for any such specific theory. Nonetheless, the overall inflationary paradigm is well grounded. Not only does it remedy the aforementioned cosmological problems, but also we shall see that some sort of inflation is indicated by diverse observational data.

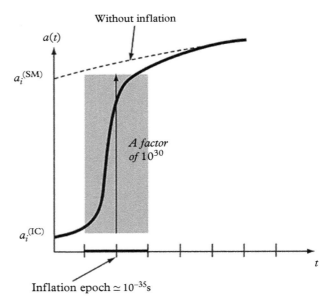

Figure 10.1 *Time evolution of the scale factor. The standard FLRW model curve is represented by the dashed line on top; the solid curve is that of the inflationary model, which coincides with the standard model curve after 10^{-35} s. The intercepts on the a axis correspond to the initial scales: $a_i^{(SM)}$ in the standard model (without inflation) and $a_i^{(IC)}$ in the inflationary cosmology.*

unified (indistinguishable from each other). Such theories appeared to imply a cosmic abundance of magnetic monopoles, in contradiction to their observed absence. He proposed (Guth 1981) that the early universe, when its thermal energy was comparable to 10^{16} GeV, could have had a large cosmological constant, which would have brought about a period of exponential expansion, thereby diluting the monopoles to a vanishingly small density. He also realized immediately that besides inflating away the monopoles, this idea provided a framework for solving the flatness and horizon problems discussed above.

In inflationary cosmology, the universe is assumed to have initially been endowed with a scalar field. This, at our present stage of understanding of particle physics theory, is a purely hypothetical energy-carrying quantum field. Following the particle physics convention of appending an "-on" to particle/field names, we call it the inflaton field. It is theorized that fluctuations of this field possessed for a finite time interval a large vacuum energy. This could be the origin of the cosmological constant that drove the big bang. Guth's original model[5] was quite specific: this effective cosmological constant was related to the false vacuum of spontaneous symmetry breaking in the GUT, and the inflaton field was identified with the associated Higgs field. However, we now view inflation as a general framework whose phenomenological success is quite independent of any specific theory of how the cosmological constant was actually generated. While there is no observational evidence for any specific inflation mechanism, we will present in Box 10.1 a brief account of the phase transition associated with spontaneous symmetry breakdown.[6] This provides a concrete example (in semiclassical language) of how such an effective cosmological constant might have been generated during the inflationary epoch.

[5] Guth's original idea was critically refined by the work of Andrei Linde (1948–) and others.

[6] For further discussion of the original GUT realization, see (Cheng 2010, Sections 11.2 and 11.6).

Box 10.1 Inflation via spontaneous symmetry breaking

Here we discuss the realization of an inflation-inducing cosmological constant through the spontaneous symmetry breaking (SSB) phase transition of a GUT. In this theory, the inflaton field is identified with the Higgs field that breaks the symmetry of the grand unified interaction, thereby making the strong and electroweak interactions distinct in the low-energy Standard Model regime.

In Fig. 10.2, the SSB phenomenon is illustrated with a Higgs field $\phi(x)$ with a temperature-dependent potential energy having a simple reflection symmetry $V(\phi) = V(-\phi)$. As the temperature drops, the form of the potential transitions from (a) to (b). During such a phase transition, the original vacuum state (i.e., the lowest-energy state) of the field system goes from the minimum of (a) into a local maximum of (b) for a finite time interval before rolling down into the minimum of (b), the final true vacuum state. During this rollover, the system is in a false vacuum; all space temporarily has a large energy density (with respect to the true vacuum). Thus the system (i.e., the universe) possesses an effective cosmological constant that can drive the exponential expansion as detailed in Section 8.4.2.

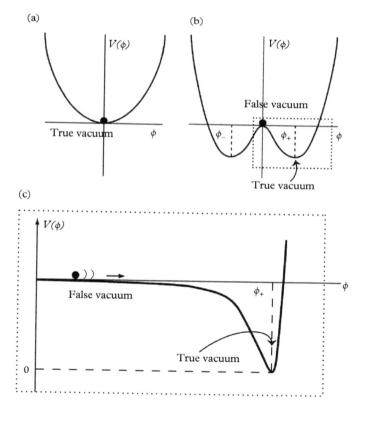

Figure 10.2 *The potential energy function of a Higgs field is illustrated by the simple case of $V(\phi) = \alpha\phi^2 + \lambda\phi^4$, possessing a discrete symmetry $V(-\phi) = V(\phi)$. The parameter α is temperature-dependent; let us assume $\alpha = \alpha_0(T - T_c)$, with positive constants α_0 and λ. (a) Above the critical temperature $(T > T_c$, hence $\alpha > 0)$, the lowest-energy state (the vacuum) is at $\phi_0 = 0$, which is symmetric under $\phi \rightarrow -\phi$. (b) Below T_c (hence $\alpha < 0$), the symmetric $V(\phi)$ has its minimum at $\phi_\pm = \pm\sqrt{-\alpha/2\lambda}$, while $V(\phi = 0)$ is a local maximum. Choosing a vacuum state (either ϕ_+ or ϕ_-) breaks the symmetry. The dashed box in (b) is enlarged in (c), but the potential function is changed from our simple example to a function with a larger flat portion near the $\phi = 0$ origin, allowing a slow-rollover transition. The dot represents the changing location of the system; it rolls from the high plateau of the false vacuum toward the true vacuum at the bottom of the trough.*

Inflation and the conditions it left behind

During the epoch of inflation, the scale factor grew exponentially: $a(t) = a(0) \exp(t/\tau)$, so the Hubble constant was in fact constant $H = \dot{a}/a = 1/\tau$; cf. (8.87). This expansion solves several outstanding cosmological problems.

The horizon problem solved   Two fixed points at a comoving distance D_0 are separated at time t by a proper distance $D(t) = a(t)D_0$ that grows with speed $V(t) = D_0\dot{a} = D(t)\dot{a}/a = D(t)/\tau$. This speed may well exceed c without violating the rule that nothing may travel faster than light relative to the comoving (expanding) rest frame.[7] By allowing such seemingly superluminal expansion, this inflationary scenario can solve the horizon problem, because two points further apart than the apparent horizon length (whether in the present era or at the photon-decoupling time when the CMB was created) could still have been in causal contact before the onset of the inflationary expansion. This phenomenon is not unique to inflationary cosmology—any two points separated by a distance greater than the Hubble length l_H recede faster than c. In a universe that is radiation- or matter-dominated back to time zero, the horizon lengths are two or three times larger than their naive value ct; moreover, an empty universe has no finite horizons, and those in a Λ-dominated universe grow exponentially.[8] Thus, during inflation, space could have grown so much so quickly (as in Fig. 10.1) that the entire observed universe could have originated from a much smaller region in thermal equilibrium from the beginning.

The flatness problem solved   Inflationary cosmology can also solve the flatness problem by stretching any initial curvature of space to virtual flatness. Applying the exponential expansion (8.87) to the Friedmann equation (10.1) yields the ratio

$$\frac{1 - \Omega(t_2)}{1 - \Omega(t_1)} = \left[\frac{\dot{a}(t_1)}{\dot{a}(t_2)}\right]^2 = e^{-2(t_2 - t_1)/\tau}. \tag{10.3}$$

Just as the scale factor was inflated by perhaps as much as $a(t_2)/a(t_1) = e^{(t_2 - t_1)/\tau} = 10^{30}$, the right-hand side may be as small as 10^{-60}. If we start with any reasonable value of $\Omega(t_1)$, we still have after inflation a very flat universe: $\Omega(t_2) = 1$ to high precision. Although gravitational attraction slows the expansion of the universe, enhancing its curvature while driving its density away from the critical value (hence the flatness problem), the accelerating expansion due to the vacuum repulsion pushes the universe (very rapidly) toward flatness: $\Omega = 1$. Thus a firm prediction of the inflationary scenario is that the universe left behind by inflation must have a 3D space with a $k = 0$ flat geometry. This implies a critical density equal to the value (10.1), but does not specify what energy/matter components make up such a density. This prediction appears to contradict the observed matter density (even with the inclusion of dark matter), which is short of the required $\Omega_0 = 1$. Nevertheless, as we shall discuss below, the observed CMB provided evidence that the spatial geometry was (and is) indeed flat.

[7] Namely, according to relativity, an object cannot travel through space faster than light (relative to any local observer), but there is no restriction stipulating that space itself (i.e., a proper distance between remote observers) cannot expand faster than c.

[8] We can change the time coordinate, $dT = dt/a(t)$, to make the Robertson–Walker metric (8.26) conformal: $ds^2 = a(T)(dT^2 - dl^2)$. This conformal space has the same lightcone structure as static space, so the horizon is just $cT = c\int_0^t dt'/a(t')$. You may plug in the scale functions for a radiation-dominated, matter-dominated, or Λ-dominated universe, or an empty universe, and integrate to get the given results. Cf, e.g., (10.17) and (10.18).

CMB anisotropy and evidence for a flat universe

The first strong evidence of the flat universe predicted by inflationary cosmology came from the observed CMB anisotropy. Following its first detection by the COBE satellite in the early 1990s, there were several balloon-based CMB experiments. In particular, collaborations such as Boomerang and Maxima-1 detected CMB fluctuations on smaller angular scales. They observed that the characteristic angular width of the temperature fluctuations in the CMB is just under 1°, so the power spectrum peaks at the multipole number $l \simeq 200$ (de Bernardis et al. 2000; Hanany et al. 2000). This is just the result we expect in a flat universe (estimated in Box 10.2). If the universe had any curvature, it would have distorted the observed angular anisotropy. Recall that the CMB radiation has been traveling uninterrupted since the universe was young, $t_\gamma \ll t_0$; the great distance it has traveled makes the CMB an ideal probe of the geometry of the universe—positive curvature ($k = +1$) would have focused the CMB rays, making the anisotropy scale appear larger; a negative curvature ($k = -1$) would have done the opposite, causing the CMB rays to diverge and shrinking the apparent scale of the anisotropy.

Box 10.2 Detecting curvature via the CMB's angular horizon

The inflationary epoch left density inhomogeneities, the seeds of the cosmic structure we observe today. In the early universe before the photon-decoupling time t_γ, the gravitational clumping of baryons was resisted by photon radiation pressure. This set up acoustic waves of compression and rarefaction, with gravity providing the driving force and radiation pressure the restoring force. These **baryon acoustic oscillations (BAOs)** took place against a background of dark matter fluctuations, which were not subject to electromagnetic radiation pressure and thus started to grow much earlier. The photon–baryon fluid can be idealized by ignoring the dynamical effects of gravitation and baryons (because the photon number density is much higher than that of the baryons). This leads to a sound-wave speed

$$c_s \simeq \sqrt{\frac{p}{\rho}} \simeq \frac{c}{\sqrt{3}}, \tag{10.4}$$

as the pressure and density can be approximated by those for radiation,[9] $p_R \simeq \rho_R c^2/3$. These cosmic sound waves left an imprint that is still discernible today. The compression and rarefaction were translated through gravitational redshift into a temperature inhomogeneity. By a careful analysis of this temperature anisotropy pattern in the CMB, we can garner much information about the early universe.

[9] See Sidenote 32 in Chapter 8.

continued

[10] This is just the relation between angle, arclength, and radius; see, e.g., Fig. 11.3(a). Just as linear momentum is conjugate to distance, angular momentum is conjugate to angle.

Box 10.2 *continued*

The sound horizon is the comoving distance (as measured today) that the BAOs would have traveled from the beginning of the universe $(t = 0)$ to the photon decoupling at t_γ:

$$r_{\rm sh} = c_{\rm s} \int_0^{t_\gamma} \frac{dt}{a(t)}. \qquad (10.5)$$

Now, in a universe with flat spatial geometry, such a horizon length $r_{\rm sh}$ on the surface of last scattering would appear as angular anisotropy of the size[10] $\theta_{\rm sh} \simeq r_{\rm sh}/R(t_\gamma)$, where $R(t_\gamma)$ is the comoving distance between us (t_0) and the surface of last scattering (t_γ),

$$R(t_\gamma) = c \int_{t_\gamma}^{t_0} \frac{dt}{a(t)}. \qquad (10.6)$$

As $r_{\rm sh}$ is the sound horizon distance, $\theta_{\rm sh}$ must then be the maximal angular size of the primordial density fluctuation.

When evaluating the integrals in (10.5) and (10.6), we shall assume a matter-dominated flat universe with scale factor $a(t) \propto t^{2/3}$ as given by (8.71):

$$\int \frac{dt}{a(t)} \propto \int a^{-1/2}\, da \propto a^{1/2} = (1 + z)^{-1/2}. \qquad (10.7)$$

The approximation of matter domination is plausible for the period prior to the photon-decoupling time t_γ because the radiation–matter equality time is several times smaller than the photon-decoupling time, as indicated by the redshift $z_{\rm RM} \gg z_\gamma$ (cf. Exercise 9.2); for the duration after t_γ, we ignore the effect of dark energy as the transition from matter to dark energy dominance is a "fairly recent" occurrence (cf. Exercise 10.4). Thus the angular size of the sound horizon at recombination can then be calculated

$$\theta_{\rm sh} \simeq \frac{r_{\rm sh}}{R(t_\gamma)} = \frac{c_{\rm s}(1 + z_\gamma)^{-1/2}}{c[(1 + z_0)^{-1/2} - (1 + z_\gamma)^{-1/2}]}$$

$$\simeq \frac{(1 + z_\gamma)^{-1/2}}{\sqrt{3}} \simeq 0.017\,{\rm rad} \simeq 1°, \qquad (10.8)$$

where we have used $z_0 = 0$, from (9.28) $z_\gamma \simeq 1100$, and from (10.4) a sound speed $c_{\rm s} \simeq c/\sqrt{3}$. This maximal angular separation in turn can be translated into the minimal multipole number $l_{\rm sh}$ (i.e., the first peak) of the CMB temperature anisotropy power spectrum as suggested by (9.48):

$$l_{\rm sh} \simeq \frac{\pi}{\theta_{\rm sh}} \simeq \pi \sqrt{3(1 + z_\gamma)} \simeq 180. \qquad (10.9)$$

Thus, in a flat universe, we expect the first peak of the power spectrum to be located at the multipole number[11] $\simeq 200$. The observed CMB power spectrum (shown in Fig. 9.4) has a prominent peak at this l value, corresponding to anisotropy spots just under 1° in angular size. This represents fundamental BAOs traveling from the beginning of the universe until the photon-decoupling time t_γ. Seen 14 Gyr later, a blob of this size will cover about 1° on the sky, if the spatial geometry of the universe is flat. As mentioned in the main text, in an universe with positive/negative curvature the observed angular size would be greater/smaller than this value.

[11] We quote the rounded-off value of 200 because (10.4)–(10.9) are rough estimates representing a simplified model. The realistic calculation in fact yields $l_{sh} = 225$.

That $k = 0$ of course implies, via the Friedmann equation, a total density $\Omega_0 = 1$. A careful matching of the measured power spectrum yields

$$\Omega_0 = 1.01 \pm 0.02. \qquad (10.10)$$

Dedicated satellite endeavors such as WMAP and subsequently Planck have produced power-spectrum results such as the one displayed in Fig. 9.4. There clearly shows that the first prominent peak is at $l \simeq 200$, indicating that the we do have a $k = 0$ flat universe. The higher-l part of the angular spectrum features peaks corresponding to acoustic oscillations. The positions and magnitudes of these peaks contain fundamental properties about the geometry and structure of the universe. Another influential cosmological project has been the survey of galaxy distributions by the Sloan Digital Sky Survey (SDSS). Their high resolution allowed them to extract[12] many important cosmological parameters: H_0, Ω_0, $\Omega_{M,0}$, Ω_B, the deceleration parameter q_0, etc.

[12] The approach in the analysis of galactic distribution is similar to that of the CMB temperature anisotropy.

The origin of matter/energy and structure in the universe

Besides the flatness and horizon problems, another issue with the standard FLRW cosmology is that it requires as an initial condition that all the energy and matter of the universe be present at the very beginning. Furthermore, this hot plasma of particles and radiation should have just the right amount of initial density inhomogeneity (density perturbation) to form, through subsequent gravitational clumping, the cosmic structure of galaxies, clusters of galaxies, voids, etc. that we observe today. Remarkably, inflationary cosmology can explain the origin of matter/energy, as well as the structure of the universe. In fact, many would argue that, while it is impressive that inflation can solve the horizon and flatness problems, its greatest success is in furnishing the mechanisms for particle production and for seeding the inhomogeneities required for the subsequent formation of cosmic structure.

Particle creation via quantum fluctuation In this inflationary scenario, all the matter and energy could have been created virtually from nothing; furthermore, the phenomenon of particle creation in an expanding universe may be qualitatively understood. In terms of the semiclassical potential energy picture of Fig. 10.2, particles were created when the system fell into and oscillated around the true vacuum.[13] Particle creation may be understood more directly in terms of quantum fluctuations of the field system. These fluctuations can take the form of the appearance and disappearance of particle–antiparticle pairs in the vacuum. Such energy-nonconserving processes are permitted as long as they take place on a sufficiently short timescale Δt that the uncertainty relation $\Delta E \Delta t \leq \hbar$ is not violated. In a static space, such virtual processes do not create real particles. However, in an expanding space, when the expansion rate exceeds the annihilation rate (recall the Gamow condition (9.12)), real particles may be created.[14] Thus, inflation in conjunction with quantum field theory naturally gives rise to particle production. This hot, dense, uniform collection of particles is just the postulated initial state of the standard big bang model. Furthermore, inflation could stretch the subatomic-sized fluctuations of a quantum field into astrophysical-sized density perturbations to seed the subsequent formation of cosmic structure. The resulting density fluctuations would be Gaussian (maximally random) and scale-invariant (i.e., the same fluctuations, of the order of 10^{-5}, on all length-scales), in agreement with the observed CMB anisotropy power spectrum and large-scale galaxy surveys.

Inflation and primordial gravitational waves Inflationary expansion would necessarily have created great ripples in spacetime, i.e., gravitational waves. Thus the detection of such primordial gravitational waves would constitute definitive evidence of cosmic inflation. These waves would have impacted upon all the matter in the universe. In particular, the spin-2 tensor nature of such perturbations, cf. (6.44), would have led to anisotropies of the plasma and its coupled radiation at recombination. This anisotropic radiation would have become polarized by scattering off electrons (Thomson scattering); cf. Box 10.3. From the presence of such a polarization pattern (the so-called B-mode), we can in principle deduce the existence of the quadrupole anisotropy imprinted by (and only by) primordial gravitational waves on the plasma just before photon decoupling. These CMB signals are expected to be very weak, because such polarization could only have been produced during the ending period of the recombination era, even weaker than the temperature anisotropy discussed previously. Thus, it came as a great surprise that the BICEP2 Collaboration (2014) announced their discovery of B-mode polarization in the CMB. This discovery still awaits confirmation by other groups, because of the concerns regarding possible contamination by foreground dust from our own galaxy.[15] Regardless of the outcome of this particular observation, such a test of the inflationary theory of the big bang is an important, albeit advanced, topic of which a student of cosmology should be cognizant.

[13] Thus all particles originated from the conversion of the inflaton field's false vacuum energy, which came about through a phase transition.

[14] This way of seeding the cosmic structure can be viewed as Hawking radiation from inflation. Recall our discussion in Section 7.3 of Hawking radiation from a black hole, in which virtual particles are turned into real ones because of the black hole event horizon. Here the production of real particles from quantum fluctuations comes about because of the effective horizon created by the hyper-accelerating expansion of the universe.

[15] As this manuscript was being readied for publication, a new announcement (BICEP2/Keck, Planck collaborations, 2015) informed us that the latest analysis showed that the 2014 optimistic conclusion was indeed unwarranted. However, with several collaborations having ever more powerful instruments pursuing this study, it is hoped that the primordial gravitational wave signal could be discovered with more sensitive analyses.

Box 10.3 CMB polarization: Thomson scattering and E- and B-mode decomposition

The low-energy ($\ll m_e c^2$) limit of Compton scattering of a photon by an electron (with unchanged photon frequency $\omega' \simeq \omega$, cf. Exercise 3.6) is called **Thomson scattering**. The physical process can be interpreted as dipole re-radiation by the electron. Thus the cross section, which is proportional to the classical electron radius squared, r_0^2, has a characteristic dependence on the angle ($\hat{\varepsilon}_1 \cdot \hat{\mathbf{k}}' \equiv \cos\alpha$) between the final photon direction $\mathbf{k}'$ and the initial photon polarization $\hat{\varepsilon}_1$ (which lies in the plane transverse to the propagation direction, $\hat{\varepsilon}_1 \cdot \hat{\mathbf{k}} = 0$):

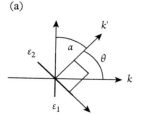

(a)

$$\left(\frac{d\sigma}{d\Omega}\right)_1 = r_0^2 \sin^2\alpha, \quad \text{and} \quad \left(\frac{d\sigma}{d\Omega}\right)_2 = r_0^2, \quad (10.11)$$

where $r_0 = e^2/m_e c^2$, and we have also displayed the cross section for an initial photon having a polarization $\hat{\varepsilon}_2$ perpendicular to $\hat{\varepsilon}_1$ and therefore (if we take $\hat{\varepsilon}_1$ to be in the plane spanned by $\hat{\mathbf{k}}$ and $\hat{\mathbf{k}}'$) perpendicular to the final radiation direction as well, $\hat{\varepsilon}_2 \cdot \mathbf{k}' = 0$; see Fig. 10.3(a). The scattering of unpolarized radiation is the average of these two cross sections:

$$\left(\frac{d\sigma}{d\Omega}\right)_{\text{unpol}} = \frac{1}{2}\left[\left(\frac{d\sigma}{d\Omega}\right)_1 + \left(\frac{d\sigma}{d\Omega}\right)_2\right] = \frac{1}{2}r_0^2(1 + \cos^2\theta), \quad (10.12)$$

where the result is expressed in term of the scattering angle $\hat{\mathbf{k}} \cdot \hat{\mathbf{k}}' \equiv \cos\theta$, as $\theta = 90° - \alpha$. Thus, as shown in (10.11), for perpendicular ($\theta = 90°$) scattering, the final-state photon is 100% polarized, with $\hat{\varepsilon}' = \hat{\varepsilon}_2$.

Above, we considered a single unpolarized photon producing a polarized photon through Thomson scattering. It is clear that for isotropic incoming radiation, with the same intensity incoming from all directions, the final radiation, being a sum of all these contributions, will again be unpolarized. As a corollary, one will have **polarized** light (in some direction) through Thomson scattering of unpolarized radiation if the initial radiation is **anisotropic**. We illustrate this point by a simple example in Fig. 10.3(b), where polarized light is obtained by scattering two unpolarized light beams of different intensities coming from perpendicular directions (as a quadrupole anisotropy).

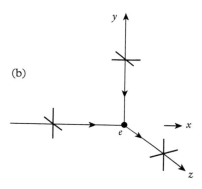

(b)

Figure 10.3 *Thomson scattering of radiation by an electron. (a) Kinematics of photon scattering $\gamma(\mathbf{k}) \to \gamma(\mathbf{k}')$, with the initial photon having polarizations $\varepsilon_1 \perp \varepsilon_2$. (b) Anisotropic unpolarized initial radiation can produce polarization in scattered radiation. Unpolarized ($\varepsilon_y^{(1)} = \varepsilon_z^{(1)}$) incoming light moving along the x axis is scattered into fully polarized light ($\varepsilon_y'^{(1)}$) moving in the z direction. Similarly, unpolarized incoming light ($\varepsilon_x^{(2)} = \varepsilon_z^{(2)}$) moving in the y direction can be scattered into polarized light ($\epsilon_x'^{(2)}$) moving in the z direction. The thicknesses of the polarization bars indicate their intensities. Thus Thomson scattering of anisotropic radiation (with unequal intensities of incoming light in the x and y directions) will produce polarized ($\varepsilon_x'^{(2)} \neq \varepsilon_y'^{(1)}$) outgoing light in the z direction.*

continued

Box 10.3 *continued*

How can we characterize the polarization state of light? The electric (or magnetic) field of radiation is perpendicular to the direction of radiation propagation. There are two independent polarization directions (in the transverse plane), each with an amplitude and phase angle. Thus, for a given light intensity, there are two independent quantities characterizing such a polarization state. Because the sum of the two squared polarization amplitudes is fixed by the light intensity, there is only one independent magnitude of the amplitudes; the second quantity is the relative phase of these amplitudes. A common characterization scheme uses Stokes' Q and U parameters. However, under a rotation in the transverse polarization plane, Q and U do not transform as the two components of a vector, but rather as the two independent elements of a 2×2 symmetric traceless tensor. Linear combinations of Q and U form[16] the E and B amplitudes. The key property of these components is that E is a scalar while B is a pseudoscalar under reflection. They are called E- and B-modes of polarization, because this parity property is the same as that of scalar projections of the familiar electric and magnetic fields (e.g., $\mathbf{E} \cdot \hat{\mathbf{r}}$ and $\mathbf{B} \cdot \hat{\mathbf{r}}$) under parity. Some simple examples of E- and B-modes are shown in Fig. 10.4.

The E and B decomposition is important, because while the temperature anisotropy discussed in Section 9.3.2 is mainly due to the matter/radiation density perturbation that seeded the structure of the universe, other types of perturbation (e.g., primordial gravitational waves and topological structures) can also contribute to the anisotropy in the CMB temperature and polarization. Crucially, especially on large angular scales, only gravitational waves can bring about polarizations in the B-mode. Hence their detection would be direct evidence of gravitational waves, as well as of the inflationary theory of the big bang.

[16] This is done by taking the covariant derivatives over a 2-sphere of this polarization tensor. Covariant derivatives in a curved space will be discussed in the next chapter.

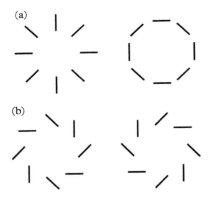

Figure 10.4 *Patterns of polarization: (a) E-mode vs. (b) B-mode. While E polarization is even under a reflection, B polarization is odd. The displayed B-modes swap under reflection, taking two reflections to get back to the same pattern.*

10.1.3   Eternal inflation and the multiverse

At our current level of understanding, inflationary cosmology is a framework, not a theory. It postulates the existence of a quantum inflaton field that momentarily transitioned to a state having a large cosmological constant. In the specific spontaneous symmetry breaking example of Box 10.1, the false vacuum brought about the exponential expansion of the space. In this semiclassical description, the system then rolled down the potential energy curve slowly enough to allow it to convert its potential energy into particles and kinetic energy. This formed our region of the cosmos, and was followed by the decelerating inertial expansion of the FLRW cosmology. From a broader perspective, one can regard the false vacuum as a quantum state subject to quantum fluctuations, and the slow-roll process as a quantum decay. From this viewpoint, such a transition was a random process. This suggests a wider view of the cosmos as a large collection of widely separated regions, each the aftermath of a portion of space that underwent inflationary explosion. One might picture such a collection as a multiverse, a huge cauldron of chaotic and eternal inflation, with causally disconnected parallel universes perpetually bubbling into existence.

10.2 The accelerating universe in the present era

We now return to the discussion of the FLRW cosmological evolution after the inflationary epoch.

Problems with a matter-dominated flat universe   By the mid to late 1990s, there was definitive evidence that the geometry of the universe is flat, as predicted by inflation. Nevertheless, there were several pieces of phenomenology that appeared to contradict such a matter-dominated flat universe.[17]

A missing energy problem   The Friedmann equation (8.48) requires a $k = 0$ flat universe to have a mass/energy density exactly equal to the critical density, $\Omega_0 = 1$, yet we observe, including both the baryonic and dark matter, only about a third of this value (radiation energy is negligibly small in the present epoch):

$$\Omega_{M,0} = \Omega_B + \Omega_{DM} \simeq 0.31. \tag{10.13}$$

Thus. if the universe is flat, it appears that we have a missing energy problem.

A cosmic age problem   From our discussion of the time evolution of the universe, we learned (see (8.71)) that a matter-dominated flat universe should have an age equal to two-thirds of the Hubble time,

$$(t_0)_{\text{flat}} = \frac{2}{3} t_H \lesssim 10\,\text{Gyr}, \tag{10.14}$$

which is shorter than the estimated age of old stars, notably the globular clusters, which have been deduced to be older than 12 Gyr (cf. Section 8.1.2). Thus it appears that we also have a cosmic age problem.

10.2.1 Dark energy and its effect

A possible resolution of these phenomenological difficulties of a flat universe (and hence of inflationary cosmology) would be to assume the presence of dark energy.

Dark energy is defined as an energy component having a negative pressure corresponding to an equation-of-state parameter[18] $w < -1/3$, so

[17] Since radiation dominance ended at the radiation–matter equality time $t_{RM} = O(10^5\,\text{yr}) \ll t_0$, we can reasonably approximate the universe as matter-dominated during its entire 14 Gyr lifetime.

[18] See (8.45) for the definition of w.

that in a dark-energy-dominated universe, the second derivative of the scale factor is positive, cf. (8.43):

$$\ddot{a} \propto -\left(p + \frac{1}{3}\rho c^2\right) = -\left(w + \frac{1}{3}\right)\rho c^2 > 0. \qquad (10.15)$$

Namely, such a negative pressure component, as shown in (8.80), gives rise to a gravitational repulsion. The simplest example of dark energy[19] is Einstein's cosmological constant, with $w = -1$. The cosmological constant assumed to be present after inflation cannot have the immense size of the one that drove the inflation. Rather, the constant dark energy density ρ_Λ should be only the missing two-thirds of the critical density required for a flat universe:

[19] One should not confuse dark energy with the energies of heavy neutrinos, black holes, etc., which are also nonluminous, but are counted as dark matter (cf. Section 8.1.2), because the associated pressure is not negative.

$$\Omega = \Omega_M + \Omega_\Lambda = 1, \qquad (10.16)$$

where $\Omega_\Lambda \equiv \rho_\Lambda/\rho_c$. A nonvanishing Λ would also provide the repulsion to accelerate the expansion of the universe. In such an accelerating universe, the expansion rate in the past, $\dot{a}(t)$, must have been smaller than the current rate H_0. This means that it would have taken a longer period to reach the present era than in a flat matter-dominated universe; hence $t_0 > 2t_H/3$. The dark energy might thereby also solve the cosmic age problem mentioned above.

10.2.2 Distant supernovae and the 1998 discovery

Because light from distant galaxies reaches us by moving with finite speed, to measure a star (or a supernova) farther out in distance is to probe the cosmos farther back in time. An accelerating expansion means that the expansion rate $\dot{a}(t)$ was smaller in the past. In a Hubble diagram, which plots luminosity distance (explained in Box 8.2) against the observed redshift (recession velocity), the Hubble curve of an accelerating universe bends upward as in Fig. 10.5. Thus objects of

Figure 10.5 *Hubble diagram: the Hubble curve for an accelerating universe bends upward. A supernova on this curve at a given redshift would be further away than anticipated (for a decelerating universe).*

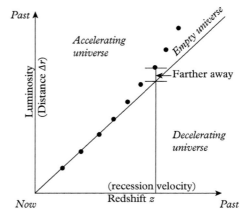

given redshift, i.e. a fixed z in (8.5), must be located farther away (a larger D to compensate the smaller expansion rate) than otherwise expected from a decelerating or empty universe; an observed light source would be dimmer than expected.

In order to observe the changing expansion rate of the universe (i.e., to measure the curvature of the Hubble curve), one must measure great cosmic distances. One needs a gauge of distance that works to over five billion light-years. Clearly, some very bright light sources are required. Since this also means that we must measure objects of an age that is a significant fraction of the age of the universe, the method must be applicable to objects present in the early cosmic era. As it turns out, a certain class of supernovae are ideally suited for this purpose.

SNe Ia as standard candles and their systematic search

Type Ia supernovae (SNe Ia) begin as white dwarfs (collapsed old stars sustained by the degeneracy pressure of their electrons) with masses comparable to the sun's. A white dwarf often has a companion star; the dwarf's powerful gravitational attraction draws matter from its companion. Its mass increases up to the Chandrasekhar limit $\simeq 1.4 M_\odot$ (cf. Section 7.2.2). When it can no longer be countered by the electron degeneracy pressure, the gravitational contraction proceeds, and the resultant heating of the core triggers a thermonuclear blast, resulting in a supernova explosion. The supernova eventually explodes and collapses. Because they start with masses in a narrow range, such supernovae have comparable intrinsic brightness. Furthermore, their brightness exhibits a characteristic decline from its maximum, which can be used to improve the calibration of their luminosity (light-curve shape analysis), making SNe Ia standardizable candles.[20] Supernovae are rare events in a galaxy. The last time a supernova explosion occurred in our own Milky Way was about 400 years ago. However, using new technology (large mosaic CCD[21] cameras), astronomers overcame this problem by simultaneously monitoring thousands of galaxies, so that on average some 10–20 supernovae can be observed in a year.

The discovery of an accelerating universe

By 1998, two collaborations,[22] the Supernova Cosmological Project, led by Saul Perlmutter (1959–), and the High-Z Supernova Search Team, led by Adam Riess (1969–) and Brian Schmidt (1967–), each had accumulated some 50 SNe Ia at high redshifts ($0.4 < z < 0.7$) corresponding to times five to eight billion years ago. They made the astonishing discovery (Fig. 10.6) that the expansion of the universe was actually accelerating, as indicated by the facts that the measured luminosities were on average 25% less than anticipated, and that the Hubble curve bent upward as in Fig. 10.5.

Extracting Ω_M and Ω_Λ from the measured Hubble curve

The Hubble curve traces the SNe's (luminosity) distances[23] as a function of their redshifts $D_L(z)$. Since Ω_M tends to bend it downward and Ω_Λ upward, by fitting

[20] Another actively investigated theory of the origin of SNe Ia posits that they are mergers of white dwarfs. Namely, there are several means by which a supernova of this type can form, but they share a common underlying mechanism.

[21] CCDs (charge-coupled devices) have the ability to transfer charges along the surface of a semiconductor from one storage capacitor to the next, so that incoming photons can be registered digitally.

[22] The discovery papers are (Riess et al. 1998) and (Perlmutter et al. 1999).

[23] In Box 8.2, we discussed the relation between luminosity distance and the more familiar proper distance, deriving a simple result (8.40) to be used in (10.19).

the observed Hubble curve, one can extract the values of Ω_M and Ω_Λ. To do so, one must express $D_L(z)$ in such a way that its dependence on the energy densities is explicit.

To observe distant galaxies' redshifts, one must receive light signals, $ds^2 = 0 = -c^2 \, dt^2 + a^2 \, dD_p^2$. The present proper distance from the light-emitting source is thus given by

$$D_p = \int_{t_{em}}^{t_0} \frac{c \, dt}{a(t)}. \tag{10.17}$$

We can replace the time integration variable by the scale factor, $dt = da/aH$ (as $H = \dot{a}/a$), and then by the redshift, $dz = -da/a^2$, because $(1+z) = a^{-1}$. This integral can thereby be expressed in terms of the redshift z of the source galaxy (using the minus sign in dz to swap the limits of integration):

$$D_p(z) = \int_0^z \frac{c \, dz'}{H(z')}. \tag{10.18}$$

Substituting this into (8.40), $D_L = (1 + z)D_p$, then yields the integral expression for the luminosity distance as a function of redshift:

$$D_L(z) = c(1 + z) \int_0^z \frac{dz'}{H(z')}. \tag{10.19}$$

The explicit dependence on the density parameters, $\Omega_{M,0}$ and Ω_Λ, enters through the Hubble constant in the above denominator via the Friedmann equation (8.47).

Exercise 10.2 Redshift dependence of the Hubble constant via the energy densities:

(a) Use the Friedmann equation to express the epoch-dependent Hubble constant $H(a) = \dot{a}/a$ in terms of the density parameter $\Omega(a) = \rho/\rho_{c,0}$.

Hint: You can replace the curvature parameter k term in (8.42) by the expression shown in (8.47), and then the $\dot{a}^2(t_0)$ factor by H_0^2. Finally, a combination of factors can be identified through (8.14) as the present critical density $\rho_{c,0}$.
(b) Put the scale dependence of the matter density in $\rho(a) = \rho_M(a) + \rho_\Lambda$ (neglect the radiation density, which is negligible long after t_{RM}) to find

$$H(a) = H_0 \left(\frac{\Omega_{M,0}}{a^3} + \Omega_\Lambda + \frac{1 - \Omega_0}{a^2} \right)^{1/2}, \tag{10.20}$$

or, in terms of the redshift in (8.35),

$$H(z) = H_0 [\Omega_{M,0}(1 + z)^3 + \Omega_\Lambda + (1 - \Omega_{M,0} - \Omega_\Lambda)(1 + z)^2]^{1/2}. \tag{10.21}$$

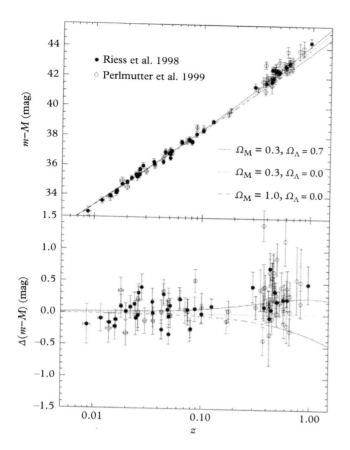

Figure 10.6 *Discovery of an accelerating universe. The Hubble plot shows data points lying above the (matter-dominated flat universe) dotted line. The solid curve is the best fit of the observational data. The vertical axes represent the distance modulus $m - M$, directly related to the luminosity distance of (8.37). In the lower panel $\Delta(m - M)$ is the difference between the measured distance modulus and its expected value for a decelerating universe with $\Omega_M = 0.3$ and $\Omega_\Lambda = 0$. The figure is reproduced with permission from (Riess 2000). ©2000, U. Chicago Press.*

With the expression (10.21) for $H(z)$ in the integrand of (10.19), one can then calculate the luminosity distance $D_L(z)$, plotted in Fig. 10.6. The observed upward bend in the Hubble curve roughly fixes $\Omega_\Lambda - \Omega_{M,0}$, as Ω_Λ lifts it up while $\Omega_{M,0}$ bends it downward. The values of $\Omega_{M,0}$ and Ω_Λ that best fit the Hubble curve to the supernova data and that yield a flat universe (consistent with CMB anisotropy data) are shown in Fig. 10.7. From these and other subsequent observations, we get

$$\Omega_{M,0} = 0.315 \pm 0.017 \quad \text{and} \quad \Omega_\Lambda = 0.692 \pm 0.010, \tag{10.22}$$

suggesting that most of the energy in our universe now resides in this mysterious dark energy, which behaves like the cosmological constant.

The cosmic time at scale factor a

Matter content shortens our estimate of the universe's age t_0 (from the Hubble time t_H, the age of an empty universe), while dark energy lengthens it; given the

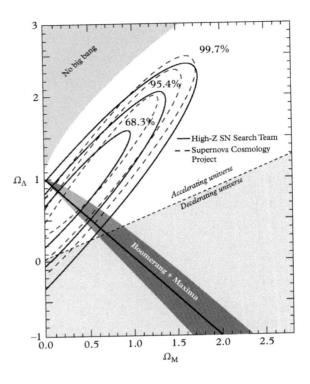

Figure 10.7 *Fitting Ω_Λ and Ω_M to the data. The favored values of CMB anisotropy experiments follow from the central values for a flat universe, $\Omega_\Lambda + \Omega_M \simeq 1$ (the straight line); those of the SNe data are represented by confidence contours (ellipses) around $\Omega_\Lambda - \Omega_M \simeq 0.5$. The graph is reproduced with permission from (Riess 2000). ©2000, U. Chicago Press.*

observed values for $\Omega_{M,0}$ and Ω_Λ, we can calculate t_0. In fact, we shall derive a more general result for $t(a)$, the age of the universe at a particular scale factor a. Of course t_0, the age of the universe now, corresponds to the cosmic time at $a = 1$. Again, the procedure is just like that in (10.17) and (10.18): one first expresses time as an integral over the scale factor, with the integrand in terms of the Hubble constant:

$$t(a) = \int_0^t dt' = \int_0^a \frac{da'}{a' H(a')}. \tag{10.23}$$

Inserting (10.20) for the scale-dependent Hubble constant yields an expression for the age in terms of the density parameters:

$$t(a) = t_{\mathrm{H}} \int_0^a \frac{da'}{[\Omega_{M,0}/a' + \Omega_\Lambda a'^2 + (1 - \Omega_0)]^{1/2}}. \tag{10.24}$$

Assuming a spatially flat universe, $\Omega_0 = \Omega_{M,0} + \Omega_\Lambda = 1$, simplifies the expression:

$$\begin{aligned} \frac{t(a)}{t_{\mathrm{H}}} &= \int_0^a (\Omega_{M,0}/a' + \Omega_\Lambda a'^2)^{-1/2} \, da' \\ &= \frac{2}{3\sqrt{\Omega_\Lambda}} \ln\left(\sqrt{\frac{\Omega_\Lambda}{\Omega_{M,0}} a^3} + \sqrt{1 + \frac{\Omega_\Lambda}{\Omega_{M,0}} a^3} \right). \end{aligned} \tag{10.25}$$

The age of the universe now   We use this result to the calculate the present age of the universe $t(a = 1)$:

$$\frac{t_0}{t_H} = \frac{2}{3\sqrt{\Omega_\Lambda}} \ln\left(\frac{\sqrt{\Omega_\Lambda} + \sqrt{\Omega_{M,0} + \Omega_\Lambda}}{\sqrt{\Omega_{M,0}}} \right) = 0.955. \qquad (10.26)$$

Thus, for the density values given in (10.22), the right-hand side comes very close to unity. Thus the decelerating effect of $\Omega_{M,0}$ and the accelerating effect of Ω_Λ coincidentally almost cancel each other. The age is very close to that of an empty universe, $[t_0]_{empty} = t_H = 14.46\,\text{Gyr}$;

$$t_0 = 0.955\,t_H = 13.8\,\text{Gyr}. \qquad (10.27)$$

Exercise 10.3 A check on the cosmic age formula

For a matter-dominated flat universe without a cosmological constant $(\Omega_0 = \Omega_{M,0} = 1$ with $\Omega_{\Lambda,0} = 0)$, check that the cosmic age formula (10.26) has the correct limit of $t_0 = \frac{2}{3}t_H$ in agreement with the result obtained in (8.71).

Transition from deceleration to acceleration

Since the immediate observational evidence from these faraway supernovae is a lower-than-anticipated luminosity, one might wonder whether there is a more mundane astrophysical explanation. There may be one or a combination of several humdrum causes that can mimic the observational effects of an accelerating universe. Maybe this luminosity diminution is brought about not because the supernovae were further away than expected, but is due to absorption by yet-unknown[24] interstellar dust and/or to some yet-unknown evolution of the supernovae themselves (i.e., supernova intrinsic luminosities were smaller in the cosmic past). However, all such scenarios would lead us to expect that the brightness of supernovae at even greater distances (hence even further back in time) should continue to diminish.

For the accelerating universe, on the other hand, this diminution of luminosity would stop, and the brightness would increase at even larger distances. This is because we expect that the accelerating epoch was proceeded by a decelerating phase. The dark energy density should be relatively insensitive to scale change (constant throughout cosmic history), while the matter and radiation energy densities, $\rho_M \sim a^{-3}(t)$ and $\rho_R \sim a^{-4}(t)$, should have been more important in earlier times. Thus, the early universe could not have been dark-energy-dominated, so it must have been decelerating. The transition from a decelerating to an accelerating phase would show up as a bulge in the Hubble curve; see Fig. 10.8. For observational support for this transition, see (Riess et al. 2004).

[24] Absorption and scattering by ordinary dust shows a characteristic frequency dependence that can in principle be subtracted out. By unknown dust we mean any possible gray dust that could absorb light in a frequency-independent manner.

The cosmic age at transition   Let us estimate the redshift at which the universe made this transition. From the second derivative of the scale factor, we define an epoch-dependent dimensionless deceleration parameter

$$q(t) \equiv \frac{-\ddot{a}(t)}{a(t)H^2(t)}.$$ (10.28)

The second Friedmann equation (8.43) can then be written with the equation-of-state parameters $w_R = \frac{1}{3}$, $w_M = 0$, and $w_\Lambda = -1$ as

$$q(t) = \frac{1}{2} \sum_i \Omega_i(1 + 3w_i)$$

$$= \Omega_R(t) + \frac{1}{2}\Omega_M(t) - \Omega_\Lambda$$ (10.29)

$$= \frac{\Omega_{M,0}}{2[a(t)]^3} - \Omega_\Lambda,$$ (10.30)

where in the last line we have dropped (at $t \gg t_{RM}$) the unimportant $\Omega_{R,0}$ term. The transition from deceleration $(q > 0)$ to acceleration $(q < 0)$ occurred at time t_{tr} when the deceleration parameter vanished: $q(t_{tr}) \equiv 0$, or

$$[a(t_{tr})]^3 = \frac{\Omega_{M,0}}{2\Omega_\Lambda}.$$ (10.31)

Plugging this into (10.25) yields

$$\frac{t_{tr}}{t_H} = \frac{2}{3\sqrt{\Omega_\Lambda}} \ln\left(\sqrt{\tfrac{1}{2}} + \sqrt{\tfrac{3}{2}}\right) = 0.53.$$ (10.32)

Thus the universe made the transition from deceleration to accelerated expansion at a cosmic age $t_{tr} \simeq 7.6$ Gyr. That is, the cosmic expansion, after spending over eight billion years slowing down, started speeding up again and has been

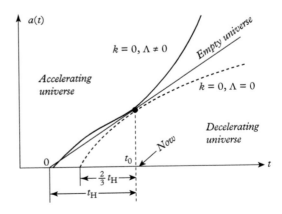

Figure 10.8 *Time evolution of an accelerating universe. It started out in a decelerating phase (the initial bulge), before expanding exponentially. The transition to an accelerating phase shows up as a point of inflection. This way, the universe has an age longer than $\frac{2}{3}t_H$, the age of a $\Lambda = 0$ flat matter-dominated universe. Compare this a(t) behavior with those without the cosmological constant, depicted in Fig. 8.7.*

accelerating for the past six billion years. The transition took place, in cosmic terms, only recently! This is sometimes referred to as the cosmological coincidence problem.

Exercise 10.4 Estimate of matter and dark energy equality time

---

Closely related to the deceleration/acceleration transition time t_{tr} is the epoch t_{MA} when the matter and dark energy components were equal, $\Omega_M(t_{MA}) = \Omega_\Lambda(t_{MA})$. Show that t_{MA} is comparable to and somewhat greater than $t_{tr} \simeq 7.6$ Gyr obtained above.

Further evidence for dark energy

We calculated the cosmic time $t_{tr} \simeq 8$ Gyr when the deceleration phase gave way to acceleration. Namely, only in the last 6 Gyr or so did the dark energy driving this accelerated expansion become the dominant force in the universe. The repulsive gravity of dark energy also slows the growth of the largest conglomerations of matter in the universe, the galaxy clusters. A group of researchers (Vikhlinin et al. 2009) used the Chandra X-ray satellite telescope to study the intensities and spectra of 86 clusters filled with hot gas that emitted X-rays. A set of 37 clusters more than five billion light-years away was compared with another set of 49 that are closer than half a billion light-years away (i.e., they are on average about five billion years younger). Theoretical models were used to calculate how the numbers of clusters with different masses would change during this timespan with different amounts of dark energy with different values of the equation-of-state parameter w. A good fit to the data clearly required the presence of dark energy with $w \simeq -1$.

10.2.3 The mysterious physical origin of dark energy

The introduction of the cosmological constant in the GR field equation does not explain its physical origin. In the inflationary framework, one could postulate that the false vacuum energy of a GUT Higgs field acts like an effective cosmological constant driving the inflationary expansion. While there is plenty of evidence for the correctness of the inflationary paradigm, there is no observational support for any particular mechanism as the source of the required cosmological constant.

What is the physical origin of the dark energy that brings about the accelerating expansion of the present epoch? Even though the observational evidence for dark energy is strong, its physical origin remains mysterious. Although the quantum vacuum energy does have constant density with respect to volume changes and hence a negative pressure, its estimated magnitude is way off—something like 120 orders of magnitude too large to account for the observed $\Omega_\Lambda = O(1)$; see the discussion in Box 10.4.

Box 10.4 Quantum vacuum energy as cosmological constant

Quantum field as a collection of quantum oscillators

Fourier components (normal modes) of waves obey simple harmonic oscillator equations. An electromagnetic radiation field (a solution to Maxwell's wave equation) can be thought of as a collection of oscillators. A quantized radiation field is a collection of quantum oscillators—simple harmonic oscillators described by quantum mechanics. Quantum field theory is usually presented as the union of quantum mechanics and special relativity. This is so for quantum electrodynamics, as Maxwell's wave equation satisfies special relativity. When quantum field theory is generalized to other particles, one works with other relativistic wave equations such as the Dirac equation, the Klein–Gordon equation, and the linearized GR wave equation. But the basic features of a quantized field discussed below remain the same.

Quantum vacuum energy has just the correct property of a cosmological constant

The simplest way to see that a quantum vacuum state has nonvanishing energy is to start with the observation that the normal modes (labeled by index i) of a field are simply a set of harmonic oscillators. Quantum mechanics predicts a discrete energy spectrum for such systems. Summing the quantized oscillator energies of all the modes, we have

$$E = \sum_i \left(\frac{1}{2} + n_i \right) \hbar \omega_i, \qquad \text{with } n_i = 0, 1, 2, 3 \ldots . \qquad (10.33)$$

From this, we can identify the vacuum energy (also called the zero-point energy) as

$$E_\Lambda = \sum_i \frac{1}{2} \hbar \omega_i. \qquad (10.34)$$

Thus, from the viewpoint of quantum field theory, a vacuum state (defined as the lowest-energy state) is not simply a void. The uncertainty principle informs us that the vacuum has an energy, since any localization (in field space) has an associated spread in the conjugate momentum value. In fact, quantum field theory portrays the vacuum as a sea of sizzling activity, with constant creation and annihilation of particles.[25] The zero-point energy has the key property of having a density that is unchanged with respect to any volume changes (i.e., an energy density that is constant). The summation of the mode degrees of freedom in (10.34) involves the enumeration of the phase-space volume in units of Planck's constant: $\Sigma_i = \int d^3x \, d^3 \, p(2\pi\hbar)^{-3}$. Since the

[25] At the atomic and subatomic levels, there is abundant empirical evidence for the reality of such a zero-point energy. For macroscopic physics, a notable manifestation of the vacuum energy is the Casimir effect, which has been verified experimentally.

zero-point energy $\hbar \omega_i / 2 = \epsilon(p_i)/2$ has no dependence on position, one obtains a simple volume factor $\int d^3 x = V$, so that the corresponding energy per unit volume, E_Λ / V, is constant with respect to changes in volume. As explained in Section 8.4.1, this constant positive energy density corresponds to a negative pressure $p_\Lambda = -E_\Lambda / V$.

Quantum vacuum energy is 10^{120}-fold too large to be dark energy

Nevertheless, a fundamental problem exists: the natural size of the quantum vacuum energy density is enormous. Here is a simple estimate of the sum in (10.34). The energy of a particle with momentum p is $\epsilon(p) = pc$ if the particle mass is negligible, $mc^2 \ll pc$, which it is over the main integration range. From this, we can calculate the sum by integrating over the momentum states to obtain the vacuum energy/mass density,

$$\rho_\Lambda c^2 = \frac{E_\Lambda}{V} = \int_0^{E_{\mathrm{Pl}}/c} \frac{4\pi p^2 \, dp}{(2\pi\hbar)^3} \left(\frac{1}{2} pc \right), \tag{10.35}$$

where $4\pi p^2 \, dp$ is the usual momentum phase-space volume factor. The integral in (10.35) would have diverged had we carried the integration to infinity. Infinite-momentum physics means zero-distance-scale physics. Since we expect spacetime to be quantized at the Planck scale (cf. quantum gravity in Section 7.3.2), it seems natural that we should cut off the integral at the Planck momentum $p_{\mathrm{Pl}} = E_{\mathrm{Pl}}/c$, since any GR singularities are expected to be modified at this short distance. In this way, the integral (10.35) yields

$$\rho_\Lambda c^2 \simeq \frac{1}{16\pi^2} \frac{E_{\mathrm{Pl}}^4}{(\hbar c)^3} \simeq \frac{(3.4 \times 10^{27} \, \mathrm{eV})^4}{(\hbar c)^3}. \tag{10.36}$$

Since the critical density (8.17) may be written in such natural units as[26]

$$\rho_c c^2 \simeq \frac{(2.5 \times 10^{-3} \, \mathrm{eV})^4}{(\hbar c)^3}, \tag{10.37}$$

we have a quantum vacuum energy density $\Omega_\Lambda \equiv \rho_\Lambda / \rho_c$ that is more than 10^{120} times larger than the observed dark energy density, which is comparable to the critical density:

$$(\Omega_\Lambda)_{\mathrm{qv}} = O\left(10^{120}\right) \quad \text{vs.} \quad (\Omega_\Lambda)_{\mathrm{obs}} = O(1). \tag{10.38}$$

Thus, if the observed dark energy originates from quantum vacuum energy, there must be some mechanism to reduce its enormous natural density to the critical value.[27] This *cosmological constant problem*, in this context, is the puzzle of why such a fantastic cancellation takes place—a cancellation of the first 120 significant figures (which stops at the 121st place)!

[26] Since $\hbar c$ has units of length times energy, the combination $(\mathrm{energy})^4/(\hbar c)^3$ has the correct units of energy per volume.

[27] Here we have considered only the zero-point energy of boson fields, i.e., particles with integer spin such as photons and gravitons. However, for half-integer-spin fermions such as electrons and quarks, the field is a collection of Fermi oscillators with negative zero-point energy. Had our universe obeyed supersymmetry exactly, with a strict degeneracy of bosonic and fermionic degrees of freedom, their respective contributions to the vacuum energy would have exactly canceled, leading to a vanishing vacuum energy. A more realistic broken supersymmetry would reduce $(\Omega_\Lambda)_{\mathrm{qv}}$ somewhat, but nowhere near the required cancellation.

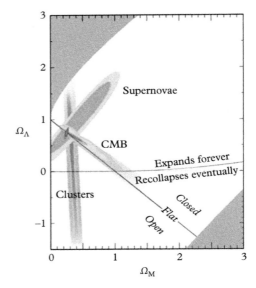

Figure 10.9 *Three independent categories of cosmological observations (supernovae, CMB, and galactic clusters) converge on the same Ω_M and Ω_Λ values. The graph is reproduced with permission from (Perlmutter 2003). ©2003 American Institute of Physics.*

10.3 ΛCDM cosmology as the standard model

After the SN Ia redshift discovery, the presence of dark energy $\Omega_\Lambda \simeq 0.7$ was further confirmed by analysis of the CMB anisotropy power spectrum, as well as of surveys of galactic distributions;[28] see Fig. 10.9. It is remarkable that we could arrive at the same conclusion by such totally different observational methods. Cosmology has seen such major achievements over the past decade that something like a standard model for the origin and development of the universe is now in place: the FLRW cosmology with a dark energy component preceded by an inflationary epoch. Many of the basic cosmological parameters have been deduced in several independent ways, arriving at a consistent set of results. These data are compatible with an infinite and spatially flat universe, having critical matter/energy density, $\Omega_0 = 1$. The largest energy component is consistent with Einstein's cosmological constant $\Omega_\Lambda \simeq 0.69$. In the present epoch, this dark energy content is comparable in size to the matter density $\Omega_M \simeq 0.31$. The matter content is made up mostly of cold dark matter $\Omega_{DM} \simeq 0.26$, more than five times as prevalent as the familiar atomic matter $\Omega_B \simeq 0.05$. Thus this standard model is often called the ΛCDM cosmology model. The expansion of the universe will never stop. In fact, having entered the accelerating phase, the expansion will keep getting faster and faster.

The cosmology presented in this book is based on the basic premise that our galaxy, the Milky Way, does not occupy a special spot in our universe. Modern cosmology with its inflationary theory of the origin of the universe now

[28] Cf. also the discussion at the end of Section 10.2.2.

suggests an even wider perspective. The inflationary multiverse theory indicates that we can entertain the possibility of the ultimate extension of the Copernican principle—even our universe could be just one of many in the cosmos!

Review questions

1. What is the flatness problem in FLRW cosmology?

2. What is the horizon problem in FLRW cosmology?

3. Give a simple physical justification of the rate equation obeyed by the Robertson–Walker scale factor $\dot{a}(t) \propto a(t)$ in a vacuum-energy-dominated universe. Explain (in qualitative terms only) how the solution $a(t)$ of such a rate equation can explain the flatness and horizon problems.

4. How can the observed temperature anisotropy of the CMB be used to deduce that the average geometry of the universe is flat?

5. What is Thomson scattering? Does the scattering of a single unpolarized photon produce a polarized photon? Can isotropic (unpolarized) radiation produce polarization? What about anisotropic radiation?

6. What is the significance of a definitive detection of a B-mode polarization pattern in the CMB?

7. The age of a flat universe without the cosmological constant is estimated to be $\frac{2}{3}t_H \simeq 9\,\mathrm{Gyr}$. Why can an accelerating universe increase this value?

8. What is dark energy? How is it different from dark matter? How is Einstein's cosmological constant related to such energy/matter contents? Do cosmic neutrinos contribute to dark energy?

9. Give two reasons why type Ia supernovae are ideal standard candles for large-scale cosmic measurements.

10. Why should the accelerating universe lead us to observe the galaxies at a given redshift to be dimmer than expected (in an empty or decelerating universe)?

11. Why is the observation that supernovae with high redshifts ($z > 0.7$) are in the decelerating phase taken to be convincing evidence that the accelerating universe interpretation of supernova data ($0.2 < z < 0.7$) is correct?

12. What is the standard ΛCDM cosmology? What is the spatial geometry in this cosmological model? How old is the universe? What is the energy/matter content of the universe?

Tensor Formalism for General Relativity

- This chapter may be viewed as the mathematical appendix of the book. While some important GR results have been stated previously without proof, we now introduce the basics of the tensor formalism needed to properly formulate GR. Specific topics are introduced with an emphasis on their mathematical content, so the reader should refer back to the previous chapters for their physics context.

- While the tensors used in GR are basically the same as those in SR, differentiation of tensor components in a curved space must be handled with extra care, because basis vectors in a curved spacetime are position-dependent.

- By adding extra terms (involving a combination of the metric's first derivatives called Christoffel symbols) to the ordinary derivative operator, we can form a covariant derivative, which acts on tensor components to yield components of a new tensor. Covariant differentiation has a clear geometric meaning in terms of parallel transport of tensors.

- The Riemann tensor reflects multiple aspects of curvature. Its expression (11.40) can be derived from
  - the change of a vector parallel-transported around a closed path (which is related to the noncommutivity of covariant derivatives of a vector);
  - the deviation of geodesics (tidal forces).

- We use the Bianchi identity to show that the Einstein tensor has no co-variant divergence, qualifying it to be the geometric term in the GR field equation. The metric tensor itself also satisfies this criterion, thereby allowing the cosmological constant term.

- The approach to Einstein field equation via the principle of least action is sketched. The relevant mathematics of its Schwarzschild solution is outlined.

A College Course on Relativity and Cosmology. First Edition. Ta-Pei Cheng.
© Ta-Pei Cheng 2015. Published in 2015 by Oxford University Press.

General relativity requires that physics equations be covariant under any general coordinate transformation that leaves invariant the infinitesimal interval

$$ds^2 = g_{\mu\nu}\, dx^\mu\, dx^\nu. \tag{11.1}$$

Just as SR requires physics equations to be tensor equations with respect to Lorentz transformations, GR equations must be tensor equations with respect to general coordinate transformations. In this way, the principle of GR can be fulfilled automatically.

General coordinate transformations

Recall our discussion in Section 3.2 that tensor components are the expansion coefficients of a tensor in terms of the basis vectors. Under a coordinate transformation, a tensor does not itself change, but its components transform because of the changed bases. The transformation rules of tensor components are listed in (3.25)–(3.27). Because repeated reference to tensor components can be cumbersome, we often simply refer to tensor components as tensors.

In Chapter 5 and in particular (5.10), we suggested that coordinate transformations can be written in terms of partial derivatives.[1] We now discuss these general coordinate transformations further. From the basic chain rule of differentiation, we have

[1] This also applies to position-independent coordinate transformations such as ordinary rotations and Lorentz transformations; see Exercise 2.4.

$$dx'^\mu = \frac{\partial x'^\mu}{\partial x^\nu}\, dx^\nu, \qquad \partial'_\mu = \frac{\partial x^\nu}{\partial x'^\mu}\, \partial_\nu. \tag{11.2}$$

We can interpret these relations as the transformations $(dx^\nu, \partial_\nu) \rightarrow (dx'^\mu, \partial'_\mu)$ by the respective transformation matrices $(\partial x'^\mu/\partial x^\nu, \partial x^\nu/\partial x'^\mu)$. Recall our Chapter 3 definitions of contravariant and covariant vector components (A^μ, A_μ); they transform in the same way as (dx^ν, ∂_ν). Thus we can write the respective transformations of the contravariant and covariant components of a vector as

$$A^\mu \longrightarrow A'^\mu = \frac{\partial x'^\mu}{\partial x^\nu} A^\nu, \tag{11.3}$$

$$A_\mu \longrightarrow A'_\mu = \frac{\partial x^\nu}{\partial x'^\mu} A_\nu. \tag{11.4}$$

We display the contravariant transformation in 4D spacetime:

$$\begin{pmatrix} A'^0 \\ A'^1 \\ A'^2 \\ A'^3 \end{pmatrix} = \begin{pmatrix} \dfrac{\partial x'^0}{\partial x^0} & \dfrac{\partial x'^0}{\partial x^1} & \dfrac{\partial x'^0}{\partial x^2} & \dfrac{\partial x'^0}{\partial x^3} \\[2ex] \dfrac{\partial x'^1}{\partial x^0} & \dfrac{\partial x'^1}{\partial x^1} & \dfrac{\partial x'^1}{\partial x^2} & \dfrac{\partial x'^1}{\partial x^3} \\[2ex] \dfrac{\partial x'^2}{\partial x^0} & \dfrac{\partial x'^2}{\partial x^1} & \dfrac{\partial x'^2}{\partial x^2} & \dfrac{\partial x'^2}{\partial x^3} \\[2ex] \dfrac{\partial x'^3}{\partial x^0} & \dfrac{\partial x'^3}{\partial x^1} & \dfrac{\partial x'^3}{\partial x^2} & \dfrac{\partial x'^3}{\partial x^3} \end{pmatrix} \begin{pmatrix} A^0 \\ A^1 \\ A^2 \\ A^3 \end{pmatrix}. \tag{11.5}$$

This way of writing a transformation also has the advantage of preventing us from misidentifying the transformation $[...]^\mu_\nu$ as a tensor.

Because any vector **A** is coordinate-independent and may be expanded as $\mathbf{A} = A^\mu \mathbf{e}_\mu = A_\mu \mathbf{e}^\mu$, the transformations of its expansion coefficients A^μ and A_μ must cancel out (i.e., be the inverse of) those of the corresponding bases:

$$\mathbf{e}'_\mu = \frac{\partial x^\nu}{\partial x'^\mu}\mathbf{e}_\nu \quad \text{and} \quad \mathbf{e}'^\nu = \frac{\partial x'^\nu}{\partial x^\rho}\mathbf{e}^\rho. \tag{11.6}$$

This is the reason why $\{A_\mu\}$ are called the covariant components: they transform in the same way as the basis vectors $\{\mathbf{e}_\mu\}$, while the contravariant components $\{A^\mu\}$ transform oppositely, like the inverse bases $\{\mathbf{e}^\mu\}$.

A tensor **T** of higher rank is likewise coordinate-independent and can be similarly expanded in terms of basis elements that are (direct) products of the vector bases:

$$\mathbf{T} = T^{\mu\nu\cdots}{}_{\lambda\cdots}\,\mathbf{e}_\mu \otimes \mathbf{e}_\nu \cdots \otimes \mathbf{e}^\lambda \otimes \cdots . \tag{11.7}$$

Therefore, tensor components $T^{\mu\nu\cdots}{}_{\lambda\cdots}$ transform like products of vector components $A^\mu B^\nu \cdots C_\lambda \cdots$. For example, mixed tensor components $T_\nu{}^\mu$ transform as

$$T_\nu{}^\mu \longrightarrow T'_\nu{}^\mu = \frac{\partial x^\lambda}{\partial x'^\nu}\frac{\partial x'^\mu}{\partial x^\rho}T_\lambda{}^\rho. \tag{11.8}$$

In particular, the metric tensor components change as

$$g_{\mu\nu} \longrightarrow g'_{\mu\nu} = \frac{\partial x^\lambda}{\partial x'^\mu}\frac{\partial x^\rho}{\partial x'^\nu}g_{\lambda\rho}. \tag{11.9}$$

Recall that the metric tensor components are related to the basis vectors $\{\mathbf{e}_\mu\}$ (and components of the inverse metric to the inverse basis vectors $\{\mathbf{e}^\mu\}$) by

$$g_{\mu\nu} = \mathbf{e}_\mu \cdot \mathbf{e}_\nu \quad \text{and} \quad g^{\mu\nu} = \mathbf{e}^\mu \cdot \mathbf{e}^\nu. \tag{11.10}$$

Since the basis vectors of a curved space are position-dependent, so must be the associated metric. Relations such as (11.9) with position-dependent $g_{\lambda\rho}$ and $g'_{\mu\nu}$ imply that general coordinate transformations must themselves vary over spacetime.

11.1 Covariant derivatives and parallel transport

Physics equations usually involve differentiation. While tensors in GR are basically the same as SR tensors, the derivative operators in a curved space require considerable care. General coordinate transformations are position-dependent, so ordinary derivatives of tensor components, except for the trivial case of a scalar tensor, are not components of tensors. Nevertheless, we shall construct covariant differentiation operations that do result in tensor component derivatives.

11.1.1 Derivatives in a curved space and Christoffel symbols

We first demonstrate that ordinary derivatives spoil the transformation properties of tensor components. We then construct covariant derivatives that correct this problem.

Ordinary derivatives of tensor components are not tensors

In a curved space, the derivative $\partial_\nu A^\mu$ is not a tensor. Namely, even though A^μ and ∂_ν transform like vectors, as indicated by (11.3) and (11.2), their combination $\partial_\nu A^\mu$ does not transform as required by (11.8):

$$\partial_\nu A^\mu \longrightarrow \partial'_\nu A'^\mu \neq \frac{\partial x^\lambda}{\partial x'^\nu} \frac{\partial x'^\mu}{\partial x^\rho} \partial_\lambda A^\rho . \tag{11.11}$$

We can find the full expression for $\partial'_\nu A'^\mu$ by differentiating ($\partial'_\nu \equiv \partial/\partial x'^\nu$) both sides of (11.3):

$$\partial'_\nu A'^\mu = \frac{\partial}{\partial x'^\nu} \left(\frac{\partial x'^\mu}{\partial x^\rho} A^\rho \right)$$

$$= \frac{\partial x^\lambda}{\partial x'^\nu} \frac{\partial x'^\mu}{\partial x^\rho} (\partial_\lambda A^\rho) + \frac{\partial^2 x'^\mu}{\partial x'^\nu \partial x^\rho} A^\rho , \tag{11.12}$$

where the chain rule (11.2) has been used. Compared with the right-hand side of (11.11), there is an extra term, the second term on the right-hand side, because

$$\frac{\partial}{\partial x'^\nu} \left(\frac{\partial x'^\mu}{\partial x^\rho} \right) \neq 0; \tag{11.13}$$

the transformations are position-dependent, which follows from the position dependence of the metric. We see that the fundamental problem lies in the changing bases, $e^\mu = e^\mu(x)$, of the curved space. More explicitly, because the vector components are the projections of the vector onto the basis vectors $A^\mu = e^\mu \cdot \mathbf{A}$, the changing bases $\partial_\nu e^\mu \neq 0$ produce an extra (second) term in the derivative:

$$\partial_\nu A^\mu = e^\mu \cdot (\partial_\nu \mathbf{A}) + \mathbf{A} \cdot (\partial_\nu e^\mu). \tag{11.14}$$

The properties of the two terms on the right-hand side will be studied separately below.

Covariant derivatives as expansion coefficients of $\partial_\nu A$

In order for an equation to be manifestly relativistic, we must be able to cast it as a tensor equation, whose form is unchanged under coordinate transformations.

Thus, we seek a covariant derivative D_ν to be used in covariant physics equations. Such a differentiation is constructed so that it acts on tensor components to yield a new tensor of rank one greater, which transforms per (11.8):

$$D_\nu A^\mu \longrightarrow D'_\nu A'^\mu = \frac{\partial x^\lambda}{\partial x'^\nu} \frac{\partial x'^\mu}{\partial x^\rho} D_\lambda A^\rho. \tag{11.15}$$

As will be demonstrated below, the first term on the right-hand side of (11.14) is just this desired covariant derivative term.

We have suggested that the difficulty with differentiating vector components A^μ is their coordinate dependence. By this reasoning, derivatives of a scalar function Φ should not have this complication, because a scalar tensor does not depend on the bases: $\Phi' = \Phi$, so

$$\partial_\mu \Phi \longrightarrow \partial'_\mu \Phi' = \frac{\partial x^\lambda}{\partial x'^\mu} \partial_\lambda \Phi. \tag{11.16}$$

Similarly, the derivatives of the vector **A** itself (not its components) transform properly, because **A** is coordinate-independent:

$$\partial_\mu \mathbf{A} \longrightarrow \partial'_\mu \mathbf{A} = \frac{\partial x^\lambda}{\partial x'^\mu} \partial_\lambda \mathbf{A}. \tag{11.17}$$

Both (11.16) and (11.17) merely reflect the transformation of the del operator (11.2). If we dot both sides of (11.17) by the inverse basis vectors, $\mathbf{e}'^\nu = (\partial x'^\nu / \partial x^\rho)\mathbf{e}^\rho$, we obtain

$$\mathbf{e}'^\nu \cdot \partial'_\mu \mathbf{A} = \frac{\partial x^\lambda}{\partial x'^\mu} \frac{\partial x'^\nu}{\partial x^\rho} \mathbf{e}^\rho \cdot \partial_\lambda \mathbf{A}. \tag{11.18}$$

This shows that $\mathbf{e}^\nu \cdot \partial_\mu \mathbf{A}$ is a proper mixed tensor[2] as required by (11.8), and this is the covariant derivative (11.15) we have been seeking:

$$D_\mu A^\nu = \mathbf{e}^\nu \cdot \partial_\mu \mathbf{A}. \tag{11.19}$$

This relation implies that $D_\mu A^\nu$ can be viewed as the projection[3] of the vectors $\{\partial_\mu \mathbf{A}\}$ along the direction of $\mathbf{e}^\nu$; we can then interpret $D_\mu A^\nu$ as the coefficient of expansion of $\{\partial_\mu \mathbf{A}\}$ in terms of the basis vectors:

$$\partial_\mu \mathbf{A} = (D_\mu A^\nu)\mathbf{e}_\nu. \tag{11.20}$$

Christoffel symbols as expansion coefficients of $\partial_\nu \mathbf{e}^\mu$

On the other hand, we do not have a similarly simple transformation relation like (11.17) when the coordinate-independent **A** is replaced by one of the coordinate basis vectors $\mathbf{e}_\mu$, which by definition change under coordinate transformations. Still, by mimicking (11.20), we can expand $\partial_\nu \mathbf{e}^\mu$ as

[2] We can reach the same conclusion by applying the quotient theorem (see Exercise 3.2) to (11.20), with the observation that since both $\partial_\mu \mathbf{A}$ and $\mathbf{e}_\nu$ are good tensors, so must be their quotient $(D_\mu A^\nu)$.

[3] We are treating $\{\partial_\mu \mathbf{A}\}$ as a set of vectors, each labeled by an index μ. The combination $D_\mu A^\nu$ is a projection of $\partial_\mu \mathbf{A}$, in the same way that $A^\nu = \mathbf{e}^\nu \cdot \mathbf{A}$ is a projection of the vector **A**.

$$\partial_\nu \mathbf{e}^\mu = -\Gamma^\mu_{\nu\lambda}\mathbf{e}^\lambda \quad \text{or} \quad \mathbf{A} \cdot (\partial_\nu \mathbf{e}^\mu) = -\Gamma^\mu_{\nu\lambda}A^\lambda. \qquad (11.21)$$

Similarly, we have[4]

$$\partial_\nu \mathbf{e}_\mu = +\Gamma^\lambda_{\nu\mu}\mathbf{e}_\lambda \quad \text{or} \quad \mathbf{A} \cdot (\partial_\nu \mathbf{e}_\mu) = +\Gamma^\lambda_{\nu\mu}A_\lambda. \qquad (11.22)$$

[4] $\mathbf{e}_\mu \cdot \mathbf{e}^\nu = [\mathbb{I}]^\nu_\mu$, so $\partial_\lambda(\mathbf{e}_\mu \cdot \mathbf{e}^\nu) = 0$. One can then apply the derivative product rule and plug in the expansion of $\partial_\nu \mathbf{e}^\mu$ to solve for $\partial_\nu \mathbf{e}_\mu$.

But the expansion coefficients $\{\Gamma^\mu_{\nu\lambda}\}$ are not tensors. Anticipating the result, we have here used the same notation for these expansion coefficients as for the Christoffel symbols introduced in Chapter 5 (cf. (5.30))—also called the affine connection (connection, for short).

Plugging (11.19) and (11.21) into (11.14), we have

$$D_\nu A^\mu = \partial_\nu A^\mu + \Gamma^\mu_{\nu\lambda}A^\lambda. \qquad (11.23)$$

Thus, in order to produce the covariant derivative, the ordinary derivative $\partial_\nu A^\mu$ must be supplemented by another term. This second term directly reflects the position dependence of the basis vectors, shown in (11.21). Even though neither $\partial_\nu A^\mu$ nor $\Gamma^\mu_{\nu\lambda}A^\lambda$ has the correct tensor transformation properties, the transformation of $\Gamma^\mu_{\nu\lambda}A^\lambda$ cancels the unwanted term in the transformation of $\partial_\nu A^\mu$ (11.12), so that their sum $D_\nu A^\mu$ is a good tensor. Further insight into the structure of the covariant derivative can be gleaned by invoking the basic geometric concept of parallel displacement of a vector, to be presented in Section 11.1.2.

One can easily show that the covariant derivative of a covariant vector A_μ takes on a form similar to (11.23) for the contravariant vector A^μ:

$$D_\nu A_\mu = \partial_\nu A_\mu - \Gamma^\lambda_{\nu\mu}A_\lambda. \qquad (11.24)$$

A mixed tensor such as T^μ_ν, which transforms in the same way as the product $A^\mu B_\nu$, will have a covariant derivative

$$D_\nu T^\rho_\mu = \partial_\nu T^\rho_\mu - \Gamma^\lambda_{\nu\mu} T^\rho_\lambda + \Gamma^\rho_{\nu\sigma} T^\sigma_\mu. \qquad (11.25)$$

There should be a Christoffel term for each index of the tensor—a $(+\Gamma T)$ for each contravariant index and a $(-\Gamma T)$ for each covariant index. A specific example is the covariant differentiation of the (covariant) metric tensor $g_{\mu\nu}$:

$$D_\lambda g_{\mu\nu} = \partial_\lambda g_{\mu\nu} - \Gamma^\rho_{\lambda\mu} g_{\rho\nu} - \Gamma^\rho_{\lambda\nu} g_{\mu\rho}. \qquad (11.26)$$

Christoffel symbols and metric tensor

We have introduced the Christoffel symbols $\Gamma^\mu_{\nu\lambda}$ as the coefficients of expansion of $\partial_\nu \mathbf{e}^\mu$ as in (11.21). In this section, we shall relate these $\Gamma^\mu_{\nu\lambda}$ to the first derivatives of the metric tensor. This will justify the identification with the symbols first defined in (5.30).

The metric tensor is covariantly constant While the metric tensor is position-dependent, $\partial[g] \neq 0$, its components are constant with respect to covariant differentiation, $D[g] = 0$ (we say that $g_{\mu\nu}$ is *covariantly constant*):

$$D_\lambda g_{\mu\nu} = 0. \tag{11.27}$$

One way to prove this is to express the metric in terms of the basis vectors, $g_{\mu\nu} = \mathbf{e}_\mu \cdot \mathbf{e}_\nu$, and apply the definition of the affine connection, $\partial_\nu \mathbf{e}_\mu = +\Gamma^\rho_{\nu\mu}\mathbf{e}_\rho$, given in (11.22):

$$\partial_\lambda(\mathbf{e}_\mu \cdot \mathbf{e}_\nu) = (\partial_\lambda \mathbf{e}_\mu)\cdot\mathbf{e}_\nu + \mathbf{e}_\mu\cdot(\partial_\lambda\mathbf{e}_\nu)$$
$$= \Gamma^\rho_{\lambda\mu}\mathbf{e}_\rho\cdot\mathbf{e}_\nu + \Gamma^\rho_{\lambda\nu}\mathbf{e}_\mu\cdot\mathbf{e}_\rho. \tag{11.28}$$

Reverting back to the metric tensors, this relation becomes

$$\partial_\lambda g_{\mu\nu} - \Gamma^\rho_{\lambda\mu}g_{\rho\nu} - \Gamma^\rho_{\lambda\nu}g_{\mu\rho} = D_\lambda g_{\mu\nu} = 0, \tag{11.29}$$

where we have applied the definition of the covariant derivative of a covariant tensor $g_{\mu\nu}$ as in (11.26). As we shall see, the covariant constancy of the metric tensor is the key property that allowed Einstein to introduce his cosmological constant term in the GR field equation.

Exercise 11.1 Christoffel symbols as the metric tensor derivative

(a) The geometry in which we are working has the property that two covariant differentiation operations on a scalar tensor commute: $D_\mu D_\nu \Phi = D_\nu D_\mu \Phi$ (we call such derivatives torsion-free). From this, prove that Christoffel symbols are symmetric with respect to interchange of their lower indices: $\Gamma^\mu_{\nu\lambda} = \Gamma^\mu_{\lambda\nu}$.
(b) Using the definition (11.22) of Christoffel symbols as the coefficients of expansion of the derivative $\partial_\nu \mathbf{e}_\mu$, we showed that the metric is covariantly constant as in (11.29). After this, derive the expression for Christoffel symbols, as the first derivatives of the metric tensor, shown in (5.30). To signify its importance, this relation is called the fundamental theorem of Riemannian geometry.

Suggestion: One can obtain the result by taking the linear combination of three equations expressing (Dg = 0) with indices cyclically permuted and by using $\Gamma^\mu_{\nu\lambda} = \Gamma^\mu_{\lambda\nu}$ as shown in (a).

Once the connection of the first derivative of the metric and the Christoffel symbols is established, we can better understand the result in (11.29). Since Christoffel symbols vanish in the local Euclidean frame, $0 = \partial_\mu g_{\nu\lambda} = D_\mu g_{\nu\lambda}$. The first equality follows from the flatness theorem (discussed in Section 5.1.3); the second follows from $\Gamma^\mu_{\nu\lambda} = 0$, hence $\partial = D$, in the local Euclidean frame. The last expression is covariant, so it must equal zero in every frame of reference, thus proving (11.27).

11.1.2 Parallel transport and geodesics as straight lines

Parallel transport is a fundamental concept in differential geometry. It illuminates the meaning of covariant differentiation and the associated Christoffel symbols. Furthermore, we can use this operation to clearly portray the geodesic as the straightest possible curve,[5] the curve traced out by the parallel transport of its tangent vector. In Section 11.3, we shall derive the Riemann curvature tensor by way of parallel-transporting a vector around a closed path.

[5] This is to be compared with our previous discussion in Box 5.2, using only ordinary derivatives.

Component changes under parallel transport

Equation (11.23) follows from (11.14). It expresses the relation between ordinary and covariant derivatives. Writing $DA^\mu = (D_\nu A^\mu)\, dx^\nu$ and $dA^\mu = (\partial_\nu A^\mu)\, dx^\nu$, (11.14) becomes

$$dA^\mu = DA^\mu - \Gamma^\mu_{\nu\lambda} A^\nu\, dx^\lambda. \qquad (11.30)$$

We will show that the Christoffel symbols in the derivatives of vector components reflect the effects of parallel transport of a vector by a displacement of $d\mathbf{x}$. First, what is a parallel transport? Why does one need to perform such an operation? Recall the definition of the derivative of a scalar function $\Phi(\mathbf{x})$,

$$\partial_\mu \Phi = \frac{d\Phi(\mathbf{x})}{dx^\mu} = \lim_{h\to 0} \frac{\Phi(\mathbf{x} + h\hat{\mathbf{e}}_\mu) - \Phi(\mathbf{x})}{h}. \qquad (11.31)$$

Its numerator involves the difference of the function's values at two different positions. Evaluating the coordinate-independent scalar function $\Phi(\mathbf{x})$ at two locations does not introduce any complication. This is not so for vector components. The differential dA^μ on the left-hand side of (11.30) is the difference $A^\mu(\mathbf{x} + d\mathbf{x}) - A^\mu(\mathbf{x}) \equiv A^\mu_{(2)} - A^\mu_{(1)}$ between the vector components $A^\mu = \mathbf{e}^\mu \cdot \mathbf{A}$, evaluated at two nearby positions, (1) and (2), separated by $d\mathbf{x}$. There are two sources of this difference: the change in the vector itself, $\mathbf{A}_{(2)} \neq \mathbf{A}_{(1)}$, and a coordinate change, $\mathbf{e}^\mu_{(2)} \neq \mathbf{e}^\mu_{(1)}$; they correspond to the two terms on the right-hand side of (11.14). Thus the total change is the sum of two terms:

$$dA^\mu = [dA^\mu]_{\text{total}} = [dA^\mu]_{\text{true}} + [dA^\mu]_{\text{coord}}. \qquad (11.32)$$

The term representing the change in the vector itself may be called the true change,

$$[dA^\mu]_{\text{true}} = \mathbf{e}^\mu \cdot d\mathbf{A} = DA^\mu. \qquad (11.33)$$

The other term represents the projection of $\mathbf{A}$ onto the change in the (inverse) basis vector between the two points separated by $d\mathbf{x}$. This change is a linear combination of the products of components of the vector A^ν with the separation dx^λ, with the Christoffel symbols as coefficients:

$$[dA^\mu]_{\text{coord}} = \mathbf{A} \cdot d\mathbf{e}^\mu = -\Gamma^\mu_{\nu\lambda} A^\nu\, dx^\lambda. \qquad (11.34)$$

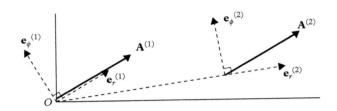

Figure 11.1 *Parallel transport of a vector A in a flat plane with polar coordinates: from position-1 at the origin, where $A^{(1)} = (A_\phi^{(1)}, A_r^{(1)})$, to position-2, $A^{(2)} = (A_\phi^{(2)}, A_r^{(2)})$. The differences in the basis vectors at these two positions, $(e_\phi^{(1)}, e_r^{(1)}) \neq (e_\phi^{(2)}, e_r^{(2)})$, bring about component changes. In particular, $A_\phi^{(1)} = 0$ while $A_\phi^{(2)} \neq 0$, and $A_r^{(1)} = \{[A_\phi^{(2)}]^2 + [A_r^{(2)}]^2\}^{1/2}$.*

[6] Strictly speaking, this statement is meaningful only in flat spaces. A tensor at one point in a curved space cannot be compared with a tensor at another point; they are different entities. For an obvious example, consider what a north-pointing vector on the equator would equal at the north pole—it is not defined. However, a curved space is locally flat (to first order), so we can parallel-transport a vector through a curved space, while keeping it constant in its local tangent space. We will see that this does induce second-order changes to a tensor after it has been parallel-transported in a closed path to its starting point, where it can be compared with its original state.

This discussion motivates us to introduce the geometric concept of **parallel transport**. It is the process of moving a tensor without changing the tensor itself (in its local tangent space).[6] The only change in the tensor components under parallel displacement is due to coordinate changes, $dA^\mu = [dA^\mu]_{\text{coord}}$. In a flat space with a Cartesian coordinate system, this is trivial, since there is no coordinate change from point to point. But in a flat space with a curvilinear coordinate system such as polar coordinates, this parallel transport itself induces component changes, as shown in Fig. 11.1.

For the vector example discussed here, we have $[dA^\mu]_{\text{true}} = e^\mu \cdot dA = DA^\mu = 0$. Thus the mathematical expression for a **parallel transport** of vector components is

$$DA^\mu = dA^\mu + \Gamma_{\nu\lambda}^\mu A^\nu \, dx^\lambda = 0. \tag{11.35}$$

Recall that we have shown that the metric tensor is covariantly constant: $D_\mu g_{\nu\lambda} = 0$. We now understand that covariant constancy of a tensor means that any change in its tensor components is due to coordinate change only. But a change in the metric, by definition, is a pure coordinate change. Hence, it must have a vanishing covariant derivative.

The geodesic as the straightest possible curve

The process of parallel-transporting a vector A^μ along a curve $x^\mu(\sigma)$ can be expressed, according to (11.35), as

$$\frac{DA^\mu}{d\sigma} = \frac{dA^\mu}{d\sigma} + \Gamma_{\nu\lambda}^\mu A^\nu \frac{dx^\lambda}{d\sigma} = 0. \tag{11.36}$$

From this, we can define the geodesic line as the straightest possible curve, because it is the line constructed by parallel transport of its tangent vector. See Fig. 11.2(a) for an illustration of such an operation in flat space. In this way, the geodesic condition can be formulated by setting $A^\mu = dx^\mu/d\sigma$ in (11.36):

$$\frac{D}{d\sigma}\left(\frac{dx^\mu}{d\sigma}\right) = 0, \tag{11.37}$$

or, more explicitly,

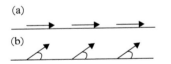

(a)

(b)

Figure 11.2 *(a) A straight line in a flat plane is a geodesic, the curve traced out by parallel transport of its tangents. (b) When a vector is parallel-transported along a straight line, the angle between the vector and the line is unchanged.*

$$\frac{d}{d\sigma}\frac{dx^\mu}{d\sigma} + \Gamma_{\nu\lambda}^\mu \frac{dx^\nu}{d\sigma}\frac{dx^\lambda}{d\sigma} = 0. \tag{11.38}$$

This agrees with the geodesic equation (5.29).

Exercise 11.2 Parallel transport of a vector along a geodesic

Show that when a vector A_μ is parallel-transported along a geodesic, the angle between the vector and the geodesic (i.e., the tangent of the geodesic) is unchanged as in Fig. 11.2(b). Namely, prove the following relation:

$$\frac{D}{d\sigma}\left(A_\mu \frac{dx^\mu}{d\sigma}\right) = 0. \tag{11.39}$$

11.2 Riemann curvature tensor

As stated in Chapter 6, the Riemann curvature (6.20) is a tensor of rank 4:

$$R^\mu_{\ \lambda\alpha\beta} = \partial_\alpha \Gamma^\mu_{\lambda\beta} - \partial_\beta \Gamma^\mu_{\lambda\alpha} + \Gamma^\mu_{\nu\alpha}\Gamma^\nu_{\lambda\beta} - \Gamma^\mu_{\nu\beta}\Gamma^\nu_{\lambda\alpha}. \tag{11.40}$$

We shall demonstrate below that although the terms on the right-hand side are ordinary derivatives and Christoffel symbols, and thus are not tensors, their combination is nevertheless a proper tensor.

$R^\mu_{\ \lambda\alpha\beta}$ determines, independently of the coordinate choice, whether a space is curved. At any point in any curved space, one can always find a coordinate system (the local Euclidean frame) in which the metric's first derivatives vanish, $\partial g = 0$. However, the second derivatives of the metric, $\partial^2 g = 0$, vanish only for a flat space. Hence, in a flat space, $\partial^2 g + (\partial g)^2 \propto R^\mu_{\ \lambda\alpha\beta} = 0$. Since the Riemann curvature is a good tensor, if it vanishes for one set of coordinates, it vanishes for all coordinates.[7] In fact, we can also show that this is a sufficient condition for a space to be flat; i.e., $R^\mu_{\ \lambda\alpha\beta} = 0$ implies a flat space.

Two separate derivations of the Riemann tensor Having learned the formalism of covariant derivatives and the concept of parallel transport, we are now ready to derive the curvature tensor expression (11.40) in a space with arbitrary dimensions.

- The first derivation uses the feature that curvature measures the deviation of geometric relations from their corresponding Euclidean versions. We discussed in Section 6.1 the particular relation (6.19) that for a 2D curved surface the angular excess ϵ of an infinitesimal polygon (the sum of the interior angles over its Euclidean value) is proportional to the Gaussian curvature K at its location:

$$\epsilon = K\sigma, \tag{11.41}$$

where σ is the area of the polygon. We will generalize this relation (11.41) for a 2D curvature K to an n-dimensional curved space. In this extension (Section 11.2.1), the concept of parallel transport plays a central role.

[7] This is exactly like the simpler 2D situation with the Gaussian curvature K of (6.7). The problem of reducing the Riemann curvature tensor of (11.40) in 2D space to the Gaussian curvature of (6.7) is worked out in Problem 13.11 in (Cheng 2010).

- The second derivation uses the feature that curvature causes relativistic tidal forces; we generalize the Newtonian deviation equation (6.24) to the GR equation of geodesic deviation. In this generalization (Section 11.2.2), covariant differentiation plays an indispensable role in finding the covariant equation that describes the relative motion of geodesic paths.

11.2.1 Parallel transport of a vector around a closed path

To extend the relation (11.41) to higher dimensions, one must first generalize the 2D quantities of angular excess ϵ and area σ to a higher-dimensional space.

Angular excess ϵ and directional change of a vector

How can an angular excess be measured in a higher-dimensional space? We first use the concept of parallel transport to cast the relation (11.41) in a form that can be generalized to n dimensions. The angular excess ϵ of a polygon is equal to the directional change in a vector after it has been parallel-transported around the perimeter. The simplest example of such a polygon is a spherical triangle with three 90° interior angles. Figure 6.2 shows that a vector parallel-transported around the triangle changes its direction by 90°, which is the angular excess. The generalization of (11.41) to an arbitrary triangle, and hence to any polygon, can be found in Sections 5.3.2 and 13.3 in (Cheng 2010). The key observation is that when a vector is parallel-transported along a geodesic, the angle it forms with the geodesic is unchanged; cf. Exercise 11.2. Recall the definition that an angle is the ratio of arclength to radius as shown in Fig. 11.3(a). Thus, the directional angular change, and hence the angular excess, can be written as the ratio of the change in a vector to its magnitude: $d\theta = \epsilon = dA/A$. Substituting this into (11.41), we obtain

$$dA = KA\sigma. \tag{11.42}$$

Namely, the change in a vector after a round-trip parallel transport is proportional to the vector itself and the area of the closed path. The coefficient of proportionality is identified as the curvature.

The area tensor

We will use (11.42) as a model for the curvature relation for a higher-dimensional curved space. We first need to write the 2D equation (11.42) in a proper index form that can be generalized to an n-dimensional space. Recall that the 2D area of a parallelogram spanned by two vectors **A** and **B** can be expressed as a vector product as in Fig. 11.3(b): $\sigma = \mathbf{A} \times \mathbf{B}$. Using the antisymmetric Levi-Civita symbol in index notation,[8] we can write this as

$$\sigma_k = \epsilon_{ijk} A^i B^j. \tag{11.43}$$

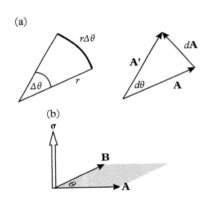

(a)

(b)

Figure 11.3 *(a) The directional change in a vector can be expressed as a fractional change in the vector: $d\theta = dA/A$. (b) The area vector of a parallelogram is the cross product of its two sides, $\sigma = A \times B$.*

---

[8] Levi-Civita symbols are discussed in Sidenote 19 in Chapter 3.

The area vector **σ** has magnitude $AB\sin\theta$ and direction given by the right-hand rule. But (11.43) is not a convenient form to use in a higher-dimensional space: (i) It refers to a 3D embedding space, even though the parallelogram resides in a 2D space. (ii) For a different number of dimensions, we would need to use the antisymmetric tensor with a different number of indices—e.g., in a 4D embedding space, $\epsilon_{\mu\nu\lambda\rho}$. We will instead use a two-index object σ^{ij} to represent the area:[9]

$$\sigma^{ij} \equiv \epsilon^{ijk}\sigma_k = \epsilon^{ijk}\epsilon_{mnk}A^m B^n = A^i B^j - A^j B^i. \tag{11.44}$$

[9] In the second equality of (11.44), we use the identity $\epsilon^{ijk}\epsilon_{mnk} = \delta^i_m\delta^j_n - \delta^i_n\delta^j_m$.

In a 2D space, we can write σ^{ij} entirely with 2D indices, without reference to any embedding space. (Recall the distinction between intrinsic vs. extrinsic geometric descriptions discussed in Chapter 5.) In an n-dimensional space, we can represent the area spanned by a^λ and b^ρ by the antisymmetric combination

$$\sigma^{\lambda\rho} = a^\lambda b^\rho - b^\lambda a^\rho, \tag{11.45}$$

with the indices ranging over the dimensions of the space: $\lambda, \rho = \{1, 2, ..., n\}$.

The curvature tensor in an *n*-dimensional space

Now we have the proper area tensor (11.45), we can cast (11.42) in tensor form to represent[10] the change dA^μ in a vector due to parallel transport around a parallelogram spanned by two infinitesimal vectors a^λ and b^ρ:

$$dA^\mu = -R^\mu_{\;\nu\lambda\rho}A^\nu a^\lambda b^\rho. \tag{11.46}$$

[10] The minus sign is required so as to be compatible with the curvature definition given in (11.40), if the direction of the parallel-transport loop is in accord with the area direction (11.45), i.e., given by the right-hand rule around σ in the 2D case (counterclockwise in Fig. 11.4).

Namely, the change is proportional to the vector A^ν itself and to the two vectors a^λ and b^ρ spanning the parallelogram. The coefficient of proportionality $R^\mu_{\;\nu\lambda\rho}$ is a quantity with four indices, antisymmetric in λ and ρ so as to pick up both terms on the right-hand side of (11.45). We shall take this to be the definition of the curvature (called the **Riemann curvature tensor**) of an n-dimensional space.[11] Explicit calculation in Box 11.1 of the change in a vector parallel-transported around an infinitesimal parallelogram then leads to the expression (11.40).

[11] We can plausibly expect this coefficient $R^\mu_{\;\nu\lambda\rho}$ to be a tensor, because the differential dA^μ (taken at a given position), a^λ, b^ρ, and A^ν are tensors, so the quotient theorem (Exercise 3.2) tells us that $R^\mu_{\;\nu\lambda\rho}$ should be a good tensor.

Box 11.1 Deriving the Riemann tensor by parallel-transporting a vector around a closed path

---

Here we shall parallel-transport a vector around an infinitesimal parallelogram $PQP'Q'$ spanned by two infinitesimal vectors a^α and b^β, shown in Fig. 11.4. Recall that under parallel transport of a vector, $DA^\mu = 0$, so the total vectorial change in (11.35) is due entirely to coordinate change:

$$dA^\mu = -\Gamma^\mu_{\nu\lambda}A^\nu dx^\lambda. \tag{11.47}$$

continued

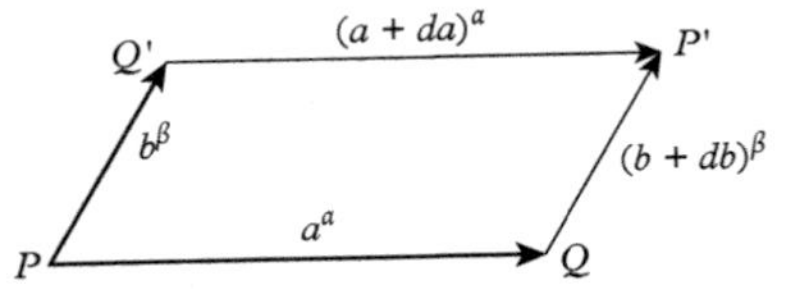

Figure 11.4 *The parallelogram $PQP'Q'$ is spanned by two vectors a^α and b^β. The opposite sides $(a+da)^\alpha$ and $(b+db)^\beta$ are obtained by parallel transport of a^α and b^β by b^ν and a^μ, respectively.*

Box 11.1 *continued*

The opposite sides of the parallelogram in Fig. 11.4, $(a+da)^\alpha$ and $(b+db)^\beta$, are obtained by parallel transport of a^α and b^β by b^ν and a^μ, respectively. The expression for parallel transport (11.47) gives the relations

$$(a+da)^\alpha = a^\alpha - \Gamma^\alpha_{\mu\nu}a^\mu b^\nu,$$
$$(b+db)^\beta = b^\beta - \Gamma^\beta_{\mu\nu}a^\mu b^\nu. \tag{11.48}$$

Using (11.47) again, we now calculate the change in a vector A^μ due to parallel transport from $P \to Q \to P'$:

$$dA^\mu_{PQP'} = dA^\mu_{PQ} + dA^\mu_{QP'} \tag{11.49}$$
$$= -(\Gamma^\mu_{\nu\alpha}A^\nu)_P a^\alpha - (\Gamma^\mu_{\nu\beta}A^\nu)_Q (b+db)^\beta.$$

The subscripts P and Q on the last line denote the respective positions where these functions are to be evaluated. Since eventually we shall compare all quantities at one position, say P, we will Taylor-expand the quantities $(...)_Q$ around the point P:

$$(\Gamma^\mu_{\nu\beta})_Q = (\Gamma^\mu_{\nu\beta})_P + a^\alpha(\partial_\alpha \Gamma^\mu_{\nu\beta})_P, \tag{11.50}$$
$$(A^\nu)_Q = (A^\nu)_P + a^\alpha(\partial_\alpha A^\nu)_P = (A^\nu)_P - a^\alpha(\Gamma^\nu_{\lambda\alpha}A^\lambda)_P,$$

where we have used (11.47) to reach the last expression. From now on, we shall drop the subscript P. We substitute into (11.49) the expansions (11.48) and (11.50):

$$dA^\mu_{PQP'} = -\Gamma^\mu_{\nu\alpha}A^\nu a^\alpha$$
$$-(\Gamma^\mu_{\nu\beta} + a^\alpha\partial_\alpha\Gamma^\mu_{\nu\beta})(A^\nu - a^\alpha\Gamma^\nu_{\lambda\alpha}A^\lambda)(b^\beta - \Gamma^\beta_{\rho\sigma}a^\rho b^\sigma). \tag{11.51}$$

We multiply this out and keep terms up to $O(ab)$:

$$dA^\mu_{PQP'} = -\Gamma^\mu_{\nu\alpha}A^\nu a^\alpha - \Gamma^\mu_{\nu\beta}A^\nu b^\beta + A^\nu\Gamma^\mu_{\nu\beta}\Gamma^\beta_{\rho\sigma}a^\rho b^\sigma$$
$$-\partial_\alpha\Gamma^\mu_{\lambda\beta}A^\lambda a^\alpha b^\beta + \Gamma^\mu_{\nu\beta}\Gamma^\nu_{\lambda\alpha}A^\lambda a^\alpha b^\beta. \tag{11.52}$$

The vectorial change due to parallel transport along the other sides, $P \to Q' \to P'$, can be obtained from this expression by the simple interchange $a \leftrightarrow b$:

$$dA^\mu_{PQ'P'} = -\Gamma^\mu_{\nu\alpha}A^\nu b^\alpha - \Gamma^\mu_{\nu\beta}A^\nu a^\beta + A^\nu\Gamma^\mu_{\nu\beta}\Gamma^\beta_{\rho\sigma}a^\rho b^\sigma$$
$$-\partial_\beta\Gamma^\mu_{\lambda\alpha}A^\lambda a^\alpha b^\beta + \Gamma^\mu_{\nu\alpha}\Gamma^\nu_{\lambda\beta}A^\lambda a^\alpha b^\beta. \tag{11.53}$$

For a round-trip parallel transport[12] from P back to P, the vectorial change dA^μ corresponds to the difference of the above two equations (which results in cancellation of the first three terms on the right-hand sides):

$$dA^\mu = dA^\mu_{PQP'} - dA^\mu_{PQ'P'} \tag{11.54}$$
$$= -(\partial_\alpha \Gamma^\mu_{\lambda\beta} - \partial_\beta \Gamma^\mu_{\lambda\alpha} + \Gamma^\mu_{\nu\alpha}\Gamma^\nu_{\lambda\beta} - \Gamma^\mu_{\nu\beta}\Gamma^\nu_{\lambda\alpha})A^\lambda a^\alpha b^\beta.$$

We conclude, after comparing (11.54) with (11.46), that the sought-after Riemann curvature tensor in terms of Christoffel symbols is just the quoted result (11.40).

[12] The order of the difference in (11.54) corresponds to parallel transport in the counterclockwise direction (on Fig. 11.4) in accordance with the area direction as defined in (11.43) and (11.45).

Exercise 11.3 Riemann curvature tensor as the commutator of covariant derivatives

---

We can obtain the same result as in Box 11.1 somewhat more efficiently by calculating the double covariant derivative

$$D_\alpha D_\beta A^\mu = D_\alpha(\partial_\beta A^\mu + \Gamma^\mu_{\beta\lambda}A^\lambda) = ..., \tag{11.55}$$

as well as the reverse order $D_\beta D_\alpha A^\mu = D_\beta(\partial_\alpha A^\mu + \Gamma^\mu_{\alpha\lambda}A^\lambda) = ...$ Show that their difference (expressed here as a commutator) is just the expression for the Riemann tensor given by (11.40):

$$[D_\alpha, D_\beta]A^\mu = R^\mu_{\lambda\alpha\beta}A^\lambda. \tag{11.56}$$

Comments: (i) At first sight, one may question this approach to the problem of parallel transport of a vector around a closed path—wouldn't parallel transport mean that $DA = 0$? But the calculation in Box 11.1 shows that $D(DA) \neq 0$; calculating the vectorial change requires consistency in keeping the higher-order terms in Taylor expansions. See Sidenote 6 for a related comment.

(ii) It is also straightforward to show that the covariant derivative commutator acting on a mixed tensor (instead of on a contravariant vector) will lead to

$$[D_\alpha, D_\beta]T^\mu_\nu = R^\mu_{\lambda\alpha\beta}T^\lambda_\nu - R^\lambda_{\nu\alpha\beta}T^\mu_\lambda; \tag{11.57}$$

i.e., for each contravariant index, there will be a $+RT$ term on the right-hand side, and, for each covariant index, a $-RT$ term.

11.2.2 Equation of geodesic deviation

In Section 11.2.1, we have derived an expression for the curvature (11.40) by a purely geometric method. A more physical approach would be to seek the GR generalization of the tidal forces discussed in Section 6.2.2. Following exactly the same steps used to derive the Newtonian deviation equation (6.24), let us consider two particles: one follows the spacetime trajectory $x^\mu(\tau)$, and the other follows

$x^\mu(\tau) + s^\mu(\tau)$. These two particles, separated by the displacement vector s^μ, obey their respective GR equations of motion, the geodesic equations, cf. (5.29):

$$\frac{d^2 x^\mu}{d\tau^2} + \Gamma^\mu_{\alpha\beta}(x)\frac{dx^\alpha}{d\tau}\frac{dx^\beta}{d\tau} = 0 \qquad (11.58)$$

and

$$\left(\frac{d^2 x^\mu}{d\tau^2} + \frac{d^2 s^\mu}{d\tau^2}\right) + \Gamma^\mu_{\alpha\beta}(x + s)\left(\frac{dx^\alpha}{d\tau} + \frac{ds^\alpha}{d\tau}\right)\left(\frac{dx^\beta}{d\tau} + \frac{ds^\beta}{d\tau}\right) = 0. \qquad (11.59)$$

When the separation s^μ is small, we can approximate the Christoffel symbols $\Gamma^\mu_{\alpha\beta}(x + s)$ by a Taylor expansion

$$\Gamma^\mu_{\alpha\beta}(x + s) = \Gamma^\mu_{\alpha\beta}(x) + \partial_\lambda \Gamma^\mu_{\alpha\beta} s^\lambda + \cdots . \qquad (11.60)$$

From the difference of the two geodesic equations, we obtain, to first order in s^μ,

$$\frac{d^2 s^\mu}{d\tau^2} = -2\Gamma^\mu_{\alpha\beta}\frac{ds^\alpha}{d\tau}\frac{dx^\beta}{d\tau} - \partial_\lambda \Gamma^\mu_{\alpha\beta} s^\lambda \frac{dx^\alpha}{d\tau}\frac{dx^\beta}{d\tau}. \qquad (11.61)$$

We are seeking the relative acceleration (the second derivative of the separation s^μ) along the worldline. So far we have only written down ordinary derivatives. In GR equations, we must use covariant derivatives. From (11.36), we have the first covariant derivative,

$$\frac{Ds^\mu}{d\tau} = \frac{ds^\mu}{d\tau} + \Gamma^\mu_{\alpha\beta} s^\alpha \frac{dx^\beta}{d\tau}, \qquad (11.62)$$

and the second covariant derivative,

$$\begin{aligned}
\frac{D^2 s^\mu}{d\tau^2} &= \frac{D}{d\tau}\left(\frac{Ds^\mu}{d\tau}\right) = \frac{d}{d\tau}\left(\frac{Ds^\mu}{d\tau}\right) + \Gamma^\mu_{\alpha\beta}\left(\frac{Ds^\alpha}{d\tau}\right)\frac{dx^\beta}{d\tau} \\
&= \frac{d}{d\tau}\left(\frac{ds^\mu}{d\tau} + \Gamma^\mu_{\alpha\beta} s^\alpha \frac{dx^\beta}{d\tau}\right) + \Gamma^\mu_{\alpha\beta}\left(\frac{ds^\alpha}{d\tau} + \Gamma^\alpha_{\lambda\rho} s^\lambda \frac{dx^\rho}{d\tau}\right)\frac{dx^\beta}{d\tau} \\
&= \frac{d^2 s^\mu}{d\tau^2} + \partial_\lambda \Gamma^\mu_{\alpha\beta}\frac{dx^\lambda}{d\tau} s^\alpha \frac{dx^\beta}{d\tau} + \Gamma^\mu_{\alpha\beta}\frac{ds^\alpha}{d\tau}\frac{dx^\beta}{d\tau} + \Gamma^\mu_{\alpha\beta} s^\alpha \frac{d^2 x^\beta}{d\tau^2} \\
&\quad + \Gamma^\mu_{\alpha\beta}\frac{ds^\alpha}{d\tau}\frac{dx^\beta}{d\tau} + \Gamma^\mu_{\alpha\beta}\Gamma^\alpha_{\lambda\rho} s^\lambda \frac{dx^\rho}{d\tau}\frac{dx^\beta}{d\tau}. \qquad (11.63)
\end{aligned}$$

For the first term on the right-hand side $(d^2 s^\mu/d\tau^2)$, we apply (11.61); for $d^2 x^\beta/d\tau^2$ in the fourth term we use the geodesic equation (11.58):

$$\frac{d^2 x^\beta}{d\tau^2} = -\Gamma^\beta_{\lambda\rho}\frac{dx^\lambda}{d\tau}\frac{dx^\rho}{d\tau}. \qquad (11.64)$$

In this way, we find

$$
\frac{D^2 s^\mu}{d\tau^2} = -2\Gamma^\mu_{\alpha\beta}\frac{ds^\alpha}{d\tau}\frac{dx^\beta}{d\tau} - \partial_\lambda\Gamma^\mu_{\alpha\beta}s^\lambda\frac{dx^\alpha}{d\tau}\frac{dx^\beta}{d\tau} + \partial_\lambda\Gamma^\mu_{\alpha\beta}\frac{dx^\lambda}{d\tau}s^\alpha\frac{dx^\beta}{d\tau}
$$

$$
+2\Gamma^\mu_{\alpha\beta}\frac{ds^\alpha}{d\tau}\frac{dx^\beta}{d\tau} - \Gamma^\mu_{\alpha\beta}s^\alpha\Gamma^\beta_{\lambda\rho}\frac{dx^\lambda}{d\tau}\frac{dx^\rho}{d\tau}
$$

$$
+\Gamma^\mu_{\alpha\beta}\Gamma^\alpha_{\lambda\rho}s^\lambda\frac{dx^\rho}{d\tau}\frac{dx^\beta}{d\tau}. \tag{11.65}
$$

After a cancellation of two terms and the relabeling of several dummy indices, this becomes

$$
\frac{D^2 s^\mu}{d\tau^2} = -\partial_\lambda\Gamma^\mu_{\alpha\beta}s^\lambda\frac{dx^\alpha}{d\tau}\frac{dx^\beta}{d\tau} + \partial_\beta\Gamma^\mu_{\lambda\alpha}\frac{dx^\alpha}{d\tau}s^\lambda\frac{dx^\beta}{d\tau} \tag{11.66}
$$

$$
-\Gamma^\mu_{\lambda\rho}s^\lambda\Gamma^\rho_{\alpha\beta}\frac{dx^\alpha}{d\tau}\frac{dx^\beta}{d\tau} + \Gamma^\mu_{\rho\beta}\Gamma^\rho_{\lambda\alpha}s^\lambda\frac{dx^\alpha}{d\tau}\frac{dx^\beta}{d\tau}.
$$

Factoring out the common $(dx^\alpha/d\tau)(dx^\beta/d\tau)$ yields the equation of geodesic deviation:

$$
\frac{D^2 s^\mu}{D\tau^2} = -R^\mu_{\ \alpha\lambda\beta}s^\lambda\frac{dx^\alpha}{d\tau}\frac{dx^\beta}{d\tau}, \tag{11.67}
$$

where $R^\mu_{\ \alpha\lambda\beta}$ is just the Riemann curvature given in (11.40). Namely, the tensor of the gravitational potential's second derivatives (tidal gravity) in (6.24) is replaced in GR by the Riemann curvature tensor (11.40).

Exercise 11.4 From geodesic deviation to nonrelativistic tidal forces

---

Show that the equation of geodesic deviation (11.67) reduces to the Newtonian deviation equation (6.24) in the Newtonian limit. In the nonrelativistic limit of slow-moving particles with 4-velocity $dx^\alpha/d\tau \simeq (c, 0, 0, 0)$, the GR equation (11.67) is reduced to

$$
\frac{d^2 s^i}{dt^2} = -c^2 R^i_{\ 0j0}s^j. \tag{11.68}
$$

We have also set $s^0 = 0$, because we are comparing the two particles' accelerations at the same time. Thus (6.24) can be recovered by showing the relation

$$
R^i_{\ 0j0} = \frac{1}{c^2}\frac{\partial^2\Phi}{\partial x^i\partial x^j} \tag{11.69}
$$

in the Newtonian limit. You are asked to prove that this expression is the limit of Riemann curvature in (11.40).

11.2.3   Bianchi identity and the Einstein tensor

We have already displayed the symmetries and contractions of the Riemann curvature tensor in Section 6.3.1. In particular, we have the symmetric Ricci tensor and the Ricci scalar:

$$R_{\mu\nu} \equiv g^{\alpha\beta} R_{\alpha\mu\beta\nu}, \qquad R \equiv g^{\alpha\beta} R_{\alpha\beta}. \tag{11.70}$$

It was suggested that some linear combination $R_{\mu\nu} + aRg_{\mu\nu}$ will enter the GR field equation as both terms are rank-2 symmetric tensors composed of the metric and its derivatives. As it turns out, the constant a can be fixed by requiring the combination to be covariantly constant: $D^\mu(R_{\mu\nu} + aRg_{\mu\nu}) = 0$ (cf. the opening discussion in Section 11.3.2). An efficient way is to use the Bianchi identity,

$$D_\lambda R_{\gamma\alpha\mu\nu} + D_\nu R_{\gamma\alpha\lambda\mu} + D_\mu R_{\gamma\alpha\nu\lambda} = 0. \tag{11.71}$$

The structure of this identity (11.71) suggests that we derive it from the Jacobi identity for the double commutators of three operators—in this case, the operators are covariant derivatives:

$$[D_\lambda, [D_\mu, D_\nu]] + [D_\nu, [D_\lambda, D_\mu]] + [D_\mu, [D_\nu, D_\lambda]] = 0. \tag{11.72}$$

Exercise 11.5 Jacobi identity and double commutator of covariant derivatives

---

(a) Prove the Jacobi identity. Namely, demonstrate explicitly that the cyclic combination of three double commutators of any three operators (in particular the differential operators in (11.72)) vanishes.
(b) Use the expression for the Riemann tensor in terms of the double commutator in (11.57) to show that

$$[D_\lambda, [D_\mu, D_\nu]]A_\alpha = -D_\lambda R^\gamma_{\alpha\mu\nu} A_\gamma + R^\gamma_{\lambda\mu\nu} D_\gamma A_\alpha. \tag{11.73}$$

Applying (11.73) to every double commutator in (11.72) acting on the covariant vector A_α, we have

$$\begin{aligned}
0 &= ([D_\lambda, [D_\mu, D_\nu]] + [D_\nu, [D_\lambda, D_\mu]] + [D_\mu, [D_\nu, D_\lambda]])A_\alpha \\
&= -D_\lambda R^\gamma_{\alpha\mu\nu} A_\gamma - D_\nu R^\gamma_{\alpha\lambda\mu} A_\gamma - D_\mu R^\gamma_{\alpha\nu\lambda} A_\gamma \\
&\quad + R^\gamma_{\lambda\mu\nu} D_\gamma A_\alpha + R^\gamma_{\nu\lambda\mu} D_\gamma A_\alpha + R^\gamma_{\mu\nu\lambda} D_\gamma A_\alpha \\
&= -(D_\lambda R^\gamma_{\alpha\mu\nu} + D_\nu R^\gamma_{\alpha\lambda\mu} + D_\mu R^\gamma_{\alpha\nu\lambda})A_\gamma \\
&\quad + (R^\gamma_{\lambda\mu\nu} + R^\gamma_{\nu\lambda\mu} + R^\gamma_{\mu\nu\lambda}) D_\gamma A_\alpha.
\end{aligned} \tag{11.74}$$

The second term on the right-hand side vanishes because of the cyclic symmetry property (6.32). It then follows that the parenthesis in the first term must also vanish, leading to the Bianchi identity (11.71).

Our next objective is to use the Bianchi identity to find the covariantly constant linear combination of $R_{\mu\nu}$ and $Rg_{\mu\nu}$, which we need for the GR field equation.

First we contract (11.71) with $g^{\mu\alpha}$; because the metric tensor is covariantly constant, $D_\lambda g^{\alpha\beta} = 0$, this metric contraction can be pushed right through the covariant differentiation:

$$-D_\lambda R_{\gamma\nu} + D_\nu R_{\gamma\lambda} + D_\mu g^{\mu\alpha} R_{\gamma\alpha\nu\lambda} = 0. \tag{11.75}$$

Contracting another time with $g^{\gamma\nu}$ yields

$$-D_\lambda R + D_\nu g^{\gamma\nu} R_{\gamma\lambda} + D_\mu g^{\mu\alpha} R_{\alpha\lambda} = 0. \tag{11.76}$$

At the last two terms, the metric just raises the indices:

$$-D_\lambda R + D_\nu R_\lambda^\nu + D_\mu R_\lambda^\mu = -D_\lambda R + 2D_\nu R_\lambda^\nu = 0. \tag{11.77}$$

Pushing through yet another $g^{\mu\lambda}$ in order to raise the λ index at the last term gives us

$$D_\lambda(-Rg^{\mu\lambda} + 2R^{\mu\lambda}) = 0. \tag{11.78}$$

Thus this combination (6.36), called the Einstein tensor,

$$G^{\mu\nu} = R^{\mu\nu} - \tfrac{1}{2} Rg^{\mu\nu}, \tag{11.79}$$

i.e., the combination $(R^{\mu\nu} + aRg^{\mu\nu})$ with the constant fixed $a = -1/2$ is covariantly constant:

$$D_\mu G^{\mu\nu} = 0. \tag{11.80}$$

$G^{\mu\nu}$ is a covariantly constant, rank-2, symmetric tensor involving the second derivatives of the metric $\partial^2 g$ as well as terms quadratic in ∂g—exactly what we were seeking for the GR field equation; cf. (6.5).

11.3   GR tensor equations

According to the strong EP, gravity can always be transformed away locally. As first discussed in Section 5.2, Einstein suggested an elegant formulation of the new theory of gravity based on a curved spacetime. The EP is thus fundamentally built into the theory. Local flatness (a metric structure of spacetime) means that SR (the theory of flat spacetime with no gravity) is automatically incorporated into the new theory. Gravity is modeled not as a force but as the structure of spacetime; free particles just follow geodesics in a curved spacetime.

11.3.1   The principle of general covariance

The field equation for the relativistic potential, the metric function $g_{\mu\nu}(x)$, must have the same form no matter what generalized coordinates are used to label worldpoints (events) in spacetime. One expresses this by the requirement that the physics equations must satisfy the principle of general covariance. This is a two-part statement:

1. Physics equations must be covariant under the general coordinate transformations that leave the infinitesimal spacetime line element interval ds^2 invariant.

2. Physics equations should reduce to the correct SR form in the local inertial frames. Namely, we must have the correct SR equations in free-fall frames, in which gravity is transformed away. Additionally, gravitational equations reduce to Newtonian equations in the limit of low-velocity particles in a weak and static field.

The minimal substitution rule

This general principle provides us with a well-defined path to go from SR equations (which are valid in local inertial frames with no gravity) to GR equations that are valid in every coordinate system in curved spacetime (curved because of the presence of gravity). GR equations must be covariant under general local transformations. The key feature of a general coordinate transformation, in contrast to the (Lorentz) transformation in flat spacetime, is its spacetime dependence—hence the requirement for covariant derivatives. To go from an SR equation to the corresponding GR equation is simple: we need to replace the ordinary derivatives (∂) in SR equations by covariant derivatives (D):

$$\partial \longrightarrow D \ (= \partial + \Gamma). \tag{11.81}$$

This is known as the minimal substitution rule, because we are assuming the absence of the (high-order) Riemann tensor $R^{\mu}_{\nu\lambda\rho}$ terms, which vanish in the flat-spacetime limit. Since the Christoffel symbols Γ are derivatives of the metric (i.e., they represent the gravitational field strength), the introduction of covariant derivatives naturally brings the gravitational field into the physics equations. In this way, we can, for example, find the equations that describe electromagnetism in the presence of a gravitational field:

$$\partial_\mu F^{\mu\nu} = -\frac{1}{c}j^\nu \quad \longrightarrow \quad D_\mu F^{\mu\nu} = -\frac{1}{c}j^\nu. \tag{11.82}$$

Namely, Gauss's and Ampère's laws of electromagnetism in the presence of a gravitational field take on the form

$$\partial_\mu F^{\mu\nu} + \Gamma^{\mu}_{\mu\lambda}F^{\lambda\nu} + \Gamma^{\nu}_{\mu\lambda}F^{\mu\lambda} = -\frac{1}{c}\,j^\nu, \tag{11.83}$$

with gravity entering through the Christoffel symbols.

GR equation of motion

Gravity in Einstein's formulation is not a force, but the structure of spacetime. Thus the motion of a particle under the sole influence of gravity is the motion of a free particle in spacetime. At this point, instead of arguing heuristically (as in Section 5.3) that the geodesic equation should be the equation of motion in GR, the minimal substitution rule actually provides us with a formal way to arrive at this conclusion. The SR equation of motion for a free particle is $d\dot{x}^{\mu}/d\tau = 0$, where τ is the proper time and $\dot{x}^{\mu} = dx^{\mu}/d\tau$ the 4-velocity of the particle. According to the minimal substitution rule, the corresponding GR equation should then be

$$\frac{D\dot{x}^{\mu}}{d\tau} = 0, \tag{11.84}$$

which we recognize from (11.37) and (11.38) to be simply the geodesic equation.

11.3.2 Einstein field equation

We note that the above minimal substitution procedure cannot be used to obtain the GR field equation, because there is no SR gravitational field equation. The search for the GR field equation must start all the way back at Newton's equation as discussed at the beginning of Chapter 6, cf. (6.4). Since the energy–momentum right-hand side of the field equation (6.4) must satisfy the GR conservation condition,

$$D_{\mu} T^{\mu\nu} = 0, \tag{11.85}$$

we must have for the geometric left-hand side a covariantly constant, symmetric, rank-2 tensor involving metric derivatives $(\partial^2 g), (\partial g)^2 \sim \partial\Gamma, \Gamma^2$, which are just the properties of the Einstein curvature tensor $G_{\mu\nu}$. This then leads to the Einstein field equation:

$$G_{\mu\nu} = R_{\mu\nu} - \tfrac{1}{2} R g_{\mu\nu} = \kappa T_{\mu\nu}, \tag{11.86}$$

where κ is a proportionality constant. This in general is a set of six coupled partial differential equations. The number of independent elements of the symmetric tensors $G_{\mu\nu}$ and $T_{\mu\nu}$ are reduced from ten to six by the four component equations of (11.80) and (11.85), the covariant constancy conditions. This number of independent equations just matches the number of independent elements of the metric solution $g_{\mu\nu}(x)$—whose ten components, as we explained in Sidenote 12 in Chapter 6, are reduced to six by the requirement of general coordinate transformation $x^{\mu} \rightarrow x'^{\mu}$ so that the metric $g_{\mu\nu}(x')$ must still be a solution.

This field equation can be written in an alternative form. Taking the trace of the above equation, we have

$$-R = \kappa T, \tag{11.87}$$

where T is the trace of the energy–momentum tensor, $T = g^{\mu\nu} T_{\mu\nu}$. In this way, we can rewrite (11.86) in an equivalent form by replacing $R g_{\mu\nu}$ by $-\kappa T g_{\mu\nu}$:

$$R_{\mu\nu} = \kappa (T_{\mu\nu} - \tfrac{1}{2} T g_{\mu\nu}).\tag{11.88}$$

Exercise 11.6 Newtonian limit of Einstein's field equation

Show that Newton's gravitational field law, written in differential form (4.7), is the leading-order approximation to the Einstein field equation (11.88) in the Newtonian limit (cf. Section 5.3.1) for a slow ($v \ll c$) source particle producing a static and weak gravitational field. In this way, one can also establish the connection between the proportionality constant κ and Newton's constant as shown in (6.39).

Einstein equation via the principle of least action In the above derivation of the GR field equation, we worked directly with the rank-2 Einstein tensor, which must satisfy a bunch of necessary conditions. $T^{\mu\nu}$ and hence $G^{\mu\nu}$ had to be covariantly constant to satisfy the constraint of energy–momentum conservation in GR, which is rather complicated conceptually. The same field equation can be obtained more systematically via the principle of least action (cf. Box 5.1). The action for any field, instead of being a time integral of the Lagrangian L for a particle as in (5.18), is a 4D integral of the Lagrangian density $\mathcal{L}$. The invariant 4D volume differential product[13] is $\sqrt{-g}\, d^4x$, where g is the determinant of the metric tensor $g_{\mu\nu}$. The relevant action for the GR metric field in vacuum is called the Einstein–Hilbert action,

$$S_g = \int \mathcal{L}_g \sqrt{-g}\, d^4x = \int g^{\mu\nu} R_{\mu\nu} \sqrt{-g}\, d^4x.\tag{11.89}$$

[13] Coordinate transformation involves the Jacobian $\mathcal{J}$ so that $d^4x' = d^4x \mathcal{J}$ and $g' = g \mathcal{J}^{-2}$ as the metric has two lower indices hence two inverse transformation factors. Thus $\sqrt{-g'}\, d^4x' = \sqrt{-g}\, d^4x$.

For the invariant source-free GR Lagrangian density, one makes the natural identification with the Ricci scalar: $\mathcal{L}_g = R$. We can then derive the $T_{\mu\nu} = 0$ version of (11.86) as the Euler–Lagrange equation resulting from minimization of this action. The variation of the action δS_g has three pieces involving $\delta g^{\mu\nu}$, $\delta R_{\mu\nu}$, and $\delta \sqrt{-g}$. The integral containing the $\delta R_{\mu\nu}$ factor, after an integration by parts, turns into a vanishing surface term. The variation $\delta \sqrt{-g}$ is easiest to compute for a diagonal metric. The determinant is the product of its elements $g = \prod_\mu g_{\mu\mu}$, so its variation (using the inverse metric $g^{\mu\nu}$) is simplified to $\delta g = g \sum_\mu \delta g_{\mu\mu}/g_{\mu\mu} = g g^{\mu\nu} \delta g_{\mu\nu} = -g g_{\mu\nu} \delta g^{\mu\nu}$. As the metric matrix is symmetric, we can diagonalize it by coordinate transformation; this expression for δg must be valid for all coordinates. Thus $\delta \sqrt{-g} = -\tfrac{1}{2}\sqrt{-g} g_{\mu\nu} \delta g^{\mu\nu}$. Consequently, the variational principle requires

$$\delta S_g = \int \sqrt{-g}\, d^4x\, (R_{\mu\nu} - \tfrac{1}{2} R g_{\mu\nu}) \delta g^{\mu\nu} = 0,\tag{11.90}$$

[14] It is possible to incorporate terms for other fields into the Lagrangian density to derive the field equations that describe the interaction of gravity with other sources.

which implies the Einstein equation (11.86).[14]

The Schwarzschild solution The Einstein equation (11.88), in the exterior region (where $T_{\mu\nu} = 0$) of a spherical source, merely states that the Ricci tensor vanishes: $R_{\mu\nu} = 0$. From Box 6.2, we learnt that a spherically symmetric metric involves only two scalar functions, g_{00} and g_{rr}. We first express in terms of g_{00} and g_{rr} the Christoffel symbols $\Gamma^{\mu}_{\nu\lambda}$, and then the Riemann curvature tensor $R^{\mu}_{\nu\lambda\rho}$, which is then contracted to give the Ricci tensor $R_{\mu\nu}$. The field equations ($R_{00} = R_{rr} = 0$) are two coupled differential equations for g_{00} and g_{rr}. One of them simply yields the relation $g_{rr} = -1/g_{00}$ of (6.57); the other is an ordinary differential equation for g_{00},

$$\frac{dg_{00}}{dr} + \frac{g_{00}}{r} = -\frac{1}{r}. \tag{11.91}$$

The solution to the homogeneous part ($d\bar{g}_{00}/dr + \bar{g}_{00}/r = 0$) is $\bar{g}_{00} = r^*/r$ for some constant r^*. By adding to the easily checked particular solution, $g_{00,\text{part}} = -1$, we obtain the general result shown in (6.57): $g_{00} = -1 + r^*/r$.

Einstein's cosmological constant The energy–momentum tensor $T_{\mu\nu}$ and Einstein curvature tensor $G_{\mu\nu}$ are covariantly constant; clearly, any term to be added to the GR field equation must also have this property. This requirement for mathematical consistency allowed Einstein to insert another metric term as in (8.74),[15] because the metric tensor itself is covariantly constant, as shown in (11.27).

[15] The gravitational Lagrangian density is modified as $\mathcal{L}_g = R + 2\Lambda$; the variation of the resulting action then leads to (8.74).

Review questions

1. What is the fundamental difference between coordinate transformations in a curved space and those in flat space (e.g., Lorentz transformations in flat Minkowski space)?

2. Writing the coordinate transformation as a partial derivative matrix, give the transformation laws for a contravariant vector $A^{\mu} \longrightarrow A'^{\mu}$ and a covariant vector $A_{\mu} \longrightarrow A'_{\mu}$, as well as for a mixed tensor $T^{\mu}_{\nu} \longrightarrow T'^{\mu}_{\nu}$.

3. From the transformation of $A_{\mu} \longrightarrow A'_{\mu}$ in the answer to the previous question, work out the coordinate transformation of the derivatives $\partial_{\mu} A_{\nu}$. Why do we say that $\partial_{\mu} A_{\nu}$ is not a tensor? What is the underlying reason why $\partial_{\mu} A_{\nu}$ is not a tensor? How do the covariant derivatives $D_{\mu} A_{\nu}$ transform? Why is it important to have differentiations that result in tensors?

4. Write out the covariant derivative $D_{\mu} T^{\lambda\rho}_{\nu}$ (in terms of the connection symbols) of a mixed tensor $T^{\lambda\rho}_{\nu}$.

5. The relation between the Christoffel symbol and the metric tensor is called "the fundamental theorem of Riemannian geometry." Write out this relation.

6. What is the flatness theorem? Use this theorem to show that the metric tensor is covariantly constant: $D_{\mu} g_{\nu\lambda} = 0$.

7. As the Christoffel symbols $\Gamma^{\mu}_{\alpha\beta}$ are not components of a tensor (and ordinary derivatives generally do not yield tensors), how do we know that $R^{\mu}_{\lambda\alpha\beta} = \partial_{\alpha}\Gamma^{\mu}_{\lambda\beta} - \partial_{\beta}\Gamma^{\mu}_{\lambda\alpha} + \Gamma^{\mu}_{\nu\alpha}\Gamma^{\nu}_{\lambda\beta} - \Gamma^{\mu}_{\nu\beta}\Gamma^{\nu}_{\lambda\alpha}$ is really a tensor?

8. The quantitative description of tidal force is the Newtonian deviation equation. What is its GR analog?

What geometric quantity of a curved spacetime replaces the second derivatives of the gravitational potential?

9. In Einstein's search for a linear combination of $R^{\mu\nu}$ and $Rg^{\mu\nu}$ that is covariantly constant, why is such a constraint required? What is an efficient way to find this combination?

10. What is the principle of general covariance?

11. If a physics equation is known in the SR limit, how does one form its GR analog? Since SR equations are valid only in the absence of gravity, turning them into their GR versions implies the introduction of a gravitational field into the relativistic equations. How does gravity enter into this alteration of the equations?

12. How can one determine the GR equation of motion from that of SR? Why can we not find the GR field equation in the same way?

Appendix A
Keys to Review Questions

Here we provide keys in the form of a few phrases by which readers are directed to the relevant passages in the text to formulate their proper answers.

Chapter 1: Introduction

1. Relativity is coordinate symmetry. SR is the symmetry with respect to inertial frames, and GR the symmetry with respect to general coordinate frames, including accelerating frames and frames with gravity.

2. Symmetry is the invariance of physics, and hence the covariance of the physics equations (i.e., physics is not changed) under symmetry transformations. Inability to detect a particular physical feature means that physics is unchanged. It is a symmetry. If one cannot detect the effect after changing the orientation, this means that the physics equation is covariant under rotation—thus we have rotational symmetry. With boost symmetry, we then have full symmetry of SR; physics is the same in all inertial frames.

3. A tensor is a mathematical object that transforms in a definite manner under a change of coordinates. Every term in a tensor equation transforms in the same way, so the equation is covariant under coordinate transformation—it is symmetric.

4. The frames in which Newton's first law applies. The frames move at constant velocity with respect to the fixed (distant) stars. GR treats as inertial frames those frames without gravity, such as free-falling frames.

5. Either light travels with speed c with respect to a particular frame of reference (the ether rest frame) or with speed c with respect to all observers, which requires new kinematics (i.e., a new relation among inertial frames).

6. The equations of Newtonian physics are unchanged under the Galilean transformation, electrodynamics under the Lorentz transformation. The Galilean transformation is the low-velocity limit ($v \ll c$) of the Lorentz transformation.

Chapter 2: Special Relativity: The New Kinematics

1. Two events that are simultaneous to one observer may not be simultaneous to another. That SR has relativity of simultaneity along with the Galilean concept of relativity of equilocality suggests that time and space are linked in SR rather than being absolutely distinct as in Newtonian mechanics. Thus time and space coordinates must be treated on an equal footing in SR.

2. See the first two paragraphs of Section 2.1.

3. See (2.13) and (2.18).

4. For time dilation, we impose the condition $x' = 0$ in the Lorentz transformation (i.e., the clock is at rest in the O' frame; hence there is no displacement, $\Delta x' = 0$). For length contraction, we impose the condition $t = 0$ (i.e., the object is moving in the O frame, and hence we must specify the times when its two ends are to be measured—the choice $\Delta t = 0$ is the simplest).

5. See Section 2.2.1.

6. Using light signals naturally incorporates the SR postulate that the speed of light is constant.

7. See Section 2.2.2, in particular (2.31) and Exercise 2.7.

8. A moving charge induces a magnetic field. Therefore a purely electric field in the charge's rest frame must transform under a boost to an electromagnetic field with a magnetic component, as seen by an observer moving relative to the charge.

Chapter 3: Special Relativity: Flat Spacetime

1. (a) $g_{\mu\nu} \equiv \mathbf{e}_\mu \cdot \mathbf{e}_\nu$. (b) Diagonal elements of $g_{\mu\nu}$ are squared magnitudes of basis vectors. Off-diagonal elements describe the non-orthogonality. (c) $g_{\mu\nu} = \eta_{\mu\nu} = \mathrm{diag}(-1, 1, 1, 1)$, so $ds^2 = \eta_{\mu\nu}\, dx^\mu\, dv^\nu = -c^2\, dt^2 + dx^2 + dy^2 + dz^2$.

2. A Lorentz transformation preserves the invariant interval Δs^2 in Minkowski space, just as a rotation preserves length in Euclidean space.

3. $A^\mu \to A'^\mu = [L]^\mu_{\ \nu} A^\nu$, $A_\mu \to A'_\mu = [L^{-1}]^{\ \nu}_\mu A_\nu$, $T^\mu_{\ \nu} \to T'^\mu_{\ \nu} = [L]^\mu_{\ \alpha} [L^{-1}]^\beta_{\ \nu} T^\alpha_{\ \beta}$, and the product $A^\mu B_\mu \to A'^\mu B'_\mu = A^\mu B_\mu$.

4. $x^\mu = (x^0, x^1, x^2, x^3) = (ct, x, y, z)$. The derivative dx^μ/dt is not a proper tensor, because coordinate time t is not a tensor. $\dot{x}^\mu = dx^\mu/d\tau = \gamma\, dx^\mu/dt = \gamma(c, \mathbf{v})$. The scalar product $\dot{x}^\mu \dot{x}_\mu = -c^2$ for particles with mass. The 4-velocity definition is not valid for massless particles, because the concept of proper time does not apply here.

5. $E = \gamma mc^2 \xrightarrow{\mathrm{NR}} mc^2 + \frac{1}{2}mv^2$ and $\mathbf{p} = \gamma m\mathbf{v} \xrightarrow{\mathrm{NR}} m\mathbf{v}$. The 4-momentum $p^\mu = (p^0, \mathbf{p}) = (E/c, \mathbf{p})$. For particles with mass, $E^2 - (pc)^2 = (mc^2)^2$. For massless particles, $E = pc$; since we still have the ratio $E/p = c^2/v$, they always move with $v = c$, and there is no nonrelativistic limit and no rest frame to define a proper time.

6. Given $\partial_\mu F^{\mu\nu} + j^\nu/c = 0$, we can show that $\partial'_\mu F'^{\mu\nu} + j'^\nu/c = 0$: $\partial' F' + j'/c = [L^{-1}]\partial[L][L]F + [L]j/c = [L](\partial F + j/c) = 0$. That is, because every term is a 4-vector, the form of the equation is unchanged under Lorentz transformation. This equation includes the statement of electric charge conservation, $\partial_\mu j^\mu = 0$, because $\partial_\mu \partial_\nu F^{\mu\nu} = -\partial_\mu \partial_\nu F^{\mu\nu} = 0$ since $F^{\mu\nu} = -F^{\nu\mu}$ and $\partial_\mu \partial_\nu = +\partial_\nu \partial_\mu$.

7. $j^\mu = (c\rho, \mathbf{j}) = c\Delta q/\Delta S^\mu$, where ΔS^μ is the Minkowski volume with x^μ held fixed (i.e., it is a Minkowski 3-surface). For example,

$\Delta S^0 = (\Delta x^1 \Delta x^2 \Delta x^3) = \Delta V$, $\Delta S^1 = (c\Delta t \Delta x^2 \Delta x^3)$, etc. The energy–momentum tensor is the 4-current for the 4-momentum field: $T^{\mu\nu} = c\Delta p^\mu / \Delta S^\nu$, so T^{00} = energy density, $T^{0i} = T^{i0}$ = momentum density or energy current density, and T^{ij} = normal force per unit area (pressure) for $i = j$ and shear force per unit area for $i \neq j$.

8. See Figs. 3.4 and 3.5

9. See Fig. 3.6 and the related discussion.

10. See Fig. 3.7(a).

11. See Fig. 3.7(b) and the related discussion. The condition is $t < vx/c^2$. Since A is not inside the forward lightcone of B (i.e., the separation between A and B is spacelike), it is not causally related to B. In order for B to influence A, a signal traveling faster than light would be required.

Chapter 4: Equivalence of Gravitation and Inertia

1. Newton's field equation is (4.7). The equation of motion is (4.9), which is totally independent of any properties of the test particle.

2. Equations (4.10) and (4.11). There is evidence of $(\ddot{\mathbf{r}})_A = (\ddot{\mathbf{r}})_B$ for any two objects A and B in experiments such as dropping objects from the Leaning Tower of Pisa, Galilean inclined planes, and Newton's pendulums.

3. The EP is the statement that the physics in a freely falling frame in a gravitational field is indistinguishable from the physics in an inertial frame without gravity. The weak EP is the EP restricted to gravity in Newtonian mechanics; the strong EP is the EP applied to gravity in situations involving all physics, including electrodynamics.

4. See Fig. 4.3.

5. (a) See the discussion leading up to (4.20); (b) (4.53).

6. (a) (4.25); (b) (4.27).

7. The satellites are moving with high speed and at high altitude. High-speed motion causes the satellite clock to run slower; high altitude (larger gravitational potential) causes the satellite clock to run faster.

8. See (4.42). The speed of light is absolute as long as it is measured with respect to the proper time and proper distance of the observer. It may vary if measured by the coordinate time, for example by an observer far away from the gravitational source. To such an observer, the light speed is no longer a constant; free space acquires an effective index of refraction. This can bring about the physical manifestation of light deflection in a gravitational field. In his 1911 EP paper, Einstein did not consider the possibility that coordinate distance may differ from proper distance, as gravitational length contraction could not be rigorously derived from the EP.

9. Plug into (4.25) position-1 as $r = \infty$ (hence $\Phi_1 = 0$ and $d\tau_1 = dt$) and position-2 as r (hence $\Phi_2 = \Phi$ and $d\tau_2 = d\tau$). The light speed according to the faraway observer is $c(r) = dr/dt$, while the constant light speed is $c = dr/d\tau$.

Chapter 5: General Relativity as a Geometric Theory of Gravity

1. An intrinsic geometric description of a space involves only measurements made in the space; no reference is made to any external embedding space. The metric elements can be found by distance measurement once the coordinates have been fixed. See (5.5) and (5.6).

2. The geodesic equation is the Euler–Lagrange equation resulting from extremization of the length integral $\int g_{ab}\, \dot{x}^a \dot{x}^b\, d\tau$.

3. Coordinate transformations in a flat space can be written in position-independent form; in a curved space, they are inherently position-dependent.

4. A small region of any curved space can be approximated by a flat space: one can always find a coordinate transformation at a given point so that the new metric is a constant up to second-order corrections.

5. In a geometric theory of a physical phenomenon, the results of physical measurements can be attributed directly to the underlying geometry of space and time. For example, the decrease in latitudinal distances as they approach either of the poles (see Fig. 5.4) reflects the geometry of the spherical surface (rather than the physics of ruler lengths). See Section 5.2.1.

6. The gravitational time dilation (4.40) can be interpreted as a measurement of the time–time metric element g_{00}, with the result $g_{00} = -(1+2\Phi/c^2)$ as shown in (5.41). Thus the geometric description of gravitational time dilation can be visualized this way: higher clocks run faster because the coordinate seconds up there are closer together in spacetime, just as the latitudinal lines on a globe are closer together at higher latitudes, cf. Fig. 5.2

7. See Fig. 5.3. In a rotating cylinder, tangential distances are length-contracted per SR, but radial distances are not. Thus the relation between circumference and radius is no longer Euclidean, $S \neq 2\pi R$.

8. This indicated to Einstein that he should identify gravity as the structure of spacetime. Curved spacetime is the gravitational field.

9. If gravity is the structure of spacetime (rather than a force), we expect objects to follow straight trajectories (geodesics) in spacetime warped by gravity.

10. The Newtonian limit corresponds to $v \ll c$ particles moving in a weak and static gravitational field. See (5.43) and (5.45) for a geometric derivation of the gravitational redshift. Namely, the metric is such that it implies that clocks run faster at higher gravitational potential. This in turn means that because $\omega \sim 1/d\tau$, frequencies measured at higher-potential locations will be redshifted.

Chapter 6: Einstein Equation and its Spherical Solution

1. Recall the familiar case of a flat plane with position-dependent polar coordinates. K vanishes only for a flat surface, independently of coordinate choice. Take a spherical surface as an example: whether it is described in polar or cylindrical coordinates, we have $K = 1/R^2$.

2. A sphere, pseudosphere, and flat plane are surfaces of constant curvature. In 4D Euclidean space, $W^2 + X^2 + Y^2 + Z^2 = R^2$ describes a 3-sphere; in 4D Minkowski space, $-W^2 + X^2 + Y^2 + Z^2 = -R^2$ describes a 3-pseudosphere.

3. Angular excess ϵ is the difference of a polygon's (interior) angular sum from its Euclidean value. Figure 6.2 shows an example where the curvature $K = 1/R^2$ is proportional to ϵ, as in the relation $\epsilon = K\sigma$ (cf. (6.19)), with σ being the area of the polygon. Another example is the deviation of the circumference of a circle (radius r) from its Euclidean value: $S - 2\pi r = -\pi r^3 K/3$ for small r, cf. (6.18).

4. Tidal forces cause relative acceleration between nearby test particles in a gravitational field. They are the gradient of the gravitational field and hence second derivatives of the gravitational potential. As the relativistic gravitational potential is the metric, the tidal forces must be second derivatives of the metric, and hence the curvature. In the GR deviation equation, the Riemann curvature takes place of the tidal force matrix in the Newtonian deviation equation.

5. The GR field equation has the structure of a curvature term (a nonlinear second-derivative function of the metric in the form of a covariantly constant, rank-2, symmetric tensor $G_{\mu\nu}$) proportional to the source term (the rank-2 energy–momentum tensor): $G_{\mu\nu} = \kappa T_{\mu\nu}$, with the proportionality constant $\kappa \sim G_{\rm N}$. Its solution is the spacetime metric.

6. See Box 6.2, in particular (6.55). Space is curved because $d\rho = \sqrt{g_{rr}}\, dr$ and time because $d\tau = \sqrt{-g_{00}}\, dt$, cf. (6.69) and (6.70).

7. Newtonian gravity for a spherical source is the same as if all the mass were concentrated at the center. Thus the time dependence of the source does not show up in the exterior field, and there is no monopole radiation.

8. The Schwarzschild solution yields $g_{00} = -1 + r^*/r = -1 + 2G_{\rm N}M/(rc^2)$. Comparing this with $g_{00} = -(1 + 2\Phi/c^2)$, we have $\Phi = -G_{\rm N}M/r$.

9. See (6.74) and (6.76). From the metric element g_{00}, we obtain $d\tau = (1 + \Phi/c^2)\, dt$. Similarly, from $g_{rr} = -1/g_{00}$, we get $d\rho = (1 - \Phi/c^2)\, dr$. By taking their ratio dr/dt, we can find $c(r)$ in terms of the potential Φ.

10. The $1/r$ Newtonian gravitational potential has the distinctive property of yielding a closed elliptical orbit. The GR correction can be thought of as adding a perturbing $1/r^3$ term in the potential, cf. (6.88). This causes the orbital ellipse to precess.

11. The metric's independence of ϕ means that $\partial L/\partial\phi = 0$, with $L = g_{\mu\nu}\dot{x}^\mu\dot{x}^\nu$. This can be translated, through the geodesic equation (6.78), into a conservation statement, $d(\partial L/\partial\dot\phi)/d\tau = 0$. The constant of motion can then be written as $\partial L/\partial\dot\phi \sim g_{\mu\phi}\dot{x}^\mu$.

Chapter 7: Black Holes

1. Physical measurement at $r = r^*$ is not singular. The Schwarzschild geometry is singular at $r = r^*$ in the Schwarzschild coordinates, but this singularity disappears in coordinates such as the Eddington–Finkelstein system. One notes

that coordinate-independent quantities such as the Ricci scalar R and the Riemann curvature product $R^{\mu\nu\lambda\rho} R_{\mu\nu\lambda\rho}$ are not singular at $r = r^*$.

2. An event horizon is an one-way barrier that particles and light can traverse in only one direction. The surface $r = r^*$ is an event horizon of the Schwarzschild geometry: this null surface allows timelike worldlines to pass inward but never outward.

3. The proper time is the time measured by an observer either (1) on the surface of a collapsing star as the stellar radius decreases beyond $r = r^*$ or (2) in a spaceship falling across the $r = r^*$ surface. The coordinate time is the time measured by a faraway observer where the spacetime approaches a flat geometry. Infinite coordinate time means that, to the faraway observer, it would take an infinite amount of time for a particle or light to come over the horizon. Alternatively, we can express this in terms of light coming over the horizon suffering an infinite redshift.

4. A null surface is a 3D subspace in spacetime made up of lightlike null lines (i.e., one of the basis vectors of the 3D subspace is null). Light can travel along the null surface, so it is everywhere tangent to lightcones on one side. Thus a timelike path (being contained inside a lightcone) can cross the null surface in only one direction, cf. Fig. 7.4.

5. In advanced EF coordinates, the time coordinate $c\,d\bar{t}$ is the distance traversed by infalling light. Consequently, $dr/d(c\bar{t}) = -1$ for infalling light. The outgoing side of lightcones changes from a slope of $+1$ in distant flat space to a slope of -1 at the $r = 0$ singularity. The lightcones never clam up, and they turn over smoothly from pointing toward the ever-larger time direction to pointing toward the ever-smaller radial distance. In retarded EF coordinates, $c\,d\bar{t}$ is the distance traversed by outgoing light. Consequently, $dr/d(c\bar{t}) = +1$ for outgoing light. The lightcones tilt over smoothly outward, cf. Fig. 7.3. In advanced coordinates, the space inside $r = r^*$ is a black hole; all worldlines move in toward the future singularity at $r = 0$. In retarded coordinates, all worldlines move out away from the past singularity at $r = 0$; the space inside $r = r^*$ is a white hole.

6. The GR correction term, $-C l^2 r^3$, dominates the effective potential in the small-r region. Consequently, in GR, a particle can go over the centrifugal barrier and fall into the black hole.

7. Gravitational binding energy powers very energetic phenomena in the vicinity of black holes. When particles fall in, a lot of energy can be released. For instance, a particle falling into the ISCO of a Schwarzschild black hole can radiate 6% of its rest energy.

8. AGN (active galactic nuclei), quasars, and some X-ray binaries.

9. When stars burn out, if their final mass is over $3M_\odot$, they are too heavy to be supported by neutron degeneracy pressure, so they must collapse into black holes.

10. One finds the following correspondence surprising: gravity on the horizon surface behaves like temperature, and horizon area like entropy.

11. QFT in a curved spacetime is a partial union of GR with quantum mechanics, because it does not treat gravity itself quantum mechanically. GR just provides its classical curved spacetime as a stage for quantum mechanical phenomena such as Hawking radiation.

Chapter 8: The General Relativistic Framework for Cosmology

1. $v = H_0 r$ is linear if H_0 is independent of v and r. This means that every galaxy sees all other galaxies as rushing away according to Hubble's law (cf. the discussion relating to Fig. 8.1).

2. $t_H = H_0^{-1} \simeq 14.4\,\text{Gyr}$; $t_0 = t_H$ for an empty universe. In a universe full of matter and energy, we expect $t_0 < t_H$, because gravitational attraction slows down the expansion. This means that the expansion rate was faster than H_0 in the past. From globular clusters, we deduced $[t_0]_{gc} \gtrsim 12\,\text{Gyr}$.

3. The rotation curves are plots of matter's rotational speed as a function of radial distance to the center of the mass distribution (e.g., a galaxy). Gravitational theory would lead us to expect a rotational curve to drop as $v \sim r^{-1/2}$ outside the matter distribution. However, rotation curves are observed to stay flat, $v \sim r^0$, way beyond the luminous matter distribution.

4. Baryonic matter is the particle physics name for matter composed of ordinary (neutral or ionized) atoms. The IGM can be detected by its electromagnetic absorption lines. It is not counted as part of dark matter, because we define dark matter as being composed of particles having no electromagnetic interactions.

5. $\Omega_M \simeq 0.32$, $\Omega_{lum} \lesssim 0.005$, and $\Omega_B \simeq 0.05$. Thus $\Omega_{DM} = \Omega_M - \Omega_B \simeq 0.27$.

6. The cosmological principle: at any given instance of cosmic time, the universe appears the same at every point: space is homogeneous and isotropic. Comoving coordinates are a system where the time coordinate is chosen to be the proper time of each cosmic fluid element; the spatial coordinates are coordinate labels carried along by each fluid element (thus each fluid element has a fixed and unchanging comoving spatial coordinate).

7. See (8.26), (8.28), and (8.29) for the Robertson–Walker metric in spherical polar and cylindrical coordinate systems. The input used in the derivation is the cosmological principle. $a(t)$ is the dimensionless scale, which changes along with the cosmic time, and it contains all the time dependence in the cosmic metric. $k = \pm 1, 0$ is the curvature parameter.

8. The Hubble constant is related to the scale factor by $H(t) = \dot{a}(t)/a(t)$. Because the wavelength scales as $\lambda_{rec}/\lambda_{em} = a(t_{rec})/a(t_{em})$, the definition of redshift $z \equiv (\lambda_{rec} - \lambda_{em})/\lambda_{em}$ leads to the relation (8.35).

9. The Friedmann equations are the Einstein equations subject to the cosmological principle, i.e., the Robertson–Walker metric and the ideal-fluid $T_{\mu\nu}$. The Newtonian interpretation of the first Friedmann equation is the energy

balance equation, with E_{tot} being the sum of kinetic and potential energies. A Newtonian interpretation is possible because of the cosmological principle—large regions behave similarly to small regions. It is only quasi-Newtonian, because we still need to supplement it with geometric concepts such as curvature and scale factor.

10. Both the critical density and the escape velocity are used to compare the kinetic and potential terms to determine whether the total energy is positive (unbound system) or negative (bound system).

11. Radiation energy, being proportional to frequency (hence inverse wavelength), scales as a^{-1}. After dividing by the volume (a^3), the density $\sim a^{-4}$. Since matter density scales as a^{-3}, radiation dominates in the early universe.

12. $p = w\rho c^2$, with $w_R = \frac{1}{3}$ and $w_M = 0$. For a flat RDU, $a \sim t^{1/2}$ and $t_0 = \frac{1}{2}t_H$; for an MDU, $a \sim t^{2/3}$ and $t_0 = \frac{2}{3}t_H$. Since the radiation–matter equality time $t_{RM} \ll t_0$, an MDU should be a good approximation.

13. See Fig. 8.7.

14. Moving the $+\Lambda g_{\mu\nu}$ term to the source side of the equation, we get

$$G_{\mu\nu} = \kappa(T_{\mu\nu} - \kappa^{-1}\Lambda g_{\mu\nu}) = \kappa(T_{\mu\nu} + T^{\Lambda}_{\mu\nu}).$$

Thus, even in the absence of a matter/energy source, $T_{\mu\nu} = 0$ (i.e., a vacuum), space can still be curved by the Λ term.

15. Constant density means $dE = \varepsilon\,dV$, with ε being the constant energy density. The first law, $dE = -p\,dV$, leads to $p = -\varepsilon$.

16. Cf. Section 8.4.1, especially (8.80). In a Λ-dominated system, since the energy density is constant, the more the space expands, the greater is the energy and negative pressure; thus $\ddot{a} \propto a$, which has a exponential solution for $a(t)$.

Chapter 9: Big Bang Thermal Relics

1. The Stefan–Boltzmann law is $\rho_R \sim T^4$ and the radiation density scaling law is $\rho_R \sim a^{-4}$. Therefore $T \sim a^{-1}$. Blackbody radiation involves only scale-invariant combinations of (volume)$\times E^2\,dE$ (recall that radiation energy $\sim a^{-1}$) and E/T.

2. Reaction rate faster than expansion rate. Cf. (9.12) and (9.13).

3. From a big bang nucleosynthesis energy E_{bbn} of the order of MeV, we have $T_{bbn} \simeq 10^9$ K. The Boltzmann distribution yields $n_n/n_p \simeq \exp(-\Delta mc^2/k_b T_{bbn})$, with $\Delta m = m_n - m_p$. Equation (9.24) leads to a mass fraction of $\frac{1}{4}$, if $n_n/n_p \simeq \frac{1}{7}$.

4. The theoretical prediction of deuterium abundance by big bang nucleosynthesis is sensitive to Ω_B and the number of neutrino flavors. The observed abundance can fix these quantities, cf. Fig. 9.1.

5. At t_γ, the reversible reaction e + p $\longleftrightarrow$ H + γ stopped proceeding from right to left. All charged particles turned into neutral atoms. The universe became transparent to photons. An average thermal energy of the order of

eV translates into $T(z_\gamma) \simeq 3000$ K. By the temperature scaling law,

$$\frac{T(t_\gamma)}{T(t_0)} = \frac{a(t_0)}{a(t_\gamma)} = 1 + z_\gamma, \tag{A.1}$$

leading to $T(t_0) \simeq 3$ K.

6. $t_{bbn} \simeq 10^2$ s, $t_{RM} \lesssim 10^5$ years, and $t_\gamma \simeq 4 \times 10^5$ years.

7. Neutrinos have only weak interaction. The cross sections for low-energy neutrinos are extremely small. No one has figured out a way to catch these particles, even though they are supposedly all around us.

8. Motion leads to a frequency blueshift in one direction and a redshift in the opposite direction. Frequency shift means energy change, and hence temperature change.

9. Primordial density perturbations as amplified by gravity and resisted by radiation pressure set up acoustic waves in the photon–baryon fluid. Photons leaving denser regions would be gravitationally redshifted and thus bring about CMB temperature anisotropy.

10. Because the dark matter was not affected by radiation pressure, it started its gravitational clumping much earlier than baryonic matter, which mainly fell into the potential well already built up by dark matter.

Chapter 10: Inflation and the Accelerating Universe

1. Gravitational attraction in a matter-dominated universe or a radiation-dominated universe would have enhanced any initial curvature. So, for the universe to be seen even somewhat flat today, $1 - \Omega_0 = O(1)$, it would need to have been extremely flat early on, $1 - \Omega(t_{bbn}) = O(10^{-14})$, an unnaturally fine-tuned initial condition. This is the flatness problem.

2. The large-scale homogeneity of the CMB suggests that distant regions of the universe had past contact to achieve thermal equilibrium. Yet, in the standard FLRW cosmology, no light signal could have traveled between them; they appear to be outside each other's horizons. This is the horizon problem.

3. The repulsive force from vacuum energy creates a positive-feedback loop: an expanding universe has more volume and hence more vacuum energy, which leads to more expansion. Thus we have exponential expansion, $\dot{a}(t) \propto a(t)$. Extreme expansion stretches out any initial curvature, leaving an extremely flat universe, which solves the flatness problem. Extreme expansion can separate distant points at a rate much greater than c. This explains how distant points that seem to be outside each other's horizons could in fact have been in past thermal contact, thus solving the horizon problem.

4. The characteristic size of the CMB anisotropy is related to the BAO sound horizon. Taking into account the comoving distance to the origin of the CMB photons at last scattering, we expect to observe angular anisotropies about $1°$ wide in the CMB, so the anisotropy spectrum should have a prominent peak at $l \simeq 200$. In a universe with curved space, the CMB rays would

converge or diverge on their way to us, respectively enlarging or shrinking the anisotropy scale. Thus the fact that we do observe the primary peak at $l \simeq 200$ is evidence of a flat universe.

5. Thomson scattering is the low-energy (classical) limit of the scattering of a photon by an electron. The perpendicular scattering of an unpolarized photon is 100% polarized. Isotropic unpolarized radiation is not polarized by scattering. But unpolarized anisotropic radiation can be polarized by scattering.

6. A *B*-mode polarization pattern in the CMB would constitute direct evidence of gravitational waves as well as of the cosmic inflation that produced them.

7. An accelerating universe must have been expanding more slowly in the past; hence it must be older than a decelerating matter-dominated universe, $\frac{2}{3}t_H$.

8. Dark energy has an equation-of-state parameter $w < -\frac{1}{3}$, for example, $w = -1$ for vacuum energy/cosmological constant. Its negative pressure causes a gravitational repulsion that accelerates the expansion of the universe. Dark matter is just an exotic form of nonrelativistic matter with $w = 0$; it does not have strong or electromagnetic interactions. Standard Model neutrinos are nearly massless, so they count as radiation, $w = +\frac{1}{3}$; exotic heavy neutrinos and WIMPs may constitute dark matter. Dark energy gives rise to gravitational repulsion, while dark matter gives rise to attraction.

9. Type Ia SNe have intrinsic luminosities that can be reliably calibrated, and they are extremely bright.

10. In an accelerating universe, the past expansion was slower than in an empty or decelerating universe. Consequently, a given redshift is farther in the past, and light travels farther to get to us and hence is dimmer.

11. Were there a mundane physical explanation for the fact that SNe with redshift $0.2 < z < 0.7$ are dimmer than they would be in a flat universe, the trend would be expected to continue to arbitrarily high redshifts. But if dark energy is accelerating the expansion, it would not always have done so. ρ_Λ is constant, but $\rho_M \propto a^{-3}$, so the universe would have been matter-dominated and hence decelerating earlier on.

12. See Section 10.3, as well as the summary at the beginning of the chapter.

Chapter 11: Tensor Formalism for General Relativity

1. Transformations in a curved space must necessarily be position-dependent.

2. See (11.3), (11.4), and (11.8).

3. The transformation of $\partial_\mu A_\nu$ has extra terms, cf. (11.12), because basis vectors and hence the transformation matrix in a curved space are position-dependent. One way to see these extra terms is through (11.14): because $A^\mu = \mathbf{e}^\mu \cdot \mathbf{A}$ is coordinate-dependent, its derivatives will have an extra $\mathbf{A} \cdot (\partial_\nu \mathbf{e}^\mu)$ factor. Equation (11.15) shows the transformation of covariant derivatives. Tensor equations are important because they are automatically relativistic.

4. $D_\nu T^{\lambda\rho}_\mu = \partial_\nu T^{\lambda\rho}_\mu - \Gamma^\sigma_{\nu\mu} T^{\lambda\rho}_\sigma + \Gamma^\lambda_{\nu\sigma} T^{\sigma\rho}_\mu + \Gamma^\rho_{\nu\sigma} T^{\lambda\sigma}_\mu$.

5. See (5.30).

6. At every point, we can always find a coordinate system (local Euclidean frame) in which $\partial_\mu g_{\nu\lambda} = 0$ and $\Gamma^\lambda_{\nu\sigma} = 0$. Therefore, in this local Euclidean frame, we have $D_\mu g_{\nu\lambda} = \partial_\mu g_{\nu\lambda} = 0$; but, being a tensor equation, $D_\mu g_{\nu\lambda} = 0$ must hold in every frame.

7. We can express the Riemann tensor as a commutator of covariant derivatives, (11.56). Since every term other than $R^\mu_{\lambda\alpha\beta}$ is known to be a good tensor, by the quotient theorem, $R^\mu_{\lambda\alpha\beta}$ must also be a good tensor. Similarly, the Riemann tensor is defined as in (11.46); all other terms being good tensors, so is $R^\mu_{\lambda\alpha\beta}$.

8. The GR analog is the equation of geodesic deviation, (11.67). Its basic elements are the geodesic equation for the particle trajectory and covariant differentiation so that the equation is covariant under general coordinate transformations. The Riemann curvature tensor replaces the second derivatives of the gravitational potential.

9. As energy–momentum conservation can be expressed in GR as $D_\mu T^{\mu\nu} = 0$, the geometric side of Einstein's equation (involving $R^{\mu\nu}$ and $Rg^{\mu\nu}$) must have the same property. The Bianchi identity is an efficient way to find this combination, since its contracted version is $D_\mu (R^{\mu\nu} - \frac{1}{2} Rg^{\mu\nu}) = 0$.

10. See the first part of Section 11.3.1.

11. Just replace ordinary derivatives by covariant derivatives (the minimal substitution rule). This is required because the coordinate symmetry in GR is local, involving position-dependent transformations. Covariant derivatives involve the Christoffel symbols, which, being derivatives of the gravitational potential (i.e., the metric), constitute the gravitational field.

12. The equation of motion for a free particle in SR is $d\dot{x}^\mu/d\tau = 0$, so the minimal substitution rule leads to the geodesic equation $D\dot{x}^\mu/D\tau = d\dot{x}^\mu/d\tau + \Gamma^\mu_{\nu\lambda}\dot{x}^\nu\dot{x}^\lambda = 0$. We cannot follow the same procedure to find the field equation, because there is no SR equation for the gravitational field.

Appendix B
Hints for Selected Exercises

Chapter 1: Introduction

1.2 Because the $\mathbf{v}t$ term is the same for both $\mathbf{r}'_A$ and $\mathbf{r}'_B$, we have an invariant separation: $\mathbf{r}'_{AB} = \mathbf{r}'_A - \mathbf{r}'_B = \mathbf{r}_A - \mathbf{r}_B = \mathbf{r}_{AB}$.

1.3 The conservation laws in these two frames differ by $\mathbf{v}[(m_A + m_B) - (m_C + m_D)]$.

1.4 The ratio of the two equations of Lorentz transformation for the space and time differentials is

$$\frac{dx'}{dt'} = \frac{dx - v\,dt}{dt - \dfrac{v}{c^2}\,dx}.$$

Chapter 2: Special Relativity: The New Kinematics

2.2 The obtained nonsynchronicity should be $\Delta t = \gamma^2 \dfrac{v}{c^2} L$.

2.3 The Lorentz transformation for position components (parallel and perpendicular to the relative velocity $\mathbf{v}$) may be written as $\mathbf{r}'_\parallel = \gamma(\mathbf{r}_\parallel - \mathbf{v}t)$ and $\mathbf{r}'_\perp = \mathbf{r}_\perp$.

2.4 Identify the transformation matrix elements with the partial derivatives in the chain rule:

$$\frac{\partial x}{\partial x'} = \gamma, \quad \frac{\partial x}{\partial t'} = \gamma v, \quad \frac{\partial t}{\partial x'} = \gamma \frac{v}{c^2}, \quad \frac{\partial t}{\partial t'} = \gamma.$$

2.5 The length can be derived from the round trip time. According to the stationary observer, it is

$$\Delta t'_1 + \Delta t'_2 = \frac{L'}{c - v} + \frac{L'}{c + v} = \frac{2L'}{c} \gamma^2.$$

The length contraction result is obtained by comparing this with the clock's proper time and length, and using the time dilation result.

2.7 $ds'^2 = ds^2 = 0$ corresponds to light speed in both frames.

2.8 The constancy of c allows us to conclude that the intervals evaluated in different coordinate frames must be proportional to each other $s'^2 = Ps^2$; then argue for $P = 1$.

2.10 During the outward-bound part of Al's journey, his yearly flashes are received by Bill every 3 years, while on the return leg, they are received every 4 months.

2.11 By calculating the relative velocity v_{12} of Al's inbound and outbound legs, we see that the original reading of $t' = 9$ years should be counted as $t = \gamma_{12}t' = 41$ years. Namely, in Al's frame, Bill's clock jumps 32 years at the turnaround point.

Chapter 3: Special Relativity: Flat Spacetime

3.1 Using the symmetry properties and relabeling the repeated indices, one can show that $S_{\mu\nu}A^{\mu\nu}$ equals exactly the negative of itself.

3.2 Plugging in the tensor transformations of $A_{\mu\nu}$ and B_ν^λ, one can then show that $C_{\mu\lambda}$ have the correct transformation property.

3.3 The matrix multiplication $[\eta] = [L][\eta][L^T]$ becomes $[1] = [L][L^T]$ when the Minkowski metric $[\eta]$ is replaced by $[1]$ in Euclidean space.

3.4 The gamma relation follows from the matrix multiplication of the Lorentz transformation for the 4-velocity:

$$\gamma'_u \begin{pmatrix} c \\ u' \end{pmatrix} = \gamma_u \gamma_v \begin{pmatrix} 1 & -v/c \\ -v/c & 1 \end{pmatrix} \begin{pmatrix} c \\ u \end{pmatrix}.$$

3.5 In (b), one obtains the new velocity by using the trigonometric identity

$$\tanh(\psi_1 + \psi_2) = \frac{\tanh \psi_1 + \tanh \psi_2}{1 + \tanh \psi_1 \tanh \psi_2}$$

for $\tanh \psi_1 = u/c$, $\tanh \psi_2 = -v/c$, and $\tanh(\psi_1 + \psi_2) = u'/c$.

3.6 The 4-momentum conservation relation and its result $k' \cdot p' = k \cdot p$ leads to the relation $k'_\mu (p^\mu + k^\mu) = k_\mu p^\mu$:

$$\left(-\hbar\omega'/c \; \mathbf{k}' \right) \begin{pmatrix} mc + \hbar\omega/c \\ 0 + \mathbf{k} \end{pmatrix} = \left(-\hbar\omega/c \; \mathbf{k} \right) \begin{pmatrix} mc \\ 0 \end{pmatrix}.$$

3.7 In the center-of-mass frame, the final 4-momentum is $P_F^\mu = (4mc, 0)$. The conservation relation $P_I \cdot P_I = P_F \cdot P_F$ leads to a lab-frame kinetic energy $K = E - mc^2 = 6mc^2$.

3.8 Only for the transverse momentum is the gamma factor a constant, which depends only on the longitudinal displacement, and hence the relative speed ($v = dx/dt$).

3.9 (b) The Lorentz-transformed field tensor $F'^{\mu\nu}$ is related to the unprimed $F^{\mu\nu}$ as

$$\begin{pmatrix} \gamma & -\beta\gamma & 0 & 0 \\ -\beta\gamma & \gamma & 0 & 0 \\ 0 & 0 & 1 & 0 \\ 0 & 0 & 0 & 1 \end{pmatrix} \begin{pmatrix} 0 & E_1 & E_2 & E_3 \\ -E_1 & 0 & B_3 & -B_2 \\ -E_2 & -B_3 & 0 & B_1 \\ -E_3 & B_2 & -B_1 & 0 \end{pmatrix} \begin{pmatrix} \gamma & -\beta\gamma & 0 & 0 \\ -\beta\gamma & \gamma & 0 & 0 \\ 0 & 0 & 1 & 0 \\ 0 & 0 & 0 & 1 \end{pmatrix}.$$

3.10 (b) For $\mu = i$, we have $\gamma F = (q/c)[-E_i(-\gamma c) + \epsilon_{ijk}B_k(\gamma v_j)]$, which is just the familiar Lorentz force law written in its components. $K^0 = \gamma \mathbf{F} \cdot \mathbf{v}/c$, because the dot product with the magnetic field term in the Lorentz force vanishes.

3.11 (c) Taking the 4-divergence of the inhomogeneous equation, we get $\partial_\nu j^\nu = -c\partial_\nu \partial_\mu F^{\mu\nu} = 0$, because $\partial_\nu \partial_\mu = +\partial_\mu \partial_\nu$ while $F^{\mu\nu} = -F^{\nu\mu}$.

3.12 $\partial_\mu F_{\lambda\rho}\epsilon^{\mu\nu\lambda\rho} = 0$ is a nontrivial relation only when the indices are unequal, e.g., $\partial_1 F_{23} + \partial_3 F_{12} + \partial_2 F_{31} = 0$. We can regard this as a relation in a particular coordinate frame with $\mu = 1$, $\nu = 2$, and $\lambda = 3$. Once written in Lorentz-covariant form, it must be valid in every frame. To prove the converse statement, all we need to do is to contract $\epsilon^{\mu\nu\lambda\rho}$.

3.13 $T^{10} = c\dfrac{\Delta p^1}{\Delta x \Delta y \Delta z} = \dfrac{\Delta E}{c\Delta t \Delta y \Delta z} = T^{01}$.

3.14 (a) Lorentz transformation requires $t < vx/c^2$ to yield $t' < 0$. (b) Event B is spacelike-displaced from the origin, so it cannot be causally related to the origin.

3.15 The figure caption would read: Three worldlines of the twin paradox: OQ for the stay-at-home Bill, OP for the outward-bound part ($\beta = 4/5$), and PQ for the inward-bound part ($\beta = -4/5$) of Al's journey. M is the midpoint between O and Q. These three lines define three inertial frames: O, O', and O'' systems. When Al changes from the O' to the O'' system at P, the point that is simultaneous (with P) along Bill's worldline OQ jumps from point P' to P''. From the viewpoint of Al, this is a leap of 32 years.

3.16 The figure caption would read: Spacetime diagram for the pole-and-barn paradox. The ground (barn) observer has (x, t) coordinates, while the runner (pole)'s rest frame has (x', t') coordinates. There are two sets of solid lines: (i) the worldline for the front door (F) and rear door (R) of the barn, and (ii) the front end (A) and back end (B) of the pole. Note the time-order reversal: $t_{AR} > t_{BF}$ and $t'_{AR} < t'_{BF}$. By adding a horizontal line to indicate the simultaneous slamming of the front and back doors in the barn frame, one can easily find their nonsimultaneous appearances in the runner's proper frame.

Chapter 4: Equivalence of Gravitation and Inertia

4.1 (a) $a_A = g \sin\theta \left(\dfrac{m_G}{m_I}\right)_A$;

(b) period $P_A = 2\pi \sqrt{\dfrac{L}{g}\left(\dfrac{m_I}{m_G}\right)_A}$.

4.2 (a) The buoyant force is always opposite to g_{eff}. (b) Drop the whole contraption.

4.3 (b) $r_S = \frac{3}{2}r_\oplus$;

(c) period $P = 3\pi\sqrt{\dfrac{3}{2}\dfrac{r_\oplus}{g}} \simeq 3\,\text{h}$.

Chapter 5: General Relativity as a Geometric Theory of Gravity

5.1 (b) $\int \sqrt{dx^2 + dy^2} = 4 \int_0^R dx \sqrt{1 + \left(\dfrac{dy}{dx}\right)^2} = 4R \left[\sin^{-1} \dfrac{x}{R}\right]_0^R = 2\pi R.$

5.3

$$\begin{pmatrix} d\rho \\ d\phi \end{pmatrix} = \begin{pmatrix} R\cos\theta & 0 \\ 0 & 1 \end{pmatrix} \begin{pmatrix} d\theta \\ d\phi \end{pmatrix}.$$

Chapter 6: Einstein Equation and its Spherical Solution

6.2 (d) Sphere $V_2 = \displaystyle\int ds_\theta\, ds_\phi = R^2 \int_{-1}^{1} d\cos\theta \int_0^{2\pi} d\phi = 4\pi R^2$;

hypersphere $\tilde{V}_2 = R^2 \displaystyle\int_1^{\infty} d\cosh\psi \int_0^{2\pi} d\phi = \infty.$

6.3 (b) The $L = 0$ equation becomes $\dot{r}^2 + \dfrac{j^2}{r^2}\left(1 - \dfrac{r^*}{r}\right) = \kappa^2.$

(d) Solving the first-order equation

$$\frac{d^2 u_1}{d\phi^2} + u_1 - \frac{1 - \cos 2\phi}{2r_{\min}^2} = 0,$$

we try the form $u_1 = \alpha + \beta \cos 2\phi$, leading to $\alpha = 1/(2r_{\min}^2)$ and $\beta = 1/(6r_{\min}^2)$, and the result, accurate up to first order in $r^*/r_{\min}$, of a bent trajectory, is

$$\frac{1}{r} = \frac{\sin\phi}{r_{\min}} + \frac{r^*}{r_{\min}^2} \frac{3 + \cos 2\phi}{4}.$$

Chapter 7: Black Holes

7.1 $\Delta\tau = 2r^*/3c.$ For a supermassive black hole, $\Delta\tau_{\text{SMBH}} \simeq 2\,\text{h}.$

7.2 $c\,d\tilde{t} = -dr\,\dfrac{r + r^*}{r - r^*}.$

Chapter 8: The General Relativistic Framework for Cosmology

8.2 Start by differentiating (8.42): $2\dot{a}\ddot{a} = \dfrac{8\pi G_N}{3} \dfrac{d}{dt}(\rho a^2).$

Chapter 9: Big Bang Thermal Relics

9.1   The entropy-conservation condition for the e^+e^- annihilation reaction may be written as $S'_\gamma + S'_{e^+} + S'_{e^-} = S_\gamma$. Plugging in the effective spin degrees and using $T'_\gamma = T'_{e^+} = T'_{e^-}$, we have $[2 + \frac{7}{8}(2+2)]V'T'^3_\gamma = 2VT^3_\gamma$ and hence

$$\frac{V'}{V} = \frac{4}{11}\left(\frac{T_\gamma}{T'_\gamma}\right)^3 = \left(\frac{T_\nu}{T'_\nu}\right)^3.$$

To arrive at the last equality, we have used $V'T'^3_\nu = VT^3_\nu$.

9.2   (a) $a(t_{RM}) = \rho_R(t_0)/\rho_M(t_0)$.

   (b)   From $\Omega_B(t_0)/\Omega_M(t_0) = 0.04/0.25$ and $n_\gamma/n_B = 10^9$, one gets $a(t_{RM}) \simeq 8 \times 10^{-5}$.

   (c) $t_{RM} = (a_{RM})^{3/2}t_0 \simeq 10^4$ years.

9.3 $\delta T/T = (\delta\omega/\omega) = (v/c)\cos\theta$.

Chapter 10: Inflation and the Accelerating Universe

10.1   (b) Combining $a(t_{bbn}) = O(10^{-9})$ from (a) with $a(t_{RM}) = O(10^{-4})$ yields

$$|1 - \Omega(t_{bbn})| = O(10^{-10}) \times |1 - \Omega(t_{RM})| = O(10^{-14}).$$

10.2   (a) From the Friedmann equation and the definition of $\rho_{c,0}$ we have

$$[H(a)]^2 = H_0^2\left(\frac{\rho}{\rho_{c,0}} + \frac{1-\Omega_0}{a^2(t)}\right).$$

10.3   For a vanishingly small Λ, we have the limit $\ln(\sqrt{\Omega_\Lambda} + \sqrt{1+\Omega_\Lambda}) \to \sqrt{\Omega_\Lambda}$.

10.4 $t_{M\Lambda} = t_H \dfrac{2}{3\sqrt{\Omega_\Lambda}}\ln(1+\sqrt{2}) \simeq 10.2\,\text{Gyr}$.

Chapter 11: Tensor Formalism for General Relativity

11.2

$$\frac{D}{d\sigma}\left(A_\mu\frac{dx^\mu}{d\sigma}\right) = \frac{DA_\mu}{d\sigma}\left(\frac{dx^\mu}{d\sigma}\right) + A_\mu\frac{D}{d\sigma}\left(\frac{dx^\mu}{d\sigma}\right) = 0.$$

11.3

$$D_\alpha D_\beta A_\mu = \partial_\alpha\left(D_\beta A_\mu\right) - \underbrace{\Gamma^\nu_{\alpha\beta}D_\nu A_\mu}_{\text{drop}} - \Gamma^\nu_{\alpha\mu}D_\beta A_\nu$$

$$= \partial_\alpha\partial_\beta A_\mu - \partial_\alpha\left(\Gamma^\nu_{\beta\mu}A_\nu\right) - \Gamma^\nu_{\alpha\mu}\partial_\beta A_\nu + \Gamma^\nu_{\alpha\mu}\Gamma^\lambda_{\beta\nu}A_\lambda$$

$$= -(\partial_\alpha\Gamma^\nu_{\beta\mu})A_\lambda - \underbrace{\Gamma^\nu_{\beta\mu}\partial_\alpha A_\nu}_{\text{drop}} - \Gamma^\nu_{\alpha\mu}\partial_\beta A_\nu + \Gamma^\nu_{\alpha\mu}\Gamma^\lambda_{\beta\nu}A_\lambda.$$

The underlined terms are symmetric in the indices (α, β) and will be canceled when we include the $-D_\beta D_\alpha A_\mu$ calculation.

11.4 For small and static $h_{\mu\nu}$,

$$R^i_{0j0} = \tfrac{1}{2}[\partial_j \partial_0 h_{0i} - \partial_j \partial_i h_{00} - \partial_0 \partial_0 h_{ji} + \partial_0 \partial_i h_{0j}]$$

$$= -\tfrac{1}{2}\partial_i \partial_j h_{00} = \frac{1}{c^2}\frac{\partial^2 \Phi}{\partial x^i \partial x^j}.$$

11.5 By expanding out the double commutators, one can check the validity of the Jacobi identity for any operators:

$$[A, [B, C]] + [C, [A, B]] + [B, [C, A]] = 0.$$

11.6 • Slow-moving source particle   Einstein's equation becomes

$$R_{00} = \tfrac{1}{2}\kappa\, T_{00}.$$

• Weak-field limit

$$R_{00} = g^{ij} R_{i0j0} = -\tfrac{1}{2}g^{ij}(\partial_i \partial_j g_{00} - \partial_0 \partial_j g_{i0} + \partial_0 \partial_0 g_{ij} - \partial_i \partial_0 g_{0j}).$$

• Static limit

$$R_{00} = -\tfrac{1}{2}\nabla^2 g_{00}, \quad \text{which is} \quad \nabla^2 \Phi = \tfrac{1}{2}\kappa\rho c^4.$$

Appendix C
Glossary of Symbols and Acronyms

Latin symbols

$a(t)$	scale factor in Robertson–Walker metric	g^*	effective spin multiplicity
$\mathbf{a}$	acceleration vector	$\mathbf{g}$	gravitational acceleration/field
A	area, nuclear mass number	$g_{\mu\nu}$	spacetime metric tensor
A^μ	electromagnetic 4-potential	G_N	Newton's constant
a_{lm}	spherical harmonic coefficients	$G_{\mu\nu}$	Einstein curvature tensor
a_{SB}	Stefan–Boltzmann constant	h	Planck's constant, $\hbar = h/2\pi$
$\mathbf{B}$	magnetic vector field	H	Hubble's constant $\dot{a}(t)/a(t)$
B	brightness	$h_{\mu\nu}$	perturbation of spacetime metric
c	light speed	$h_{+,\times}$	plus and cross polarizations of gravitational
$c(x)$	light speed by coordinate time		plane waves
c_s	sound speed	$[\mathbb{I}]$	identity matrix
$C(\theta)$	correlation function	$\mathcal{J}$	angular momentum, Jacobian
C_l	angular power spectrum	$\mathbf{j}$	electromagnetic current 3-vector
$[A, B]$	commutator ($= AB - BA$)	j^μ	electromagnetic current 4-vector
D	distance	K	Gaussian curvature
D_0	comoving distance now, $D(t_0)$	k	wavenumber $k = |\mathbf{k}| = 2\pi/\lambda$, curvature
D_L	luminosity distance		signature $(0, \pm 1)$
D_p	proper distance	$\mathbf{k}$	wave 3-vector
D_μ	covariant derivative	k^μ	wave 4-vector
$\text{diag}(a_1, a_2)$	diagonal matrix $\begin{pmatrix} a_1 & 0 \\ 0 & a_2 \end{pmatrix}$	k_B	Boltzmann's constant
		K^μ	covariant force 4-vector
e	eccentricity, electric charge, electron field	L	Lagrangian, length
$\mathbf{E}$	electric vector field	$\mathcal{L}$	luminosity, Lagrangian density
E	energy	$[L]^\lambda_\mu$	Lorentz transformation
E_{Pl}	Planck energy	l	length, angular momentum multipole
$\mathbf{e}_\mu$	basis vectors in a manifold		number
$\mathcal{E}$	GR analog of nonrelativistic total energy	l_H	Hubble length
F	force	l_{Pl}	Planck length
f	flux	m	mass (usually test mass), multipole node
$F_{\mu\nu}$	electromagnetic field tensor, Yang–Mills field tensor		number
		M	mass (usually source mass)
$\tilde{F}_{\mu\nu}$	electromagnetic field dual tensor	m_G	gravitational mass
g	number of states with same energy metric determinant	m_I	inertial mass
		m_{Pl}	Planck mass
$[g]$	metric matrix	M_c	critical mass

n	index of refraction, number density
N	total number, number of neutrons in a nucleus
$O(x)$	of the order of x
$[\hat{O}g]$	(derivative) operator acting on the metric
p	pressure, momentum magnitude
$\mathbf{p}$	momentum 3-vector
p^{μ}	momentum 4-vector
q	charge, deceleration parameter
r	radial coordinate
$\hat{\mathbf{r}}$	unit vector in the $\mathbf{r}$ direction
r^{*}	Schwarzschild radius
r_{sh}	comoving distance for BAO sound horizon
R	distance, radius of curvature, Ricci curvature scalar
R_0	radius now in Robertson–Walker geometry
$[\mathbf{R}]_{ij}$	element of a rotational matrix
$R^{\mu}{}_{\lambda\alpha\beta}$	Riemann curvature tensor
$R_{\mu\nu}$	Ricci curvature tensor
S	surface area, entropy, action, circumference of a circle
s	invariant spacetime interval
$\mathbf{s},\, s^{\mu}$	relative displacement vectors
ΔS^{μ}	Minkowski volume with $\Delta x^{\mu}=0$
t	time (usually coordinate time)
$\bar{t}(\tilde{t})$	advanced (retarded) EF coordinate time

t_0	cosmic time of the present epoch, age of the universe
t_{H}	Hubble time ($=1/H_0$)
t_{Pl}	Planck time ($=l_{\mathrm{Pl}}/c$)
T	temperature (on absolute scale)
T_{Pl}	Planck temperature
$T_{\mu\nu}$	energy–momentum–stress 4-tensor
T_{ij}	momentum–stress 3-tensor
u	energy density, speed, inverse orbital radius
$\mathbf{u}$	velocity 3-vector
u^{μ}	velocity 4-vector
v	magnitude of boost velocity
v_i	velocity component
v_{esc}	escape velocity
V	velocity, volume, potential energy
w	equation-of-state parameter
W	complexion, number of microstates
x	spatial or spacetime position
$\tilde{x}$	comoving position
x^{μ}	position (displacement) 4-vector
$\dot{x}^{\mu}$	4-velocity $dx^{\mu}/d\tau$
y	helium mass fraction
$Y_{lm}(\theta,\phi)$	spherical harmonics function
z	wavelength shift, redshift
Z	number of protons in nucleus

Greek symbols

α	orbital distance $\perp$ major axis
β	v/c, velocity in units of c
γ	the Lorentz factor $(1-\beta^2)^{-1/2}$
Γ	reaction rate
$\Gamma^{\nu}_{\lambda\rho}$	Christoffel symbols
$\delta x,\, \Delta x$	change in x
δ_{ij}	Kronecker delta
∂_i	3D del operator, $\equiv \partial/\partial x_i$
∂_{μ}	4D del operator, with $\partial_0 = (1/c)\partial/\partial t$
$\boldsymbol{\nabla}$	del operator with components $(\partial/\partial x_1, \partial/\partial x_2, \partial/\partial x_3)$
ε_{ijk}	3D Levi-Civita symbol
$\varepsilon_{\mu\nu\lambda\rho}$	4D Levi-Civita symbol
ϵ	energy, angular excess
ϵ_0	permittivity of free space
$\zeta(n)$	Riemann zeta function
$\eta_{\mu\nu}$	metric of (flat) Minkowski spacetime
θ	polar angle coordinate

θ_{sh}	angular size of BAO sound horizon
κ	particle energy in GR
κ	gravity strength, $8\pi G_{\mathrm{N}}/c^4$
λ	wavelength, neutron-to-proton ratio
Λ	cosmological constant
$[\Lambda]$	coordinate transformation matrix
μ_0	permeability of free space
ν	frequency
ξ	Robertson–Walker cylindrical radial coordinate
ρ	mass density, charge density, cylindrical radial coordinate, proper radial distance
ρ_c	cosmic critical density
σ	area of a polygon, reaction cross section
τ	proper time, lifetime
Φ	electromagnetic or gravitational potential
ϕ	phase of a wave
ϕ	azimuthal angle coordinate

$\phi(x)$	(generic) scalar field	Ω	$d\Omega = \sin\theta\, d\phi\, d\theta$ element of solid angle
χ	Robertson–Walker spherical radial coordinate	Ω	ratio of density to critical density (ρ/ρ_c)
ψ	rapidity of relative frames, GR effect parameter	Ω_B	baryonic matter density ratio
		Ω_M	matter density ratio
Ψ	quantum wave function	Ω_{DM}	dark matter mass density ratio
ω	angular frequency	Ω_Λ	dark energy density ratio

Acronyms

1D, ..., 4D	one-dimensional, ..., four dimensional	IGM	intergalactic medium
AGN	active galactic nuclei	ISCO	innermost stable circular orbit
B	baryon	LHS	left hand side
BAO	baryon acoustic oscillation	ΛCDM	lambda cold dark matter
BBN	big bang nucleosynthesis	ΛDU	dark-energy-dominated universe
BH	black hole	ΛM	dark energy–matter equality
BICEP	Background Imaging of Cosmic Extragalactic Polarization	MDU	matter-dominated universe
		NR	nonrelativistic
CCD	charge-coupled device	QCD	quantum chromodynamics
CDM	cold dark matter	QFT	quantum field theory
CM	center of mass	QM	quantum mechanics
CMB	cosmic microwave background	RDU	radiation-dominated universe
COBE	Cosmic Background Explorer	RM	radiation–matter equality
DM	dark matter	RW	Robertson–Walker
DOF	degrees of freedom	SDSS	Sloan Digital Sky Survey
EF	Eddington–Finkelstein	SI	International System of Units
EM	electromagnetic	SM	Standard Model
EP	equivalence principle	SN	supernova
ETH	Eidgenössische Technische Hochschule [(Swiss) Federal Polytechnic Institute]	SNe	supernovae
		SR	special relativity
		SSB	spontaneous symmetry breaking
FIRAS	Far Infrared Absolute Spectrophotometer	SSU	steady-state universe
FLRW	Friedmann–Lemaître–Robertson–Walker	SUSY	supersymmetry
GPS	Global Positioning System	WIMP	weakly interacting massive particle
GR	general relativity	WMAP	Wilkinson Microwave Anisotropy Probe
GUT	grand unified theory		

Miscellaneous symbols and units

Metric prefixes

Prefix	Symbol	10^n
nano	n	10^{-9}
micro	μ	10^{-6}
milli	m	10^{-3}
centi	c	10^{-2}
kilo	k	10^3
mega	M	10^6
giga	G	10^9

Tensor indices

Greek indices	(usually $\mu, \nu, \lambda, \rho, \alpha, \beta$): 4D spacetime
Latin indices	(usually i, j, k, l, m, n): 3D space (usually a, b, c): 2D space

Selected units

Symbol	Name of unit	Physical quantity	Definition
°	degree	plane angle	$2\pi/360 \simeq 0.0175$ rad
′	arc minute	plane angle	$1°/60 \simeq 2.91 \times 10^{-4}$ rad
″	arc second	plane angle	$1'/60 \simeq 4.85 \times 10^{-6}$ rad
Å	angstrom	length	10^{-10} m
AU	Astronomical Unit (average distance between earth and sun)	length	$\simeq 1.50 \times 10^{11}$ m
eV	electronvolt	energy	$\simeq 1.60 \times 10^{-19}$ J
g	gram	mass	
Hz	hertz	frequency	s^{-1}
J	joule	energy	$kg \cdot m^2/s^2$
Jy	jansky	flux per unit frequency	10^{-26} W$/(m^2 \cdot$ Hz$)$
K	kelvin	temperature	
lyr	light-year	length	$c \cdot yr \simeq 9.46 \times 10^{15}$ m
m	meter	length	
pc	parsec	length	$1\,AU/1'' \simeq 3.09 \times 10^{16}$ m
rad	radian	plane angle	2π around a circle
s	second	time	
sr	steradian	solid angle	4π over a sphere
W	watt	power	J/s
yr	year	time	$\simeq 3.15 \times 10^7$ s

Particle and element names

γ	photon
ν	neutrino
e	electron
μ	muon
τ	tau
q	quark
p	proton
n	neutron
H	hydrogen
D	deuterium
He	helium
Li	lithium
Fe	iron

Sign conventions

Beware of various sign conventions $[S] = \pm 1$ used in the literature:

$$\eta_{\mu\nu} = [S1] \times \mathrm{diag}(-1, 1, 1, 1),$$
$$R^{\mu}_{\lambda\alpha\beta} = [S2] \times (\partial_\alpha \Gamma^{\mu}_{\lambda\beta} - \partial_\beta \Gamma^{\mu}_{\lambda\alpha},$$
$$+ \Gamma^{\mu}_{\nu\alpha}\Gamma^{\nu}_{\lambda\alpha} - \Gamma^{\mu}_{\nu\beta}\Gamma^{\nu}_{\lambda\alpha}$$
$$G_{\mu\nu} = [S3] \times ((8\pi G)/c^4))T_{\mu\nu}.$$

Thus the convention adopted in this book is $[S1, S2, S3] = (+ + +)$. The sign in the Einstein equation $[S3]$ is related to $[S2]$ as well as the sign convention in the definition of the Ricci tensor $R_{\mu\nu} = R^{\alpha}_{\mu\alpha\nu}$.

Letter and symbol subscripts

b	boson
bh	black hole
c	critical
cosmo	the entire universe
eff	effective
em	emitter
EH	event horizon
f	fermion
h	horizon
galac	galaxy
gc	globular cluster
mo	motion
obs	observed
p	proper
qv	quantum vacuum
rec	receiver
rel	relativity
s	sound
sh	sound horizon
S	satellite
$\odot$	symbol for the sun (e.g., $M_\oplus$ = solar mass)
$\oplus$	symbol for the earth (e.g., $R_\oplus$ = earth's radius)

Cosmological component subscripts

R	radiation
gas	nonluminous gas (mostly IGM)
lum	luminous matter (stars)
B	baryonic matter (gas +lum)
DM	dark matter
M	nonrelativistic matter (B +DM)
Λ	dark energy
tot	total

Cosmological epoch subscripts

0	present epoch (now)
bbn	big bang nucleosynthesis
γ	photon decoupling time (recombination)
fr	proton–neutron freeze-out
$M\Lambda$	matter–dark energy equality
RM	radiation–matter equality
tr	transition from deceleration to acceleration

References and Bibliography

References

Alpher, R.A. and Herman, R. (1948). "Evolution of the universe," *Nature*, **162**, 774.

Bennett, C.L. et al. (2013). "Nine-year WMAP observations: final maps and results," *Astrophys. J. Suppl. Ser.*, **208**, 1.

Berger, A. (ed.) (1984). *The Big Bang and Georges Lemaître: Proceedings of a Symposium in Honour of G. Lemaître Fifty Years after His Initiation of Big-Bang Cosmology*, Reidel, Dordrecht.

Berry, M.V. (1989). *Principles of Cosmology and Gravitation*, Institute of Physics, Bristol.

BICEP2 Collaboration (2014). "Detection of *B*-mode polarization at degree angular scales by BICEP2," *Phys. Rev. Lett.*, **112**, 241101.

BICEP2/Keck, Planck Collaborations (2015). "A Joint Analysis of BICEP2/Keck Array and Planck Data" to be published in *Phys. Rev. Lett.*, **114**, 101301.

Burles, S. et al. (2001). "Big-bang nucleosynthesis predictions for precision cosmology," *Astrophys. J. Lett.*, **552**, L1.

Carroll, S. (2004). *Spacetime and Geometry: An Introduction to General Relativity*, Addison-Wesley, San Francisco.

Cheng, T.P. (2010). *Relativity, Gravitation and Cosmology: A Basic Introduction*, 2nd ed., Oxford University Press, Oxford.

Cheng, T.P. (2013). *Einstein's Physics: Atoms, Quanta, and Relativity–Derived, Explained, and Appraised*, Oxford University Press, Oxford.

Clowe, D. et al. (2006). "A direct empirical proof of the existence of dark matter," *Astrophys. J.*, **648**, L109.

CPAEe (1989). *The Collected Papers of Albert Einstein* (English translation), Vols. 2 and 6, Princeton University Press, Princeton, NJ.

de Bernardis, P. et al. (Boomerang Collaboration) (2000). "A flat universe from high-resolution maps of the cosmic microwave background radiation," *Nature*, **404**, 955.

Deprit, A. (1984). "Monsignor Georges Lemaître," in (Berger 1984), pp. 363–392.

Dicke, R.H., Peebles, P.J.E., Roll, P.G., and Wilkinson, D.T. (1965). "Cosmic blackbody radiation," *Astrophys. J.*, **412**, 414.

Dietrich J.P. et al. (2012) "A filament of dark matter between two clusters of galaxies," *Nature*, **487**, 202.

D'Inverno, R. (1992). *Introducing Einstein's Relativity*, Oxford University Press, Oxford.

Dodelson, S. (2003). *Modern Cosmology*, Academic Press, San Diego.

Einstein, A. (1905a). "On the electrodynamics of moving bodies," *Annalen der Physik*, **17**, 891 [*CPAEe*, **2**, 140]; see also (Einstein et al. 1952).

Einstein, A. (1905b). "Does the inertia of a body depend upon its energy content?," *Annalen der Physik*, **18**, 639 [*CPAEe*, **2**, 172]; see also (Einstein et al. 1952).

Einstein, A. (1916). "The foundation of the general theory of relativity," *Annalen der Physik*, **49**, 769 [*CPAEe*, **6**, 146]; see also (Einstein et al. 1952).

Einstein, A. (1917). "Cosmological considerations in the general theory of relativity," *Preussische Akademie der Wissenschaften, Sitzungsberichte*, p. 142 [*CPAEe*, **6**, 421]; see also (Einstein et al. 1952).

Einstein, A. (1946). "Elementary derivation of the equivalence of mass and energy," *Technion J.*, **5**, 16.

Einstein, A. et al. (1952). *The Principle of Relativity—A Collection of Original Papers on the Special and General Theory of Relativity*, Dover, New York.

Ellis, G.F.R. and Williams, R.M. (1988). *Flat and Curved Space-Times*, Clarendon Press, Oxford.

Fixsen, D.J. et al. (1996). "The cosmic microwave background spectrum from the full COBE FIRAS data set," *Astrophys. J.*, **473**, 576.

French, A. (ed.) (1979). *Einstein—A Centenary Volume*, Harvard University Press, Cambridge, MA.

Fruchter, A. et al. (NASA/ERO Team) (2001). "Image of galaxy cluster Abell 2218, showing gravitational lensing effects," STScI and ST-ECF.

Gamow, G. (1946). "Expanding universe and the origin of elements," *Phys. Rev.*, **70**, 572

Gamow, G. (1948). "The evolution of the universe," *Nature*, **162**, 680.

Gamow, G. (1970). *My World Line, An Informal Autobiography*, Viking, New York, p. 44.

Greene, B. (2011). *The Hidden Reality: Parallel Universes and the Deep Laws of the Cosmos*, Vintage, New York.

Guth, A.H. (1981). "The inflationary universe: a possible solution to the horizon and flatness problems," *Phys. Rev. D*, **23**, 347.

Guth, A.H. (1997). *The Inflationary Universe*, Addison-Wesley, San Francisco.

Hanany, S. et al. (2000) "Constraints on cosmological parameters from MAXIMA-1," *Astrophys. J. Lett.*, **545**, L5.

Harrison, E. (2000). *Cosmology: The Science of the Universe*, 2nd ed., Cambridge University Press, Cambridge.

Hartle, J.B. (2003). *Gravity: An Introduction to Einstein's General Relativity*. Addison-Wesley, San Francisco.

Hobson, M.P., Efstathiou, G., and Lasenby, A.N. (2006). *General Relativity: An Introduction for Physicists*, Cambridge University Press, Cambridge.

Isaacson, W. (2007). *Einstein: His Life and Universe*, Simon & Schuster, New York.

Jacobson, T. (1995). "Thermodynamics of spacetime: the Einstein equation of state," *Phys. Rev Lett.*, **75**, 1260.

Kenyon, I.R. (1990). *General Relativity*, Oxford University Press, Oxford.

Kibble, T.W.B. (1985). *Classical Mechanics*, 3rd ed., Longman, London.

Kolb, E.W. and Turner, M.S. (1990). *The Early Universe*, Addison-Wesley, San Francisco.

Landau, L.D. and Lifshitz, E.M. (1962). *The Classical Theory of Fields*, 2nd ed., Pergamon Press, Oxford.

Lambourne, R.J.A. (2010). *Relativity, Gravitation and Cosmology*. Cambridge University Press, Cambridge.

Liddle, A. (2003). *An Introduction to Modern Cosmology*, 2nd ed., Wiley, New York.

Liddle, A. and Loveday, J. (2008). *Oxford Companion to Cosmology*, Oxford University Press, Oxford.

Logunov, A.A. (2001). *On the Articles by Henri Poincaré "On the Dynamics of the Electron"* (English translation by G. Pontecorvo), 3rd ed., JINR, Dubna.

Misner, C., Thorne, K., and Wheeler, J.A. (1973). *Gravitation*, W.H. Freeman, New York.

Moore, T.A. (2012). *A General Relativity Workbook*, University Science Books, Mill Valley, CA.

Mukhanov, V. (2005). *Physical Foundations of Cosmology*, Cambridge University Press, Cambridge.

Ohanian, H. and Ruffini, R. (2013). *Gravitation and Spacetime*, 3rd ed., Cambridge University Press, Cambridge.

Ostriker, J.P. and Mitton, S. (2013). *Unraveling the Mysteries of the Invisible Universe*, Princeton University Press, Princeton, NJ.

Ostriker, J.P. and Peebles, P.J.E. (1973). "A numerical study of the stability of flattened galaxies: or, can cold galaxies survive?," *Astrophys. J.*, **186**, 467.

Padmanabhan, T. (2010). "Thermodynamical aspects of gravity: new insights," *Rep. Prog. Phys.*, **73**, 1.

Pais, A. (2005). *Subtle is the Lord ... The Science and Life of Albert Einstein*, Oxford University Press, Oxford.

Peacock, J.A. (1999). *Cosmological Physics*, Cambridge University Press, Cambridge.

Peebles, P.J.E. (1980). *The Large Scale Structure of the Universe*, Princeton University Press, Princeton, NJ.

Peebles, P.J.E. (1993). *Principles of Physical Cosmology*, Princeton University Press, Princeton, NJ.

Peebles, P.J.E. (1984). "Impact of Lemaître's ideas on modern cosmology," in (Berger 1984), pp. 23–30.

Penzias, A.A. and Wilson, R.W. (1965). "A measurement of excess antenna temperature at 4080 Mc/s," *Astrophys. J.*, **412**, 419.

Perlmutter, S. et al. (Supernova Cosmology Project) (1999). "Measurements of Ω and Λ from 42 high-redshift supernovae," *Astrophys. J.*, **517**, 565.

Perlmutter, S. (2003). "Supernovae, dark energy, and the accelerating universe," *Physics Today*, April 2003, 53.

Planck Collaboration (2014). "Planck 2013 results: XVI. Cosmological parameters," *Astron. Astrophys.*, **571**, A16.

Pound, R.V. and Rebka, G.A. (1960). "Apparent weight of photons," *Phys. Rev. Lett.*, **4**, 337.

Raine, D.J. and Thomas, E.G. (2001) *An Introduction to the Science of Cosmology*, Institute of Physics, Bristol.

Riess, A.G. et al. (High-Z Supernova Search Team) (1998). "Observational evidence from supernovae for an accelerating universe and a cosmological constant," *Astron. J.*, **116**, 1009.

Riess, A.G. (2000). "The case for an accelerating universe from supernovae," *Publ. Astron. Soc. Pac.*, **112**, 1284.

Rindler, W. (1968). "Counterexample to the Lenz–Schiff argument," *Am. J. Phys.*, **36**, 540.

Rowan-Robinson, M. (2004). *Cosmology*, 4th ed., Oxford University Press, Oxford.

Ryden, B. (2003). *Introduction to Cosmology*, Addison-Wesley, San Francisco.

Ryder, L. (2009). *Introduction to General Relativity*, Cambridge University Press, Cambridge.

Schutz, B.F. (2009). *A First Course in General Relativity*, 2nd ed., Cambridge University Press, Cambridge.

Schwinger, J. (1986). *Einstein's Legacy—The Unity of Space and Time*, Scientific American Books, New York.

Silk, J. (2000). *The Big Bang*, 3rd ed., W.H. Freeman, New York.

Stephani, H. (2004). *Relativity: An Introduction to Special and General Relativity*, 3rd ed., Cambridge University Press, Cambridge.

Taylor, E.F. and Wheeler, J.A. (2000). *Exploring Black Holes: Introduction to General Relativity*, Addison-Wesley, San Francisco.

Tegmark, M. (2014). *Our Mathematical Universe*, Knopf, New York.

Thorne, K.S. (1994). *Black Holes & Time Warps: Einstein's Outrageous Legacy*, Norton, New York.

Tolman, R.C. (1934). *Relativity, Thermodynamics and Cosmology*, Clarendon Press, Oxford.

Verlinde, R.P. (2011). "On the origin of gravity and the laws of Newton," *JHEP*, **1104**, 29.

Vikhlinin, A. et al., (2009). "Chandra Cluster Cosmology Project III: cosmological parameter constraints," *Astrophys. J.*, **692**, 1060.

Wald, R.M. (1984). *General Relativity*, University of Chicago Press, Chicago.

Weinberg, S. (1972). *Gravitation and Cosmology*, Wiley, New York.

Weinberg, S. (1977). *The First Three Minutes: A Modern View of the Origin of the Universe*, Basic Books, New York.

Weinberg, S. (2008). *Cosmology*, Oxford University Press, Oxford.

Weisberg, J.M. and Taylor, J.H. (2003). "The relativistic binary pulsar B1913+16," *Proceedings of Radio Pulsars, Chania, Crete, 2002* (eds. M. Bailes et al.) (Astronomical Society of the Pacific Conference Series), University of Chicago Press, Chicago.

Will, C. (1986). *Was Einstein Right?—Putting General Relativity to the Test*, Basic Books, New York.

Zee, A. (2013). *Einstein Gravity in a Nutshell*, Princeton University Press, Princeton, NJ.

Zee, A. (2001). *Einstein's Universe: Gravity at Work and Play*, Oxford University Press, Oxford.

Bibliography

General relativity (including cosmology)   (Zee 2013), (Ohanian and Ruffini 2013), (Moore 2012), (Cheng 2010), (Lambourne 2010), (Hobson, Efstathiou, and Lasenby 2006), (Hartle 2003), (Taylor and Wheeler 2000), (D'Inverno 1992), (Kenyon 1990), (Misner, Thorne, and Wheeler 1973), (Weinberg 1972), (Landau and Lifshitz 1962).

General relativity (more mathematically oriented)   (Ryder 2009), (Schutz 2009), (Carroll 2004), (Stephani 2004), (Wald 1984).

Cosmology and astrophysics: (Weinberg 2008), (Mukhanov 2005), (Rowan-Robinson 2004), (Dodelson 2003), (Liddle 2003), (Ryden 2003), (Raine and Thomas 2001), (Harrison 2000), (Silk 2000), (Peacock 1999), (Peebles 1993), (Kolb and Turner 1990), (Berry 1989), (Peebles 1980), (Tolman 1934).

General interest and biographical books   (Tegmark 2014), (Cheng 2013), (Ostriker and Mitton 2013), (Greene 2011), (Liddle and Loveday 2008), (Isaacson 2007), (Pais 2005), (Zee 2001), (Guth 1997), (Thorne 1994), (Will 1986), (Schwinger 1986), (French 1979), (Weinberg 1977).

Index